Beyond the Valley

How Innovators around the World Are Overcoming Inequality
and Creating the Technologies of Tomorrow

Ramesh Srinivasan

The MIT Press
Cambridge, Massachusetts
London, England

This book was set in Stone Serif and Stone Sans by Westchester Publishing Services. Printed and bound in the United States of America.

Library of Congress Cataloging-in-Publication Data

Names: Srinivasan, Ramesh, author.
Title: Beyond the Valley : how innovators around the world are overcoming
   inequality and creating the technologies of tomorrow / Ramesh Srinivasan.
Description: Cambridge, MA : MIT Press, [2019] | Includes bibliographical
   references and index.
Identifiers: LCCN 2019005607 | ISBN 9780262043137 (hardcover : alk. paper)
Subjects: LCSH: Internet—Moral and ethical aspects. | Internet industry—
   Moral and ethical aspects.
Classification: LCC TK5105.878 .S75 2019 | DDC 174/.93843—dc23
   LC record available at https://lccn.loc.gov/2019005607

10 9 8 7 6 5 4 3 2 1

Contents

Acknowledgments

A book like this, which in its first draft was over seven hundred pages, can't just write itself! It took inspirational friends, family, and colleagues to make *Beyond the Valley* happen. And to each of you I'm deeply grateful.

Thank you to my wonderful parents (the Srinivasans), my brother Mahesh and his partner Amanda, and the central figure in my life: my deepest confidante and love Syama. Thank you also to my university (UCLA) and colleagues around the world for all their support throughout the process.

MIT Press has been a joy to work with. A special thank you to Gita Manaktala, an always constructive, supportive, and fantastic editor. Gita, your mindful and positive approach toward this book has been essential in making it a real contribution.

I want to thank my reassuring, smart, and savvy agent Jeff Shreve (and Peter Tallack) of the Science Factory for being such great partners with this project. Jeff, thank you for believing in me and in this ambitious book from the start, for lending your great eye to the proposal, and for your support in giving it the global reach that it needs and deserves given the topics I discuss. Thank you also to Doug Rushkoff for your mentorship and the powerful preface. And thank you Cathy O'Neil for all your support in helping this book reach a wide audience.

I worked with a wonderful team of supporters to conduct the research and express the arguments that form the foundation of *Beyond the Valley*. Chi Truong, thank you for being so easy and efficient to work with—helping draft and edit content with your fantastic mind and editorial eye. Rene Bermudez, it's been such a pleasure—your attitude, creativity, consciousness, and wonderful writing skills have been instrumental in making this book happen. Philip Smart, thank you for your support in Africa, and afterward with the text. Shane Boris, Sarah Gavish, and Jill Giardino, thank you for your generosity in reading and commenting on this book. Thank you to my wonderful copyeditor Mary Bagg for all of your support and excellent edits.

And then there is the real glue that kept the project together on a day-to-day basis. To Leia Yen, a UCLA undergraduate way beyond your years, one of the most remarkable students I have ever come across: thank you for helping me research the arguments in this book, and for bringing them together in polished form as this book reached its final stages. Jasmine Huang, thank you so much for doing such impeccable work to help me glean insight from the interviews, and for being available for this project despite living thousands of miles away! Vanessa Wong, a hearty thanks for your support with interviews and other essential book items. Michael Lumpkin and Roey Reichert, thanks for your support in this book's initial stages.

Thank you also to Aditi Mehta for joining me to co-author chapter 17, "Keeping Network Power Local," so I could add these inspiring efforts to this book's story. Adam Reese, thank you for you your meticulous, diligent, always-on approach toward chapter 23, "Blockchain: A Crazy Free-for-All, and Maybe More?" It is a true contribution—an analytical, reflective, and important assessment of this space. In terms of blockchain community connections, thank you to Anthony Donofrio for introductions and to my dear friend Vikash Singh for your support.

The people, places, and projects featured in *Beyond the Valley* speak to the power of human creativity and the dreams we have of a more equal digital world; one that recognizes our values and rights as sovereign human beings and the communities to which we belong. I'm just sharing all of your stories; you are leading the way. Thank you to each and every one of you.

Foreword

Digital technology is the subject of our time.

It's more than just the topic of so many articles, panels, talks, and books. Our technology serves as the very focus and direction of our society. But the more we look to our devices, platforms, and networks for an understanding of our collective situation, the further untethered we become from the real world in which we live.

We tend to look at technology as the subject of our concerns, and human society—human beings ourselves—as the objects being acted upon. In part this is because we no longer use our technologies so much as our technologies use *us*. With every swipe of our fingers, our smart phone gets smarter about us, even as we get dumber about it.

Indeed, as long as we attempt to understand technology by the content pouring through its many screens, we will remain clueless about its real impacts on our cultural, economic, and planetary environment. We simply become part of a feedback loop between people and a media landscape that is programmed to keep us distracted, atomized, and powerless.

Ramesh Srinivasan breaks this cycle. He recognizes the true ethos of Silicon Valley tech development: earn enough money to insulate yourself from all the harm created earning money in this way. It's a page from the same rulebook that has guided corporations since the era of the British East India Company. The object of the game is to extract as much value as possible by any means necessary, and then externalize the impact to people and places somewhere else. Digital technology platforms, with their illusion of cleanliness and hermetically sealed purity, hide their externalities even better than traditional Industrial Age enterprises. But the human slavery, environmental destruction, social alienation, economic oppression, and civic collapse they engender are just as real. Moreover, they now happen "at scale."

The inwardly turned campuses of our tech monopolies keep their workers' focus on the code, and off the impact. (How can a company live by the credo "don't be evil" if it is not even looking beyond its webpage metrics?) Likewise, when our evaluation of power dynamics in a digital age relies on analysis of the companies themselves, the affordances they are embedding into their platforms, or even structural critique of their surveillance services and business plans, we risk losing sight of the bigger and more relevant picture. We end up trying to understand technology through the lens of the very technologies we are trying to see. And in the process, we make them even more central to the story, relegating ourselves—humanity—to the background. Mere externalities.

That's why it's so important that a scholar of Srinivasan's rigor and insight actually traveled the world to witness, first-hand, the impact of algorithms on electorates, ride-hailing apps on African cities, smart phones on indigenous Mexicans, and blockchain on the environment. In venturing beyond Silicon Valley and its user interfaces, he found rampant, underreported, and actively camouflaged devastation. Digital technology companies are involved in shattering lives and livelihoods, enslaving children, undermining democracy, ruining economies, and threatening the environment.

But he also found people fighting back, subverting the intended functions of technologies to empower themselves and their communities. From Oaxaca, Mexico, where activists built community-owned digital networks that promote the solidarity and autonomy of indigenous people, to Sub-Saharan Africa, where people and small businesses leverage the power of networks not to compete but to collaborate and exchange value. By subverting the extractive intentions of the technology companies seeking to colonize and coopt them, these people and places are staging a revolution against the dominance of global corporate capitalism.

Those of us insulated by our technologies, our privilege, or our Western bias, owe it to ourselves to learn from these efforts. If we model these same approaches, we stand a chance of restoring our own collective agency, and avoiding civilization-scale catastrophe.

By reporting from the ground up, Ramesh Srinivasan reminds us that in a world of digital domination, we are all indigenous people.

It's time we act that way.

Douglas Rushkoff

New York, February 2019

Introduction

Sergio, a friend of mine from Argentina, recently completed a trip around the world. I met him last year in southern Mexico, where I've been visiting regularly since 2016 to learn from indigenous communities who have been creating their own digital networks. My visits there were inspired by my two-decade-long fascination with the internet and digital technology's rapid development: first as an engineer and, later, as an educator and researcher exploring how digital technologies impact the lives of diverse cultures and societies.

Sergio had not traveled extensively outside of his country before. When we met at a food vendor's table on a street in Oaxaca, he was buzzing with the excitement of having visited five continents and holding various jobs (as a carpenter and farmer, for instance, and in the service industry, at hotels and restaurants) in just the past year. When I asked him about his range of experiences in parts of the world so different from where he grew up in working-class Buenos Aires, he made an offhand remark that stopped me in my tracks: "Regardless of where you go, people get pushed into the internet. That is what they want."

After we parted ways, Sergio's comment stayed with me. I was struck by the contrasting but intertwined images it contained: the first being of people thrust into some alternate universe that exists beyond our screens; the second being of people who relate to the internet as the place where we share common ground. Like many of us, Sergio had the perspective of a user, simultaneously awed yet also perplexed by the characteristic "pushiness" with which websites, mobile apps, virtual reality headsets, and myriad other digital devices have appeared and made their way into our lives.

I wrote this book as a way to explore how we, the billions of internet users, can respond to the sea change that transformed an open world of online possibility into something else altogether: a digital landscape of walled gardens and predetermined paths already programmed for us, for which we have little visibility or control. If we are going to turn over our lives to these devices and systems, shouldn't we, at the minimum, have power over how they are designed and profited from? Shouldn't technology be people-centered, not in use and addiction, but in creation and application?

As access to the internet and cell phones expands around the world, so too does the power and wealth of but a few technology companies located on the West Coast of the United States (Silicon Valley and California) and in China. These companies offer services that have, without question, created value for their billions of users, providing efficient and economical ways for us to find and disseminate information, telecommute to our jobs, socialize with one another, and buy and sell goods. Many of us love the cheap prices on Amazon and how quickly the products we purchase arrive at our doorstep. Others believe that things like surveillance cameras, GPS tracking of people using smartphones, and the use of shared data to apprehend terrorists provide great value. But we may not also recognize that in the process we give up our individual rights to privacy, put the most vulnerable members of our society at risk, threaten the viability of small businesses, and contribute to greater economic inequality and political division.

In domesticating the "wild west" of the internet, the big tech companies have provided a vibrant market of beneficial tools and services for users and amassed unimaginable wealth for their executives and stockholders. But in the process they have also supplanted the open, democratic internet with a vast network of privately owned architecture—imagine digital fences, highways, roads, bridges, and walls—whose sole purpose is to control people's movement through digital space in ways that benefit their companies' bottom lines.[1] Clinging to this idea of an open internet obscures the reality: The digital world is structured by static, hefty, and inflexible digital architectures. And like invisible borders, they enforce specific paths.

We are increasingly aware, of course, that the phones and websites linking us to a wider world keep track of us, even when we are not using them, and that our personal data is vulnerable to covert collection and monetization. We have also experienced the amplification and rapid spread of propaganda, misinformation, and hate speech on the internet, too often with tragic results, all of which distracts us from facts, contexts, and multiple points of view. Meanwhile, journalists, activists, scholars, and more of the public are raising concerns that Chinese, Western, and white male interests dominate the content and systems that power the internet rather than those who reflect the full diversity of us online. Finally, against the backdrop of profound economic inequality across the world, we have seen how the expansion of the gig economy has made work, wages, and benefits less secure than ever before, even as digital automation threatens to eliminate much of the current job market.

I, like so many of us, derive incredible value from the services and products provided by Amazon, Google, Apple, Microsoft, and Facebook. But I also know that I have agreed to their terms of service without understanding exactly what I have given up in exchange for what I am getting: I am not being given the means to think through the downstream implications of my actions even for myself, much less for others in my networks. As an example, Amazon has made it so inexpensive and easy to buy almost anything, that almost all of us use the site regularly. The Chinese Alibaba is no different. But is it okay that both these companies have monopolistically overtaken an open marketplace? What about Amazon's facial recognition technology being sold to military contractors, the police, and to support President Donald Trump's actions with the Immigration and Customs Enforcement (ICE) agency? Are we okay with these sorts of transactions, similar examples of which we can find involving every powerful tech company?

How many of us know that the internet itself largely came to be thanks to public, taxpayer-funded research? So how come the spoils of our public investments, the trillions of dollars associated with Silicon Valley, have only lined the pockets of uber-rich investors and executives? Somehow we gave away all the money and power to the 1 percent, a secretive cadre of middlemen.

Don't get me wrong, the value provided by the most powerful and ubiquitous technologies is unmistakable: Facebook, Google, Apple, Microsoft, and Amazon have created efficient, helpful, and beautiful services and objects most of us wouldn't want to live without. And these Big 5 tech companies are not the only major players. Although I don't discuss Chinese tech giants in detail in this book because I couldn't gain sufficient access to information via first-hand interviews, they are equally pervasive in their presence and influence, particularly in Asia. In 2018, China took 9 of the top-20 slots ranking internet leaders of the world, with their businesses and products reflecting the kind of design and performance values championed by Silicon Valley.[2] In the e-commerce sector, Alibaba is the No. 1 retailer in the world with profits and sales surpassing Amazon and eBay combined.[3] It too, has worked toward expanding and synthesizing its services, from cloud computing to offering virtual reality shopping experiences for its users.[4] On top of it all, China has just launched an Artificial Intelligence (AI) news anchor, according to state news agency Xinhua. The *Guardian* reported that Chinese viewers in November 2018 were greeted with a digital version of regular anchor Qiu Hao, who promised them, "Not only can I accompany you 24 hours a day, 365 days a year. I can be endlessly copied and present at different scenes to bring you the news."[5]

I wrote this book not to merely criticize or raise alarms, but to advocate for and illustrate a future where the connectivity and the online services we love don't come with the costs of surveillance, income disparity, false information, and extreme imbalances in how technologies are designed and deployed. The possibilities are vast, and we can learn a great deal from the whole range of strategies and efforts already underway. In order to put the power of technology to work for the common global good, we must identify what is dangerous about the current arrangement, and mobilize a vision for positive change.

The challenge doesn't just face the big tech companies but any who presume that their so-called neutral systems always do good, even when their troubling effects may look more like social engineering. For example, consider the company Faception's attempts to predict IQ, personality, and violent behaviors through the application of "deep" machine-learning techniques to facial features and bone structure.[6] We've made such horrible mistakes before, for example treating black people's faces and body types as supposed evidence of their inferiority. Are we about to justify continued racism thanks to "machine learning"?

In *Beyond the Valley* I show examples of technologies from around the world that balance efficiency with values of equity, democracy, and diversity. I tell stories of global innovators that point the way toward a humane and balanced internet of tomorrow. I explain the perils that occur when a narrow understanding of efficiency becomes our primary focus and ironically creates inefficiency. Efficiency on our consumer platforms can make all of our lives inefficient by disturbing our sense of security and privacy, interfering with our democracy, and furthering economic inequality. Efficient industries can create the inefficiency (and massive threat) of climate change. And similarly, the blind embrace of efficiency can result in tragic hidden costs to vulnerable humans—like when the reliance on AI-powered safety-features that could not be overridden in an emergency resulted in fatal plane crashes in 2018 and 2019.

To address these challenges I tell stories of entrepreneurs creating technologies with a social mission, users pooling resources and ideas to support grassroots politicians and ethical causes, and communities coming together to own digital platforms. In all of these cases the idea is for people, not private tech companies, to design their own networks and services based on shared values and belief systems.[7] These small-scale, environmentally conscious, user-governed, and often decentralized efforts are examples of what some call "appropriate" or "people-centered" technology. I also share examples that are less widely established, but which are on the cutting edge of innovation, such as privacy-protecting systems, universal basic income, portable benefits, union organizing, digital cooperatives, worker councils, and more.

These are not all new ideas; they exist around us and it's time to see what we can gain by paying attention to them. But we can also do more, starting with demanding that governments and big tech companies be more accountable and communicative with their users. We can also work with engineers who are not trained to analyze the social, political, or economic effects of the systems they create to develop a more reflective and inclusive design process. We have a range of positive directions in front of us, and they can drive the change needed for a connected world that is more just, more representative, and more diversely democratic.

Part of pursuing this future is pursuing the internet we were promised, but haven't yet received: an internet that acts as a "global village," bringing us all together; an internet that creates, or at least supports, equality; an internet that lifts all boats. The coming pages will look at the past and the future, around the world, and at our cultural, political, and economic lives to point toward a digital future that supports diversity, democracy, and the belief that our collective and individual welfares are interwoven.

I Intrusive Tactics: Tracked, Hijacked, and Hooked

1 The Power of Data

In November 2017 my Apple MacBook Pro was stolen. It happened on the island of Ibiza, where I was visiting my friend, a documentary filmmaker. We were thinking of making a series of short films to illustrate innovative uses of new technology around the world.

We sat together writing down examples of technological innovation, subversion, and creativity we'd already witnessed. We discussed themes I explore in this book such as surveillance, the gig economy and the future of work, artificial intelligence (AI), and cryptocurrencies. After uploading our ideas to Google's cloud-based storage platform, Google Drive, we took a short break, locking our laptops in the trunk of our rented car. When we returned, both laptops were gone.

Although upset, I was also relieved. I'd be able to recover most of my data because almost all of it was in the cloud: that was what I signed on to when I stored my files with Google (as well as with Apple, Microsoft, and Dropbox). But in the bargain I'd allowed them to profit from my data. It dawned on me—given the "terms of service" I'd agreed to—that the value of my data to the storage companies was far greater than the value of my laptop to the thief.

How many of us are fully aware that companies providing us with cloud storage have ways to monetize our data? They use targeted advertising, for instance, and charge fees, once we get hooked on the "security" and peace of mind their free services offer. We put faith in these companies to store our files securely. But our private information may no longer remain private if, let's say, the government asks to search through our data. It's also not private because all of this data is surveilled, computed, and (often) manipulated in mysterious ways to influence and shape our behaviors as these corporations see fit. By creating accounts with these companies, we give them access to ourselves. So when we consent to privacy policies, in reality we are signing on to systems of surveillance never before seen in human history, and we do so without being included or consulted. By desiring to keep our data accessible and recoverable, we lock ourselves into using their services.

As I thought about getting my files back in Ibiza, I wondered: Does access to my own property have to come with such strings attached?

In my relationship to Apple, I found the strings impossible to untie. By purchasing an Apple laptop, I had been unwittingly transformed from a customer into a part of the company's product line. I had provided Apple with private data and trusted Apple to protect it. Now, to retrieve that data, I had to purchase a new laptop with Apple-owned peripherals and devices.

So I caught the next flight to Madrid. I then took the train to the city's Apple Store located in Puerta del Sol (see figure 1.1). In many ways this central plaza still serves in its centuries-old role as a landmark and gathering place for the entire Spanish nation, a point from which numerous roads emanate.

With its distinctive aesthetic and its strategic placement in a civic space, the Apple Store stood out among the adjacent parliamentary and government buildings as if the global brand were more powerful than Spain itself. The day I was there more people entered and left the Apple Store than passed through the doors of any neighboring building.

Figure 1.1
Madrid's well-trafficked Apple Store (source: Gumcam).

In 2017, Apple—a private corporation worth more than a trillion dollars—actually began to conceptualize its retail stores as "town squares," thereby branding itself an integral part of our everyday lives.[1] Other big technology corporations describe themselves in similar terms. Facebook CEO and founder Mark Zuckerberg, for instance, refers to the company as a "social infrastructure" for the "global community."[2] So what we have in the digital world are privately owned public spheres. They are branded as public, civic, and virtuous but in reality are dominated by a single logic—extending profitability and economic value.

How can these companies' interests be the same as the "public" interest? Like many other large corporations, Apple is shareholder-owned and therefore primarily accountable to its investors and executives, rather than to its employees, customers, or the general public. I had not begun to unpack the meaning of all this until I realized how monumental the Puerta del Sol Apple Store actually was, as an icon and a structure.

As I walked into that store, an Apple employee invited me to purchase a new machine, log onto my iCloud account, and retrieve from Apple cloud storage the data that I "lost" from my stolen laptop.

These transactions, completed in less than an hour, were simple and efficient. My data was "extraterritorial," easily retrievable in Spain, and likely would be in most other countries. Purchasing a new laptop and getting my data back supported Apple's economic model. National borders were no threat or hindrance. The same was true with retrieving data I had stored on other corporate cloud servers managed by Google and Dropbox.

Who Is the Product?

In technology circles, this saying has made the rounds: *If you're not the customer, then you're the product.*

Many of today's technologies come to us for free. But in return our personal lives come "for free" to these technologies. They are accessible for the companies' profit and gain. Yes, when we search Google we look for information, but do we realize that Google is also searching us? Users create the value for other users, and thus for the companies who own the technologies. But we, their users, have become the product in another sense: the companies who provide our "free" platforms and services have figured out how to make money by using our data and capturing our attention. It took them a while to do this. That they have done it—and especially how they have done it—really matters.

In her 2016 book *Weapons of Math Destruction*, Cathy O'Neil explains that algorithms, the building blocks of most online systems we use, are merely mathematical expressions (or models) repeatedly applied to any domain or task.[3] Private corporations often create these algorithms in secret, even though such algorithms affect the lives of us all, the public. Algorithms, says O'Neil, are "opinions embedded in code … [not] objective and true and scientific." It's a marketing trick, she contends, "to intimidate you with algorithms, to make you trust and fear algorithms because you trust and fear mathematics."[4]

Disguised algorithmic power can lead to what the scholar Judith Butler has referred to as the "foreclosure of identity," the power to define a person in a way that closes off other possibilities.[5] How does that happen? The collection of our personal data occurs at all moments and places, whether or not we willingly provide it. It isn't limited to the time we spend on Facebook or YouTube. Companies of all kinds aggregate our credit card purchases, court records, mobile phone activity, and medical histories to create models of who we are and what we do. These revealing profiles, which we never see, are used to predict, influence, and shape our behavior.

At risk here is our autonomy as citizens. If our profiles determine what content, advertising, and recommendations we see, what options and choices have we lost? The ability to change over time, the ability to experiment, develop, and become "ourselves" in a way that does not map onto existing categories—and even the ability to forget something—are in the balance.

Algorithms and platforms "understand" us through our past behavior, not our future possibilities. Their tools and objectives are computation, permutation, and efficiency rather than creativity, imagination, or soulfulness. If we aren't careful, we might just find ourselves annexed by this mode of being, unable to develop in ways that set us wandering, exploring, or simply thinking in a different direction than the ones mapped out by computers. The more data we provide as we carry our mobile phones everywhere, the more the door is open to intervention over how we behave and who we will become.

As we blog, comment, and post, we practice what the political philosopher Jodi Dean describes as a "circuit" of pleasure, production, and surveillance.[6] When we go online, we feel as if we are freely self-expressing, connecting with billions of fellow users. But behind the spectacle of expression, data is captured, attention is monetized, and surveillance is legitimated. Just as labor fed the industrial revolution's inequalities, the capture of data fuels the digital capitalism of today. Dean sums it up in her description of the common digital experience: "More circuits, more loops, more spoils for the first, strongest, richest, fastest, biggest."[7]

As a result, an overwhelming amount of information begged for services like search, social media, and even GPS to help to sort through the noise. In helping us feel less overwhelmed, however, algorithms decide what to include and what to exclude, what to push to the top of a feed or a list of results, and what to bury. In doing so they tell us, invisibly and in the background, what we should see and who we should communicate with. They tell us how to understand our political, cultural, and global reality. They define how we think. They have transformed how we see ourselves. The blinders of privately owned web platforms both filter and compress our glimpse into the "open" internet. This is how the internet has become a set of privatized spaces for the capture and surveillance of our data and attention. Indeed, data today is understood by experts to be the "oil of the digital economy," something to be collected, scraped, aggregated, harvested, and speculated upon.[8]

By what key step did this happen? Foremost we allowed monitoring and surveillance on a scale never seen before; *even a network of sophisticated surveillance cameras is no match for the phones in our pockets.* Google, for example, is able to monitor us not just when we are on its homepage but in many other ways, even when we are offline. The company is generally vague about how it does so, and even quieter when it comes to what it does with this data. And this is not merely a story about Google, as monitoring and surveillance are ubiquitous when it comes to any of the largest social media, search, or e-commerce companies. We have accepted the loss of our privacy and autonomy as part of a bargain that gave us efficient, free technology.

There is no single correct answer to the question of how to balance surveillance with one's values; and besides, human values are often more diverse than similar. Whatever values hold people together—in a relationship, in a community, or in a place—these values are worth protecting. This *doesn't* have to mean giving up the efficiency, connectivity, or utility the internet has brought us—not at all. But it does mean reconfiguring the *rela-tionships* between tech companies and users so that the power of increased efficiency and connectivity promotes the social good.

As of now, billions use technologies created by a few companies located in but a sliver of the world, whether on the West Coast of the United States or the powerful cities of Shanghai, Beijing, and Hong Kong.[9] As a result, the cultures that produce popular technologies are not nearly as diverse as their users. This cuts off a majority of users from participating in building, own-ing, or designing the technologies that will shape our global future.[10] That is just one example of an imbalanced relationship between producers and consumers at the heart of the tech industry today.

But it doesn't have to be this way. I will show you examples from across the world that demonstrate how diverse cultural values influence the cre-ation of technologies and the ends they serve. This is all the more impor-tant because 54 percent of the world's internet and mobile phone users no longer live in the Northern Hemisphere. And the percentage will increase as attempts to expand connectivity and access move forward.[11]

Jaron Lanier, a pioneer in the field of virtual reality (VR), believes that we—as citizens, users, and human beings—must bear some responsibility for the predicament we face. "The solution," writes Lanier in a 2017 *Guard-ian* article, "is to double down on being human."[12] In his recent book *Who Owns the Future,* Lanier calls us to action. "The only way to say no" to the deal of participating in the technologies offered by the "Big 5" tech compa-nies, he states, is for users of the internet to "transcend the role of consumer once in a while."[13] At the TED 2018 conference Lanier elaborated on the "being human" factor in the digital age: "If we want our species to survive, we cannot have a society in which, if two people wish to communicate, the only way that can happen is if it's financed by a third person who wishes to manipulate them."[14]

Tim O'Reilly, a revered figure among technologists and the man who popularized the term "Web 2.0," has quipped that there is a master algorithm in our society: optimizing shareholder value over all else.[15] O'Reilly implies that with the handover of our data, time, and attention, the biggest tech-nology companies in the world can mostly ignore the political, social, cultural, and economic effects of their technologies. The ultimate account-ability of a publicly traded company is to its shareholders, as measured by its market cap. But is ignoring the negative consequences of financial growth the only way to pursue economic success? It need not be a zero-sum game.

I contend that human communication—our need to connect, socialize, and shape our worlds—is a resource. Like timber or water, we can use it responsibly or irresponsibly. We have fought for cleaner energy and better protections for animals and lobbied against sweatshop labor. We have seen all of these wins as "progress." So why should we passively accept the specific version of the internet that corporate imaginations have handed us? Why can't we imagine another version—an internet that provides as much value to us users as ever, but which doesn't exploit us or endanger our livelihoods?

The Manufacturing of Consent

In 1988 Noam Chomsky, a prolific scholar/author in fields ranging from linguistics and cognitive science to political activism and social criticism, co-wrote *Manufacturing Consent: The Political Economy of Mass Media* with Edward Herman. The authors argued that mass-media platforms, though seemingly pointed toward an earnest search for truth and justice, actually served as propaganda machines for the state and corporations. They claimed that the media industry's underlying profit models impacted what readers viewed and how stories were framed.[16] What does Chomsky think of these arguments today?

When I met Chomsky in March 2018, he still spoke of major media as a huge corporation whose goal is to develop "a product for the market, for other corporations."[17] Readers and viewers themselves are the product, said Chomsky. "That has not changed."[18] One lesson that *Manufacturing Consent* can offer today concerns how power operates when media systems interact with people. What happens when citizens are both the customers and the products of private, for-profit corporations whose primary allegiance is to their bottom line? An industry with this structure will tend to produce propaganda, bias, and a lack of diversity or objectivity.

On April 13, 2018, Chomsky posted the following tweet: "The most effective way to restrict democracy is to transfer decision-making from the public arena to unaccountable institutions: kings and princes, priestly castes, military juntas, party dictatorships, or modern corporations."[19] This is consistent with a point that Chomsky has long made—that private, unaccountable institutions are the most central threat to democracy, and therefore to values of political and economic equality.

But what about online news content today? Facebook, Google, Instagram, YouTube, Twitter—these are not the media giants Chomsky initially described. But they are similar to those he critiqued three decades ago because they are largely unaccountable institutions. None employ reporters; therefore the news stories they disseminate are rarely sourced or framed. Nevertheless, these corporations not only serve news to their users from many of the traditional institutions Chomsky critiques, but they also use algorithmic formulas that wield power over what we see. Because these formulas are generally optimized to confirm our existing viewpoints, they run the risk of limiting rather than opening our worldviews. Matt Taibbi, a journalist who has also analyzed the relevance of *Manufacturing Consent* for the twenty-first century, notes that Chomsky's stress on big media reportage as "narrowly constrained" is even more pointed today as frustrations with the media have become bipartisan. And, he says, "There's a whole generation of people that consider themselves to be progressive who don't understand this concept of an artificially narrowed reality."[20]

What we see online profoundly influences what we believe. Studies that confirm this also reveal other ways in which online content, such as top Google search results or Facebook newsfeed posts, affect users.[21] Research on the neuroscience of internet addiction reveals that online activity can release dopamine into the ventral tegmental area of the midbrain, which in turn releases the dopamine hormone into the brain's pleasure centers.[22] This can generate addiction and compulsive behavior, which internet companies can exploit through the design of their algorithms and user features. As a 2012 article from the *Atlantic* points out, "the release of dopamine [also] forms the basis for nicotine, cocaine, and gambling addictions."[23] Although dopamine's role in our obsessive behaviors is commonly known by now, it may be less obvious that the amount of time we spend on addiction-optimized internet platforms is just as unhealthy as junk food or cigarettes. It is easy to see technology and the internet as neutral, as a source of knowledge and a means of communication. But that doesn't accurately describe our position as users, and therefore consumers, when it comes to for-profit technology companies and their services.

Much as the media industry of the late twentieth century transformed reading and viewing customers into profit-making products, we, as digital tech consumers, remain the most valuable product online today. Our data and attention produce profits in ways that draw from traditional advertising models.[24] The concerns Chomsky and Herman raised more than thirty years ago have become magnified. As new technologies, mass media systems, and governmental systems work together to "manufacture consent" through the addiction of users, they naturalize and normalize online life as part of the "order of things," perpetuating invisible, seemingly neutral, systems of power and control.[25]

The Power of the Big 5

Today the major internet and digital technology companies are more powerful than ever. As of 2017, the Big 5—Amazon, Apple, Google, Microsoft, and Facebook—were worth more than $3 trillion, with each company's stock prices growing between 30 and 60 percent during 2016 and 2017.[26] These companies give us the benefit of experiencing the "open" digital world, but they do so on their own terms. Yes, there are many examples of how platforms like Twitter, Instagram, and Facebook have been used to support movements like #MeToo, Black Lives Matter, and Occupy Wall Street.[27] But more often we see the role of Western technology companies in society defined by their proprietary control over the dominant tools and systems we use to communicate with one another, search for information, buy, sell, watch, or read.

I recently discussed these concerns with Siva Vaidhyanathan, author of *Antisocial Media: How Facebook Disconnects Us and Undermines Democracy.* "Each of these companies is trying to become the operating systems of our lives," he said. "We've already seen the battle to be the operating system of our laptops; that's over, that's stale, not interesting. And then [the fight over] the operating systems on our mobile devices…got old, stale. Now [these companies] want to be the operating systems of our bodies."[28]

In 2016 alone, several stunning statistics involving Amazon underscored Vaidhyanathan's point: The company accounted for 43 cents per every dollar spent online in the United States, horizontally integrating to take over retail, logistics, web services, warehousing, and numerous other markets.[29] Amazon has achieved this dominance through its brilliant and unrelenting consolidation of a sprawling marketplace. The aggregation of attention online is equivalent to market share, and Amazon has accomplished this with its strategy of low prices, free shipping, fast delivery, and seemingly endless selection. Its customer base is so large and so loyal that no brand and few retailers—including Amazon's direct competitors—can afford to compete. As Brad Stone documents in his book *The Everything Store*, Amazon is ruthless and retaliatory in its treatment of these partners, who cannot afford to push back. Amazon now owns the customer base—otherwise unreachable, even by the largest companies.

Amazon is also a dominant player in the domains of cloud computing and server storage, boasting both the oldest and largest cloud infrastructure.[30] Some estimates claim that its infrastructure has more than ten times the computing capacity of the next fourteen-largest vendors combined, meaning that when one puts content online, it is often interlinked with some Amazon service with storage capabilities both online and off.[31] Think of Acme, the fictional Looney Tunes conglomerate that made and distributed products of all types imaginable to Wile E. Coyote and you'll have a comparable sense of the ubiquitous Amazon's scope.

In an interview published in April 2018, Amazon CEO and founder Jeff Bezos told a Business Insider reporter that funding his for-profit space company, Blue Origin, by liquidating a small fraction of his (then) $127 billion personal wealth—that's a billion dollars' worth of Amazon stock per year, to be precise—is "his most important work." Indeed, in a frank moment he admitted, "The only way I can see to deploy this much financial resource is by converting my Amazon winnings into space travel."[32]

Earlier that month the US senator Bernie Sanders, in conversation with CNN news anchor Jake Tapper,[33] weighed in with concerns that Amazon had gotten too big. Several months later on Twitter, he pointed out that although independent retail is declining we see Amazon entering into almost every area of commerce.[34]

Three years before Franklin Foer published his 2017 book *World Without Mind: The Existential Threat of Big Tech*, he profiled Amazon in an article for the *New Republic*: "Amazon is *the* shining representative of a new golden age of monopoly that includes Google and Walmart," Foer declared. These behemoths have become "self-styled servants of the consumer," and we enjoy their benefits without much considering the long-run dangers.[35] In Amazon's case, the company has already threatened competitors, forcing dependence or bankruptcy for independent businesses of all kinds. And this activity stands aside from Bezos's troubling ownership of the *Washington Post*, one of the most well-known US newspapers.

Low pricing and free services are a hallmark strategy of the biggest tech companies, and it's easy to see why. But it's less easy to see the effect of this strategy on employment, economic security, open markets with equal opportunities, and politics.

Ethan Zuckerman, a scholar of civic media and technology, recently told me that the striking similarities between Amazon and the massive retailer Walmart, also a major political lobbyist, have to do with their functional positions in the marketplace as part of the Big 5 tech companies. Both operate as "the intermediary," forcing customer and supplier dependencies on them, shaping an economic climate where they remain in a central position of power:

> Amazon is high-tech in its front but it's Walmart in the back. It's the mullet of tech companies. It looks like a Silicon Valley company up front, but it's really a Walmart-like, drive-it-to-the bottom ideology at the back.... [Amazon] has basically said, "We just really don't care, you can sell whatever you want. If you get into sufficient trouble, we will yank you. But what is more important to us than anything else is to be the store for literally everything.[36]

Google and Facebook have cornered the digital advertising market, controlling over 70 percent combined.[37] Each occupies an iconic place in their respective domains, with more than 2 billion users worldwide. Google processes over 40,000 search queries every second on average, or over 3.5 billion searches per day and 1.2 trillion searches per year worldwide.[38] Facebook, which owns the video and social media network Instagram, and the messaging service WhatsApp, has become the dominant platform not just for social media, but for news and current events as well.[39] Largely thanks to the iPhone, the number of Apple device users has now exceeded 1 billion across the world.[40]

These facts justify the importance of looking behind the curtain at each company. For example, in considering Facebook's ownership of Instagram, WhatsApp, and Oculus, we should scrutinize the behavior of the company beyond antitrust or regulation concerns.[41] It would be naive to trust WhatsApp's promise of "end user" encryption, knowing full well that the data still goes to Facebook, when Facebook has not shown itself to be trustworthy when it comes to personal data.[42] Overall, to understand the greater landscape in which technology corporations exist and operate, we should consider the specific technologies we use as well as the portfolio of the mother company.[43]

The Big 5—or the "frightful five," as Farhad Manjoo, the *New York Times* tech columnist refers to them—are better seen as global empires rather than visionary technology companies.[44] They exert significant control over our present, yet the true fright comes imagining a future where their systems and tools become more closely embedded in our lives, our homes, and bodies. The 24/7 listening devices called Google Home, Amazon Echo, Apple Siri are designed to maximize profit through surveillance. It is nearly impossible to turn them off, disable them, or even understand what they listen to and how they learn.

2 The Social Contract

In recent years digital technologies have reached ever deeper into our lives. That's why we must begin to think of them not as one-size-fits-all bundles of tools and systems, but rather as specific sets of agreements. When we use a given technology, we "accept" it via a legal contract clogged with terms too lengthy and complex for anyone other than trained attorneys to decipher. This contract, however, disguises that we are signing onto is a relationship in which we have no power. That's why it's important to consider the promises and perils associated with the decisions we make about digital technology.

David Ige, the governor of Hawaii, once explained that he was not able to notify his state's citizens of a missile alert (that turned out to be a false alarm) because he had forgotten his Twitter password.[1] Situations like these reveal the troubling extent to which we have delegated our individual, social, economic, and political lives to private and corporate technologies. The fate of a nation can rest in the hands of a politician's accessible Twitter account, yet what do we really know about Twitter?

What if we turn from the impact of digital technology on civic activities and look at the pervasive use of our biological data? Is the outlook any more reassuring? Recent iPhone models offer identification technology such as thumbprints and retinal scans. Apple's healthkit collects medical data as well. What about genetic identification technology itself, such as 23AndMe, a biotechnology and personal genomics company providing customers with DNA-testing results?[2] This company, founded by Esther Wojcicki, the ex-wife of Google co-founder Sergey Brin, warrants scrutiny, and not just because our DNA data has the potential to be used for wrongdoing. How comfortable can we be when our data is accessible to a private corporation, particularly one so closely connected to Google?[3] Should not the burden of proof be upon the company to disclose how the data is being (or will be) used? So far 23AndMe has established a track record for breaking promises when it comes to privacy: for example, its British subsidiary Deep-Mind Health reneged on the agreement to keep data separate and therefore private from Google accounts and services. Why would the company do this? Because more data is almost always seen as better from the perspective of the company even if it threatens previous pledges to users and their privacy.

Ray Kurzweil, the futurist and former MIT professor who now works for Google, has written and spoken about transforming life with a vision toward immortality by merging our consciousness into machines, thus overcoming the limitations of the aging human body.[4] He believes in a future era called the Singularity "in which the pace of technological change will be so rapid, its impact so deep, that human life will be irreversibly transformed.... This epoch will transform the concepts that we rely on to give meaning to our lives, from our business models to the cycle of human life, including death itself."[5] Kurzweil, like many futurists primarily concerned with advances in technological development, treats the Singularity as a given, rather than as a possibility to be questioned or reflected upon relative to the world's realities.

Lacking such discussions, we march forward myopically. With DNA and a massive trove of digital data available, has the time neared where "intelligent" android machines can be created that somehow resemble us? Are we okay with this, and are we even aware that these developments are ones we may have signed onto?

Sure, humans and machines have increasingly fused, making science fiction shows like *Westworld* or *Black Mirror*, movies like *Blade Runner*, *Minority Report*, or the *Matrix*, and books by cyberpunk authors like JG Ballard, Philip K. Dick, and William Gibson seem more realistic than not. These works present a world that has witnessed for better or worse a breakdown of traditional social orders. In that world powerful forces have blurred the lines between truth and lies, control and influence, freedom and slavery, sickness and health, even pleasure and pain. These stories show what it may mean to grapple with computational systems we don't necessarily understand. They teach us that as we delegate increasing control to these systems, we may also lose narratives of the future that are rooted in principles of equity and democracy.

The award-winning BBC filmmaker Adam Curtis recently spoke about what we can learn from the political and technological upheaval that has occurred since his 2002 documentary *The Century of Self*. In that four-part film he explored a theory popularized and practiced beginning in the 1920s by Edward Bernays, Sigmund Freud's nephew and the so-called father of public relations. Bernays believed that consumerism could be used to control people's irrational desires, leaving them susceptible to fascist and nationalist propaganda. Today, Curtis holds that Bernays's concept of humans as "very simple irrational robots...managed by...machines and behavioral psychology" has permeated everything, especially with the rise of social media that "manage you beautifully, yet at the same time...allow you to keep feeling like an individual."[6]

To paraphrase the philosopher Friedrich Nietzsche, those who fear the rise of the robots should beware not to become robots themselves.[7] As these digital systems increasingly interact with themselves and one another, they risk being even more opaque and detached from human understanding or control. As Curtis points out, "We live with the ghosts of our own decisions played back to us in slightly varied forms every day and we don't believe in anything, we have no vision of the future, or big vision of any alternative future to this—we just live with it."[8]

Perhaps the major technology companies have been able to maintain power over our current state of mind because they have presented us with a compelling, well-developed, and attractive definition of technology and a vision of the future: their brand images. Through branding they have carefully cultivated the popular perception that they are cutting-edge, innovative "disruptors" that will shape a fun, efficient, and meritocratic future. They also advertise their intentions to build the "social infrastructure" that will bring local and global communities together (Facebook), to "organize the world's information and make it universally accessible and useful," and to "[not] be evil" but rather assimilate information into something like the "mind of God" (Google). Apple promises to "think different." Amazon pledges to be "the most customer-centered company in the world," a place where one can "work hard, have fun, make history." Microsoft vows to be "what's next."[9] In these statements the big technology companies claim to be not only be the place for global community but also to serve as divine innovators and inventors of the future. But when the "fun" stops and the "evil" sneaks in, when their technologies have unintended, negative consequences—when they hurt people—these companies are largely unmoved. They do not change their business models, their penchant for secrecy, or their often short-term, profit-oriented thinking. Suddenly, when called to account for the harms of their technologies, these history-making masters of the universe are merely "providing a platform" and cannot be held accountable for the ways in which it is used.[10]

Instead of viewing these companies as intransigent and ourselves as powerless, we should fight for technologies that better reflect our values. The first step is to do away with the myth that technologies are value-neutral, or somehow inherently positive, and instead demand greater transparency around the values that drive the design, engineering, and business choices these companies make. When those values serve the well-being of executives and stockholders but fail the users, it's time we demand change, reconsider the technologies we use, and scrutinize businesses we support.

Is Technology Value-Neutral?

"Guns don't kill people, people kill people."[11]

If you live in the United States, you've likely heard this statement before. It implies that values are not inherent to a technology but exist only in how that technology is put to use. It's similar to the myth that technology is neutral. Consider the example of Alfred Nobel: His efforts to provide a safe use for highly explosive liquid nitroglycerin resulted in his invention of dynamite. But the subsequent benefits of dynamite for the mining and demolition industries could not justify fatalities that occurred during experimentation. When the "misuse" of dynamite by the military to power cannons during the Spanish American War led a French newspaper to nickname the inventor, a self-proclaimed pacifist, as the "merchant of death," Nobel felt compelled to establish the prizes in his name to improve his legacy.[12]

Nobel's invention of dynamite raises a question about the *purpose and potential driving the development of any technology*. Nobel aimed to improve the world, and therefore his own morals and values can be judged accordingly. But even so, we should ask how successful he was in achieving this goal. Substances such as dynamite may require outside agents to activate their potential. But nonetheless they are capable of great destruction and can be designed with that goal in mind. So it is safe to say that such technologies are value-laden, not neutral.

Although it is true that technology *itself* does not make moral decisions, the way it is designed can enable and encourage certain types of behavior (such as sharing life events on social media). Or it can prohibit our ability to act (for example, Apple's refusal to allow any hardware modifications to a product after it is purchased impinges upon our ability to repair an older laptop).

We should consider the ambitions, profits, and models of growth associated with various tech companies rather than assume that they are committed to socially benevolent values. Not surprisingly, many companies starting out with a social mission end up driven by revenue models antithetical to that mission. This is no less true in tech than in other industries. The incredible reach of technology into our everyday lives, however, makes it uniquely powerful.

The ideologies that govern society, the same ones encoded into algorithms, are responsible for society's inequalities, and too often they are supported by those who hold power within it. Consider examples of bias built into facial recognition systems like the one in the iPhone X, which couldn't tell multiple Asian faces apart in China. What about the Russian-built FaceApp, which turned President Barack Obama's face whiter?[13] (See figure 2.1.) And then look at what happened when Microsoft created an AI chatbot named Tay and set it up on Twitter (TayTweets @TayandYou). The company quickly shut the bot down because its tweets turned racist, homophobic, xenophobic, and sexist within a matter of a couple days, thanks to

Figure 2.1
FaceApp's "whitening" of Barack Obama (source: TechCrunch).

its ability to "learn" from the discriminatory and partisan world where it was introduced.[14]

Two principal arguments drive the technological debate. The first sees technology as central to how we live and who we are. Think of Marshall McLuhan, a philosopher and pop culture darling of recent decades, who coined the popular phrase, the "medium is the message."[15] From this perspective, our experiences are *determined by technology*, specifically how we receive information. Television differs greatly from radio, for example, in terms of how it shapes our attention and immersion. McLuhan's expression reveals that the way a technology is built or designed—its very form rather than what we understand as its function—shapes our experiences as users or viewers.

The second perspective, which I explore in this book, places greater agency in the user, viewer, and consumer of information to interpret, remake, design, and take control over the technologies and systems we use. Communities, users, and cultures indeed have significant power over shaping technology's impact on our lives. We owe it to ourselves to critically look at any technology, regardless of what it is, and ask the question: What were the values that shaped its design?

Many of the richest tech entrepreneurs in the world today assume that disastrous "end times," in which we scrabble for scarce resources on a dying planet, await us all in the near future. The *New Yorker* journalist Evan Osnos, for example, has written that some uber-rich technologists are stockpiling weapons and food, even preparing underground bunkers. Ironically, for all of their professed faith that technology "solves problems," these tech billionaires seem to question whether their inventions are improving the overall systems in which we live.

In any context, when a few people profit at the expense of the majority, with the cost being quality of life, the situation has the potential to become unjust. The fact that we blindly sign onto contracts with a tech company or an internet service provider reveals our powerful dependency on digital technologies. Acknowledging that dependency is a first step to advocate for alternatives.

3   Foreclosing the Future

As the largest technology companies shift toward greater integration into new markets, such as driverless cars, virtual reality, gaming technology, healthcare, and artificial intelligence systems, we may witness new monopolies in the making.

Monopolies are dangerous for several reasons. First, they limit freedom of choice for consumers, who cannot turn to other options if a company permits unfair price fluctuations on a product (think of the pharmaceutical industry for example) and for laborers, who suffer from unfair wage practices. A monopoly that is too large can change prices at its own discretion and leave consumers with nowhere else to turn.[1] In addition—despite antitrust laws passed in different countries to limit the abilities of monopoly development in the marketplace—monopolies hold concentrated economic and political power. For example, US Supreme Court decisions, in cases such as *Citizens United v. Federal Election Commission*, have granted corporations the power of unlimited political lobbying and the legal status of personhood.[2]

But monopolies equally threaten a democratic society because they can limit discussions about the value systems affecting technological innovation. An open public dialogue about automated cars, for instance, could address questions of how such machines would operate in the public space, how they may be regulated, who would be allowed to own them, and their effects on human labor. But such conversations seem to be either nonexistent or sidelined when it comes to the decisions being made about our future.

In the United States a process called *vertical integration*—the development of various parts of a business to allow it to extend end-to-end from production to consumption—is legally permitted. For example, Amazon produces its own television shows and then sells the services through which the public streams and watches them. But *horizontal integration*—acquiring another company that offers the same services—is typically trickier to pull off in the courts, partly because of existing laws.

Remember when Amazon just sold books? Now it is a monopoly-in-the-making, but in a distinct fashion. Rather than dominating all of a single sector, it provides retail services for everything across the board, from grocery delivery, to entertainment, to clothes, books, and toys.[3] Jeff Bezos himself claims that he wants Amazon to sell *everything*. This kind of business model moves beyond the classic premise of monopoly, in which the best competitor simply wins out in the marketplace. Now we see market competition taken to such an extreme that a single company is consuming the market entirely. Already, as *The Nation* reports, Amazon

> also produces hit television shows and movies; publishes books; designs digital devices; underwrites loans; delivers restaurant orders; sells a growing share of the Web's advertising; manages the data of US intelligence agencies; operates the world's largest streaming video-game platform; manufactures a growing array of products, from blouses to batteries; and is even venturing into health care.[4]

Amazon does have one global competitor, the Chinese e-commerce giant Alibaba. The two are currently in an arms race over investments into AI cloud-based services.[5] Amazon has re-architected its distribution not only of physical goods but also of binary digits, or bits, aiming to become the center of both worlds. Such massive distribution advantages could create a tension with historical antitrust standards. Nonetheless by moving aggressively into logistics, distribution (through automated drones), and even physical retail stores, Amazon has vertically expanded to take power over the entire pipeline: from our data in the cloud to the consumption of physical goods on the ground. Its monopolistic activity plants a foothold across the spectrum—dominance of web services, ownership of the *Washington Post*, and expansion into the physical world of grocery stores such as Whole Foods and new retail stores, as well as delivery technology and warehouses.[6]

Tech giants in the United States continue to transform antitrust laws already in place. In June 2018, the US District Court ruled that AT&T could take over Time Warner Cable, an acquisition challenged by the Justice Department because of horizontal integration antitrust concerns. But these telecommunication companies argued that this move was necessary in order to compete against other corporate giants that pose an even greater risk by monopolizing or controlling prices at will.

With questions of monopolistic power in mind, I reached out to Roger McNamee, a billionaire investor and part-time guitarist who in 2004, after decades of funding a range of technology businesses, formed the private equity firm Elevation Partners. (The music star Bono was a partner in the venture.) Because McNamee considers himself a mentor to Mark Zuckerberg, and is the one who introduced the Facebook founder to the company's current chief operating officer Sheryl Sandberg, the media have sought him out to discuss what's gone wrong with Facebook, particularly in relation to the 2016 United States election. As McNamee explained it to me, we're witnessing a "narrowing of values…a reducing of innovation because [monopolistic corporations] are choking off a lot of other viewpoints that could enter" into healthy conversations and bring to the table multiple perspectives about new technologies, and how they are to be developed and introduced.[7]

As industries from retail to banking have concentrated, similar dynamics shape the consolidation of economic and political power into a few big technology companies. Without diversity, flexibility, or room for creative new voices to enter the field, we are doomed to engineer more of the same: centralized, rarified, and controlling organizations run by a small elite with outsized power. When technologies come from such elite places of privilege, we can imagine how unlikely it is that they serve our interest as users.

As we think about the dangers of giving any private enterprise too much power, consider the 2017 firing of the respected antitrust policy researcher, Barry Lynn, from the New America Foundation. The story has an interesting backdrop: just after the EU's decision to levy its $2.7 billion fine on Google for antitrust violations, Lynn wrote a statement applauding this decision, which the New America Foundation posted on its website.[8] Yet hours later the statement disappeared.

According to an article in the *Daily Beast*, for decades Lynn had been "warning about the pernicious effect monopolies have on all facets of American life: from the food one eats, to the financial system one uses, to the forms of communications on which one depends."[9] Google, according to a *New York Times* article, is one of the foundation's major donors, and Eric Schmidt, Google's former CEO, sits on the think tank's board.[10] As a result, because Lynn was an important force in shaping discussions around the regulation of tech giants, he seems to have become a casualty.

Lynn is now the head of the Open Markets Institute, which had convened a summit in 2016 dedicated to reigning in what he saw as the excesses of the tech giants. His team had launched the Citizens Against Monopoly initiative, arguing that the tech giants, particularly Google, have developed a political influence operation that is sophisticated and impactful, perhaps even more so than many major governments.[11]

Lynn wrote a *Washington Post* op-ed piece titled, "I criticized Google. It got me fired. That's how corporate power works," in which he discussed what might happen when corporate donors overtake a think tank's agenda.[12] Corporate power can threaten any independent institution—a university, newspaper, or community organization—that we depend on to fight for equality, diversity, and justice. As Lynn points out, "by design, the private business corporation is geared to pursue its own interests. It's our job as citizens to structure a political economy that keeps corporations small enough to ensure that their actions never threaten the people's sovereignty over our nation." Although his discussion is focused on the United States, it can be applied to smaller nations that have even less ability to sway the activities of massive technology firms.

Privatization of Public Life

Some of the most powerful companies in the world are responsible for creating the technologies that define how we discover information, communicate with one another, and buy and sell products. Today, these technologies are in the hands of billions of people worldwide, meaning that the *public experiences* defining our collective lives online are often controlled for us, and profited from, by private corporations.

Much of public life has been privatized, and not just in the United States. Schools, hospitals, banks, roads, militaries, police, even prisons, have all followed this path. These institutions directly affect our public lives and are charged with serving the public interest. Yet by their very definition as for-profit institutions, the public may not be their first priority. The accountability of these private institutions is primarily to their executives and shareholders, and their mission is to make a profit—even if they are funded by public tax dollars.

Figure 3.1
In the COMPAS system the black man on the left is likely to score a higher recidivism rate than the white man on the right (source: ProPublica).

The internet has also followed this path. Some of the early spaces that defined communication online, such as Usenet forums, were nonprofit. Some operating systems, such as Linux, and browsers, such as Mozilla, were not produced for commercial use. Today, however, we see the expansion of private, for-profit power over our public lives online with specific automated and AI systems designed to take over many of the tasks of government functions and institutions.

Consider the following two examples in the field of law enforcement. Both show how public institutions are using technology developed by private corporations. The first system, called COMPAS, calculates a *recidivism* score, or the likelihood of a convicted criminal to commit another crime in the future (see figure 3.1). The second, PredPol, is a *predictive policing* system. Police departments across the nation use it to determine algorithmically where and when a crime is likely to occur, when police should monitor certain neighborhoods and zones, and whether a given crime constitutes gang activity.

COMPAS and Suspect Recidivism Rates

The COMPAS system, developed by the Northpointe Company, has received significant scrutiny, thanks to the in-depth reporting of two journalists from the nonprofit organization ProPublica.[13] They first chronicle the case of Brisha Borden, an eighteen-year-old African American girl without a criminal record. In 2014 Borden and her friend picked up a kid-sized scooter and bicycle on the sidewalk, valued at $80, and started to ride the vehicles down the street. A neighbor spotted them and called out that the bike and scooter belonged to her six-year-old son. The girls stopped riding and walked away. But another neighbor who witnessed the event had already called the police, who arrested Borden and charged her with burglary and petty theft.

Compare that to the case of a forty-one-year-old white male, Vernon Prater, who had an armed robbery conviction and had served multiple years in prison. Prater was apprehended for shoplifting merchandise worth $86.35 from a Home Depot store. The COMPAS system concluded that Borden, the teenage black girl (who had committed only a misdemeanor as a juvenile) was at higher risk for recidivism than Prater, the white man convicted of a felony.

The system's determinations suggest that race, age, and gender are more important in judgments of potential criminality than a felony conviction and prison time. COMPAS has also flagged black defendants as being twice as likely to commit crimes compared to white criminals, despite prior criminal records. Based on data from Broward County, Florida, its predictions around violent crimes have been found to be incorrect 80 percent of the time. Describing the system, Mark Boessencker, a judge in Napa County (California) Superior court, where the system has also been used, explained: "A guy who has molested a small child every day for a year could still come out as a low-risk because he probably has a job. … Meanwhile, a drunk guy will look high risk because he's homeless."[14]

The ProPublica study also included the case of Paul Zilly, a forty-eight-year-old construction worker sent to prison for stealing a lawn mower and some tools. COMPAS rated Zilly with a high score for violent recidivism, even though Zilly had been working with a Christian pastor to address problems that influenced his decision to steal. Despite this proactive attempt to reform, the COMPAS system doomed him to a more punitive sentence than the one he might have received had the judge not seen the rating. The issue, Zilly explained, extends beyond innocence or guilt to concern power over one's own life: "Not that I'm innocent, but I just believe people do change."[15]

Questions such as "Was one of your parents ever sent to jail or prison?" and "How often did you get in fights while at school?" form the basis for the system's algorithmic rating formulas.[16] Puzzlingly, many COMPAS supporters insist that the answers to such questions are not highly correlated to race, despite social science research that shows how race plays a factor in criminal determination.[17] In response to this research we often hear another question: Is this not better than a racist judge? This question presumes that our only alternatives are racist judges or racist algorithms. What if instead we demanded other technologies that support the world we aspire toward, a world where racial justice is prioritized and racial bias is eliminated? What if the most criminalized populations in the country were brought in to help design and implement these technologies?

Predictive Policing and Racism

The Los Angeles Police Department (LAPD) has incorporated the PredPol (predictive policing) system since 2014 in over a third of its divisions.[18] Instead of using a narrative crime report to determine whether a given crime is gang-related, PredPol produces a "partially generative neural network," in other words, an algorithmically written crime report,[19] by considering quantitative, less "messy" data such as the number of suspects, the primary weapon used, and where the crime took place. A police department then draws upon these three criteria to provide a map of what it deems to be gang territory.

In an article for the journal *e-flux* Jackie Wang explains in greater detail: PredPol computes jurisdiction maps covered with red square boxes that indicate where crimes are likely to occur and at what times throughout the day. Police can then patrol these "spatial" zones in the flesh. Wang raises a number of "what happens next" questions, some already troubling to civil rights groups and critics of predictive policing:

> What is the attitude or mentality of the officers who are patrolling one of the boxes? When they enter one of the boxes, do they expect to stumble upon a crime taking place? How might the expectation of finding crime influence what the officers actually find? Will people who pass through these temporary crime zones while they are being patrolled by officers automatically be perceived as suspicious? Could merely passing through one of the red boxes constitute probable cause?[20]

Wang has also written *Carceral Capitalism*, a compilation of essays on topics related to technologies of control (both inside and outside prison) that affect perceptions about racism and crime. She notes how the PredPol system evolved from software funded by the Pentagon and used in Iraq to track insurgents and predict casualties. Jeffrey Brantingham, a UCLA anthropology professor and PredPol's co-developer, holds an evolutionary view of criminals as "modern-day urban foragers whose desires and behavioral patterns can be predicted." Such a model, says Wang, "appeals to our desire for certitude and knowledge about the future."[21] But how might satisfying that desire ultimately work against us?

Let's go back to examine PredPol's selling point, that it avoids "messy" crime scene descriptions. This implies that the system's algorithmically formulated and visualized data represents a neatly reflected (and therefore true) reality. Instead, says Wang, such data "actively *constructs* our reality."[22] The system thus ignores the subjectivities and contexts that shape specific real-world events. And those events are admittedly more difficult to process and compute using objective learning algorithms. And what if behind the system's prediction there only lies data that shows police have patrolled the corner before, perhaps because it's in a poor neighborhood?

PredPol consistently claims that it can be linked to a decrease in crime in cities that have implemented it. But crime rates have been decreasing steadily throughout the United States since the 1990s, and there is little evidence to correlate any decrease to PredPol. For example California's gang database, the very source of so-called objective information upon which the algorithms are based and modeled, was found to be full of errors. For example, it mistakenly included forty-two gang member names with birth dates of one year old or younger.[23]

When criticized about the ethical concerns the technology has raised, Hau Chan, one of seven co-authors (with Brantingham) of a paper presented at the inaugural Artificial Intelligence, Ethic and Society conference in February 2018, replied, "I'm just an engineer."[24] As we weigh that comment, we must remember that PredPol is a *for-profit business* that sells services with the potential to *socially engineer* the police system.

Both PredPol and COMPAS illustrate how private companies have developed technologies that impact our public institutions and lives. Both are "intelligent systems" that disproportionately target racial and sexual minorities as well as those of low socioeconomic status, using secretive data sets. Because they are seen as "statistical" or "mathematical," it becomes very difficult to make prejudice visible. Both systems, and therefore the courts and cops using them, do not take into account the impact of incorrect judgments or predictions. Nor do they consider the amount of pain their decisions may have on individuals and communities who are already racially profiled and placed at the margins of social and economic power. But this myopic viewpoint ignores alternative values that we must safeguard: complexity, social context, civic decision-making, and an empathic approach that sees us all as connected. And so, instead of further criminalizing vulnerable communities with technology, we can design tools in the image of compassion and restorative justice.

Reddit: Throwback or Future Path?

The internet was never solely a professional or governmental undertaking. Instead, a wide range of entrepreneurial, countercultural, and artistic figures contributed to its evolution. From its early "online community" days in the mid-1980s, we saw the internet as a way to share common interests, hobbies, professions, and cultural identities with others. But we did not anticipate how our decentralized use of the internet, with each of us on our computers and phones, could be controlled. We did not imagine how the scale and expansion of the internet could be co-opted through the capture and manipulation of data. Nor did we foresee how that data could be optimized within a capitalist economic system where tech companies have one major role: increasing their bottom line.

We still see stragglers of the internet's past living on today. Consider Reddit, a website that proclaims itself to be "the front page of the internet."[25] I recently interviewed Reddit's CEO and co-founder Steve Huffman. He described his site, with its 350 million to 400 million users, as "organic and pure," the most "human place on the internet."[26] Why? For one, the site looks like a throwback to the Usenet forums and newsgroups of the 1980s and 1990s. It's not a slick black box as much as a messy list of threads and posts, a site that provide users with opportunities to enter different conversations around topics of their choice.

Unlike Facebook or YouTube, Reddit is completely user-driven. Registered users post content, vote to support it, and recommend different communities to one another. (There are millions of such communities, called subreddits.) With Facebook and YouTube, users also contribute by posting, voting, and so on—but the decisions shaping what they see are not so user-powered. Reddit has not "updated" its original approach to take advantage of hidden algorithms based on personal data collected about users, and thus it does not control the content of sites the way Facebook and YouTube do. Reddit will suggest, or even recommend, other threads its users may wish to join, but it does not direct or otherwise manipulate the users' attention to them. This user-oriented method works for the site, well enough for Alexa (that's right, Amazon's Alexa), to rank Reddit the fifth most popular site in the United States and the eighteenth worldwide.[27]

Reddit makes money from advertising but its practices are far more transparent than those of its big tech competitors. The company doesn't offer advertisers options to microtarget users through a hidden system of data-collection, profiling, and aggregating—advertisers can simply post their ads where they wish without accessing a user's intimate personal data. Why? Because Reddit doesn't need such tactics, Huffman told me. If the skin care brand L'Oréal wants to advertise, it can simply go to the "Makeup Addiction" community. The approach is to link advertisers with users based on the communities they voluntarily joined, not in hidden and sneaky ways.

Reddit, like Wikipedia, remains dominantly English language–based and Eurocentric, a point that Huffman acknowledges. The site has also dealt with its fair share of controversy, having hosted intolerant content that forced moderators of the politics community to ban a number of posts and links. Misinformation planted on the site has gone viral. For example, after the 2014 Boston Marathon bombings, when a grainy picture of the suspected bomber circulated on social media, a user posted it on a Reddit forum alongside a photo simultaneously featured on Facebook: that of Sunil Tripathi, a student at Brown University who was missing at the time and eventually found dead by suicide. The Reddit user accompanied the set of photos with speculation about Tripathi's involvement in the bombings and the FBI's involvement in his search.[28] Thus Reddit, like other internet forum spaces, can be a place where misinformation is born, and from which it can travel across the site's user communities. From there it can get picked up by attention-optimized algorithmic aggregators—including those that shape what is visible on Facebook or Google News.

But despite Reddit's shortcomings, the site reminds us that the internet signified "community" to its early proponents and users. This was even true for the tech giants of today. For example, Facebook started as a community network in 2004—a space for Harvard students to communicate and share information. It then expanded to other universities before eventually opening itself up to the world, at least to anyone with an email address.[29] Google also described itself as a pro-community technology: within a brief ten-year span since its founding in 1998, the site operated services for "blogging, instant messaging, shopping and social networking," it challenged Microsoft's role in the workplace with a "suite of word processing, spreadsheet and other tools," and it began to build a software platform for mobile phones intended to compete with Apple's iPhone.[30]

I spoke briefly with the Harvard legal scholar Lawrence Lessig about where we and the internet have been headed considering how community oriented technologies turned into massive private corporations. In several of his influential books, Lessig has argued that the internet facilitated forms of "virtuous expression" through which the peer-to-peer experience of uploading and downloading (or streaming) supported the formation of new communities based on non-monetary interests. Lessig told me he wishes to "make obvious to policy makers and technologists the intimate connection between *technical design and public values*." He stressed how "particular values that manifest [themselves] on the internet were contingent on [the technology's] design."[31] Yet he expressed frustration that internet technologists have created a cash cow by influencing internet users to consume addictive content as if it were junk food and by allowing rumors and misinformation to spread like wildfire. Regulators should do something about this, Lessig declared, recognizing that market-dominant technologies have become addictive, monopolistic, and unhealthy for democracy. Our policymakers must do what they can to push forward a climate in which competitors and alternatives to the tech companies today are also viable.

4 Disconnection and Connection

The intelligent computer systems being designed for use in courtrooms and police departments, increasingly represent the norm rather than the exception when it comes to the troubling ways many technologies are developed and deployed today. These systems follow a pattern, as we saw in the last chapter. They arrive in our lives promising efficiency and connectivity. But they operate in ways that blur our understanding of what we are connected to and what it means to be connected. This happens because those who hold power over connecting our world to the internet are all too often disconnected from the populations they claim to serve. The PredPol developer Hau Chan (discussed in chapter 3) is "just an engineer." Mark Zuckerberg is just a kid from Harvard who had a dream. Steve Jobs is just an iconoclast who traveled the world. Jeff Bezos is just one of the wealthiest men in the world obsessed with privatized space travel.

How did this problem of disconnection arise?

Let's go back to 1997. At that time, we celebrated the internet for liberating information and democratizing access. But even then the French philosopher Paul Virilio saw the internet as a potential "information bomb." Such an explosion of information, he argued, would overwhelmingly disorient internet users, who would then relinquish total control to technologies produced by private corporations and powerful nation states.[1] Virilio cautioned that the speed at which information could travel would create "virtualization," where neither time nor space would matter to the extent it did before. We would all risk being controlled by technologies created to support the interests of the powerful.

Since 1997 Virilio has been one of many writers to raise concerns about the digital world. Even the controversial neoconservative statesman Henry

Kissinger argues today that new technology's "emphasis on speed inhibits reflection; its incentive empowers the radical over the thoughtful; its values are shaped by subgroup consensus, not by introspection."[2]

Our Digital Footprints

Because they use our personal data, many technology platforms also hold the potential to steer our economic, cultural, and political futures. Instead of serving people, however, they tend to support the agendas of a few private corporations notable for their lack of transparency.

Companies that collect, sell, and analyze our data now claim they are able to use it to predict what we will do. Several studies show how tech companies can make conclusions about a person's identity and behavior based on a simple set of Facebook "likes." But the problem goes beyond the *intentional* posts, likes, or comments we put online. Much of the data used to assess and impact our behavior comes from information we release without awareness.

For example, when we use our mobile phone or credit card we make data available to the companies who produce and control the digital applications behind them—or to brokers who buy and sell our data. As we carry around our smartphones, software technologies not only track our biometric data but also collect information about where we go and the places and people with whom we interact. This data is the "digital footprint" we leave behind when we browse online or use credit cards, or when surveillance cameras capture our movement in public spaces. We have little control or awareness regarding how much data we are giving up, and how often we do so while using the internet in our daily lives. It's concerning that so much Facebook data about us is available to be used this way. But far scarier is the ability to combine this data with all the other digital footprints we release. Yes, you'll get a glimpse of my identity from what Facebook has on me, but add to that my aggregated online data and you'll have a penetrating, invasive portrait.

The computational models analyzing our data can hugely impact our behaviors, decisions, and experiences. They do so by feeding back the data they gather about us into their *persuasion architecture*, an online environment they build for us that determines what we see and how we see it.

Consider the persuasion architecture of the supermarket checkout line: it exposes us to well-placed junk or snack foods and the ubiquitous tabloids, and it influences our decisions to purchase the items on display. Persuasion architecture exists online at a massive scale with greater immediacy: what we see is personalized based on real-time information we provide, making the "options" available to click on or view that much more powerful.

Research concerning the potential to analyze our digital footprints is not new. Michal Kosinski, a professor at Stanford, demonstrated in 2012 that with a mere 68 Facebook likes he could predict a user's skin color, sexual orientation, and political affiliation. But that's not all. He and other researchers have also been able to predict intelligence, religion, drug and alcohol use, and even whether a person's parents are divorced.[3] With 70 likes, there's data enough to evaluate a person's behavior better than a colleague ordinarily could; 150 likes would beat what parents know about their child; and with 300, what a partner or spouse knows.[4] The claim is that just by gathering and computing across enough data trails, a person's IQ becomes predictable.[5]

It's troubling that these models define terms like "sexuality," "intelligence," and "behavior" in extremely reductive ways. In doing so they relegate the manner in which we define of the most complex parts of life to the binary, yes-or-no logic of labels. Even worse, they do this in ways that are hidden from us. And that hiddenness means we might never be able to find out what details are included or excluded about us in the process. It's understandable to trust the recommendations we get online, but what if they are at odds with how we view ourselves?

Welcome to the Hyperreal

Our ability to communicate today by email, text, and video chat, still maintains continuity with the ways we've kept in touch, shared, and exchanged information in the recent past. But the speed of communication—which brings us together in immediate or near immediate timeframes (for very little cost, and often for free)—guarantees seemingly frictionless ease.

Yet such efficiency does not come without a cost. In a recent opinion column for the *New York Times*, Cal Newport suggests that the smartphone draws "us in with endless diversions, like the warm ping of social approval

delivered in the forms of likes and retweets, and the algorithmically amplified outrage of the latest 'breaking' news or controversy."[6]

Siva Vaidhyanathan also sees danger lurking behind the frictionless ease
and endless diversions. In his 2011 book, *The Googlization of Everything: And
Why We Should Worry*, he examined the downside of the company's global
embrace and its mission to "organize everything." And in his recent book,
Antisocial Media, he discusses how Facebook and its widespread use could
undermine democracies across the world.[7] When I asked him about these
concerns he replied with his own question:

> Has it made us smarter, has it made us wiser?…It has really just overwhelmed
> us with audio, video, and text. In some ways it's paralyzed us and stunted us, in
> other ways it's enlightened us and empowered us. There's no one answer to that
> question. But I think it's safe to say that what has happened to the removal of fric-
> tion is that we've opened a whole lot of new problems that we did not anticipate.[8]

Vaidhyanathan points out that we must not equate "positive" technological transformations with social empowerment. There's a danger when the
corporate or institutional entities that drive these technological shifts are
disconnected from the communities their products serve. "A queer young
person in a straight, unfriendly part of the world might get tremendous
value out of the connective power of social media," he says. "But *collectively
it's a disaster*. Because it's one part of a system that changes at scale."[9]

The scale of these platforms, Vaidhyanathan contends, gives them power
to shake our economic, political, and cultural foundations, with zero input
or accountability from any of us. Our media, financial institutions, places
of buying and selling, libraries, and, most dramatically, our democratic institutions, are all being challenged by new technologies. These institutions
deserve critique. Keeping them accountable is a tenet of any democratic
society, but we must not throw the baby out with the bathwater in our zeal
to technologize everything.

Consider the photo that appeared in the Dominican Republic's newspaper *El Nacional* in February 2017 (figure 4.1). It looks like Donald Trump but
it's actually Alec Baldwin, the actor who has been spoofing the president
on *Saturday Night Live* ever since the campaigns leading up to the 2016 US
presidential election.[10] Was this just an innocent mix-up? Most likely, but
something more significant may be going on. It brings to mind "the hyperreal," a term the philosopher Jean Baudrillard used to describe the blurring
of fiction and reality produced within our image-saturated world.[11]

Trump dice colonias de Israel no favorecen paz

Es la primera vez que el presidente de Estados Unidos asume una posición sobre el desarrollo de los conflictivos asentamientos israelíes

JERUSALÉN. AFP. El presidente estadounidense Donald Trump dijo en una entrevista publicada el viernes que el desarrollo de los colonias israelíes "no es bueno para la paz", su primera toma de posición sobre esta cuestión desde que llegó a la Casa Blanca.

"No soy alguien que piense que el desarrollo de las colonias sea bueno para la paz" dijo Trump en una entrevista publicada en hebreo por el periódico gratuito israelí Israel Hayom, pocos días antes de recibir al primer ministro Benjamin Netanyahu.

También aseguró que está estudiando "muy seriamente" el traslado de la embajada de Estados Uni-

ruptura con la política de Estados Unidos y de gran parte de la comunidad internacional, que considera que el estatuto de Jerusalén -también reivindicada por los palestinos como capital de su futuro estado-, tiene que decidirse mediante negociaciones.

En caso de traslado, el presidente Mahmud Abas dijo que la Organización por la Liberación de Palestina, considerada por la comunidad internacional como representante de todos los palestinos, podría dejar de reconocer a Israel, una medida que llevaría los esfuerzos de paz 20 años atrás.

Seguridad
WASHINGTON. EFE. El presi-

Donald Trump, presidente de EEUU.   Benjamin Netanyahu, premier israelí.

Familiares policías protestan en Brasil

RÍO DE JANEIRO. EFE. Pequeños grupos de familiares de policías militarizados se manifestaron hoy a las afueras de varios cuarteles de Río de Janeiro en demanda de mejoras salariales y de trabajo de los efectivos.

La protesta comenzó de manera pacífica a primera hora de la mañana durante el cambio de turno en algunos batallones de la Policía Militarizada, como los de Frei Caneca, en el centro de la ciudad; Olaria y Tijuca, en el norte.

Los manifestantes, en su mayoría mujeres, portaban carteles pidiendo mejoras salariales la renovación de equipamientos y beneficios laborales.

El movimiento tiene características parecidas al que comenzó el último sábado en el vecino estado de Espírito Santo y donde familiares perma-

Figure 4.1
Not the real Donald Trump (source: *El Nacional*).

We see the hyperreal at work in the "Trump photo" not only as a case of mistaken identity but also in terms of how we understand the US president himself, as a former "reality TV" star. Even the president's Twitter handle, @realdonaldtrump, suggests that his tweets are the real deal and other representations are fakes. This invokes a far more important question: How are we to discern fact from fiction, and reality from imagination, within a world of information overload, surrounded by immersive technology?

When Baudrillard discussed the hyperreal in his 1981 book, *Simulacra and Simulation*, he drew on a fable by the Argentine author Jorge Luis Borges in which cartographers were asked to create a map of an empire.[12] The resulting map embodied such a level of precision and detail that in the end it became indistinguishable from the actual territory it sought to represent. In the fable, the physical map fades and crumbles into the soil, as does the empire into the landscape, marking the final blurring between signifier and signified. As Baudrillard stressed: "It is the real, and not the map, whose vestiges persist here and there in the deserts that are no longer those of the Empire, but ours: *the desert of the real itself.*"[13]

New technologies expand our connections and disconnections. If they are created by powerful organizations or individuals who are themselves disconnected from the lives of those they claim to serve, they can take us into spaces of the hyperreal—spaces that blur our understandings of what is real, who we are, how we communicate, and how we learn.

Dependency, Addiction, and Misinformation

The internet's promise to decentralize aims to give each user power and voice. Intuitively that goal seems in harmony with a democratic society where citizens, states, or communities are part of the whole, but also have the opportunity to self-govern and participate as they see fit. But problems come when tech companies, particularly Facebook and Google, exploit this decentralization as they engineer ways of capturing and controlling data.

Google has developed numerous efficient and well-designed applications that work in tandem. Gmail and Google's search, maps, drive, and hangouts (chat) applications work perfectly with one another, even though we could explore the wilderness of the internet through search alone. According to Roger McNamee, the outspoken billionaire tech investor I quoted in chapter 3, Google twice displayed "genius insight": first when it recognized "twenty or twenty-five different functions that were common to everybody on the internet," and second when it "systematically created easy-to-use, free applications to address each of those functions."[14] Those functions and apps shaped the default infrastructure by which users navigate and engage with the internet to support their interests.

Thanks to the "genius insight" of the Silicon Valley giant, users have now become dependent on the company's products. As McNamee points out, "People would no sooner give up their Google application than cut off their right pinky, and the result is that Google has been able to centralize the internet and make the open web irrelevant." The company forced "all of the customers—whether they were users or businesses—and the content holders into their system." That, says McNamee, "is the very basis of all the problems" the company causes. "And in a world where there is weak antitrust regulation, that insight was worth hundreds of billions of dollars."[15]

As McNamee sees it, Google's move to dominate the internet with at least half a dozen products and over a billion users is equivalent to "simply walking onto the beach at Santa Monica and building a shopping mall … or taking over New York City's Central Park and charging [admission] just because they could."[16]

Facebook went on to imitate the Google strategy of centralization, McNamee adds. "[They] got essentially an 80 percent share of consumer photos [and] consumer video to Google's 20 percent," and "Google got 80 percent, maybe 90 percent of search and mail." Yet unlike Google, McNamee claims that Facebook's business model is "*manifestly* bad for [all] of its users.... People are harmed daily. The parasitic model of the media was started by Fox News, and Facebook took it to another level.... In order to make up for that, they focused on personalization, and that's when they began their whole surveillance model."[17]

Facebook also generated addiction among its users as billions grew ever more dependent on a single platform from which they could view one another's photos, videos, and posts. The smartphone, a tool that allowed content to be delivered *universally*, in theory to anyone at any time, became the catalyst for this achievement. Unlike Google, however, Facebook did not create a set of efficient applications. It opted instead to provide a centralized place for social life. By doing so it moved far away from the democratic vision of the online world as a public space, park, or square, where people can digitally come together and build community in ways over which they, and not a distant corporation, have power. As a result Facebook has become a space that accommodates various forms of surveillance and, ultimately, misinformation.

From my personal experience and perspective as a trained engineer, data scientist, and researcher of the internet, I find it hard to conclude that Facebook's intentions were sociopathic, as McNamee alleges.[18] But several Facebook executives have come out to deride the company's business practices.[19] Chamath Palihapitiya, for example, the company's former vice-president for user growth, spoke to an audience at Stanford Business School: "We have created tools that are ripping apart the social fabric of how society works. The short-term dopamine-driven feedback loops we've created are destroying how society works.... You don't realize [it], but you are being programmed."[20]

Sean Parker, Facebook's first president (who is revered in tech circles as the co-founder of Napster), has now revealed the motivation behind design choices Facebook made as it engineered and updated its platform. In an interview published on the Axios website, Parker acknowledged that the company, and Instagram, which it owns, constantly honed its objectives, none of them altruistic, by asking the following question: "How do we consume as much [user] time and attention as possible?" We were "exploiting a vulnerability in human psychology," Parker admitted. "The inventors, creators—it's me, it's Mark [Zuckerberg], it's Kevin Systrom on Instagram, it's all of these people—understood it consciously. And we did it anyway."[21]

Facebook's decision to grab its users' attention and feed their addiction is all the more disturbing because the company disseminates destructive and hateful content to achieve both goals. It is well known how fake news and conspiracy theory content has gone viral on the Facebook platform when it comes to elections, mass shootings, and neo-Nazi demonstrations. So-called intelligent algorithms have been "credited" for making these decisions, not human beings. According to McNamee, Facebook realized that "if you want to keep people in the filter bubble, ... you can show people car wrecks for a while, but after the wreck loses its salience you have to go to something more tragic than that."[22]

The use of algorithms in decision-making also implicates YouTube, the Google-owned video-sharing platform, which now has over 1.5 billion monthly users.[23] In February 2018, the journalist Paul Lewis filed a report in the *Guardian* claiming that YouTube's recommendation algorithm promotes divisive clips and conspiracy videos in the "Up next" thumbnails that entice a user to remain on the site.

To prove his point Lewis cleared his browsing history, deleted cookies, and opened a private browser to ensure that YouTube recommendations for him were not based on past, personalized choices, but rather on what the artificial intelligence–powered system selected to maintain his attention.[24] The "Up next" selections that came his way were unsettling: more fake news videos; more conspiracy theories gone viral. Macabre, hateful, and violent videos, including one associated with Peppa Pig, the animated TV character who has a dedicated channel on YouTube, were served up to toddlers. Lewis concluded that YouTube's recommendation algorithm both values and supports hysteria and sensationalism. On YouTube, he said, "Fiction is outperforming reality."[25]

We've heard of the "win-win" narrative that binds video creators to platform-holding corporations: if users watch a video and stay on the platform, the relationship is symbiotic. This feeds the popular misperception that YouTube and other social media platforms are great democratizers. So does YouTube's self-promotion when it comes to politics: prior to the 2008 US election it praised its role (with CNN) as a co-moderator of a presidential debate between Barack Obama and John McCain, claiming that it "lowers the barrier…to engage in the political process, and levels the platform for political discussion."[26]

What the win-win narrative fails to consider are the rest of us, the users. As we get addicted to these platforms, and there is increasing evidence that we do, we give into their recommendations and allow our attention to be exploited.[27] As a result, we now have billions of users stuck in siloes of manipulation and misinformation, believing things that, in McNamee's words, are "demonstrably not true.…But you can't get through to [these users].…It's like they're part of separate cults."[28]

In an effort to break the barrier, McNamee and Tristan Harris, a former Google employee and design ethicist, created the Center for Humane Technology, a coalition of tech creators studying what its co-founders believe are the most pernicious problems with technology today: one being its addictive use by youth, and another being its potential to influence elections and politics. McNamee spoke to me about how the center has been catching "lightning in a bottle," garnering attention from a wide range of journalists and warning US and international users alike about the dangers of ceding too much control to the biggest technology companies.[29] The center, however, has not yet engaged with the most pressing issue we face: the hidden capture, control, and manipulation of our lives.

Whether we're talking about massive lobbying power or accumulated wealth, McNamee explains that we shouldn't expect to see significant change as long as stockholder valuations keep skyrocketing. Accountability to shareholders is the linchpin of a publicly traded corporation. If the current, secretive model of controlling attention and data continues to work for companies such as Facebook and Google, he claims, they are likely to take little action.[30]

McNamee calls the notion that technology should be solely valued through trade the most "specious piece of bullshit imaginable."[31] Yet shareholder trade is what ultimately matters for commercial valuation, because those creating the technologies that reach the hands of billions are members of private corporations. Meanwhile we, the users, the ones whose data and attention run through the pipes to fuel this entire endeavor, are reduced to just that: a passive resource to be burned through, with no meaningful voice, value, or vision of our own to contribute as equal citizens.

The Collection of Our Data

McNamee defends Apple from some of the critiques he levels at Facebook and Google by arguing that the corporation's model is more traditional. "They made a much better product, got much higher margins...people f—ing love it."[32] He does admit, however, that Apple products and services are also addictive—in addition to its laptops and phones, users are dependent on its iTunes store, which serves as a centralized data-gathering gateway for the podcasts, music, and videos users consume. McNamee sees a distinction here: for Apple the human being is the customer rather than the product; he views the company's policy concerning data and privacy as less predatory.

What's missing from his characterization of Apple? Several issues, one being Apple's approach toward *planned obsolescence*, or its design of expensive devices intended to die after a given period of time.[33] Apple also has a troubling environmental and labor record, which involves the use of industrial sweatshops in countries like China.[34] Additionally, the company has exploited consumers by forcing them to buy expensive "new" dongle peripherals (such as chargers and cables that are only compatible with iPhones) to ensure fully functioning devices.[35]

The music business is where we see Apple's monopolistic aspirations in action. As Jonathan Prince, from rival Spotify, points out: "Apple has long used its control of iOS to squash competition in music, driving up the prices of its competitors, inappropriately forbidding us from telling our customers about lower prices, and giving itself unfair advantages across its platform through everything from the lock screen to Siri. You know there's something wrong when Apple makes more off a Spotify subscription than it does off an Apple Music subscription, and doesn't share any of that with the music industry. They want to have their cake and eat everyone else's too."[36]

Thus, in different but related ways, we see how the biggest tech companies have achieved dominance. Facebook hijacks attention and data. Google centralizes the supposedly decentralized internet by cajoling us into using their interlocking, data-gathering products and services. Apple forces us into a dependence on new products, peripherals, and their own version of data-gathering centralized media content services. Amazon vertically integrates between bits and atoms—grabbing our data, pushing us into their media networks and offerings, and centralizing their services for the distribution and consumption of retail products and groceries.

Both Amazon and Google have moved into our homes with the Amazon Echo and Google Home Mini, devices that listen to us for location information, brands we like, and other personal details. These and other companies have entered the surveillance business—the more they learn about us the more they can influence our behavior and sell us more products and services.

Facebook tracks users when they are not online, including those who have never even signed up for a Facebook account. The company is now the subject of a lawsuit in California initiated by the former startup Six4Three, which alleges that the company gathered information about non-user friends of people with Facebook accounts, reading text messages, tracking locations, and accessing photos on their phones, even if they did not have the Facebook app installed on it.[37]

Google is well known for hoarding data from users and non-users alike, and has excelled at acquiring and developing so many distinct applications that it has become difficult not to provide the company with data unknowingly.[38] As a thought experiment, the company's Google X division has considered the possibility of organizing all the data collected by a user's devices in a "ledger" and presenting it "as a bundle of information that can be passed on to other users for the betterment of society."[39]

Microsoft and Apple have been experimenting with facial recognition software, which suggests that the technology might be even more aggressively marketed for policing or vetting job candidates. Although Apple has claimed that it will not share the location data it collects about us with third parties, we know that facial and retinal scans are now acclaimed features of its newest technologies, including the iPhone X.[40]

Amazon has developed an AI system called "Rekognition" that can be adopted for government surveillance. According to a May 2018 report by the ACLU of Northern California, the system could "identify, track, and analyze people in real time and recognize up to 100 people in a single image … [by scanning information] it collects against databases featuring tens of millions of faces."[41] Rekognition has already been used in several US cities and states. But in January 2019 results of an MIT Media Lab study suggested a potential setback for Amazon and its efforts to popularize its surveillance system: Rekognition "misclassified women as men 19 percent of the time...and mistook darker-skinned women for men 31 percent of the time."[42]

Imagine how facial recognition, combined with police body cameras, can facilitate real-time surveillance of protests, vulnerable immigrants, and more. Anything that a policeman's body camera sees can now be potentially identifiable, tracked, logged, and abused.[43] These technologies, used surreptitiously by law enforcement or any other government agency, make it all the more imperative that we protect our civil liberties, which include "speaking freely and [going] about our business anonymously in public."[44]

5 Blind Solutions

Today we face a bait-and-switch when we go online. Private companies commonly develop and pitch their technologies as solutions, yet they rely on our blind trust and increased dependency to reap greater corporate gain. In his second book, *To Save Everything Click Here*, Evgeny Morozov writes that "Silicon Valley's quest is to fit us all into a digital straightjacket by promoting efficiency, transparency, certitude and perfection—and by extension eliminating…friction, opacity, ambiguity and imperfection."[1]

As he examines systems created to assess criminal policing, health tracking, politics, and much more, Morozov exposes the myth of "solutionism," that all problems can be solved through technology.[2] *Solutionism* is an apt term to describe how private companies that dominate the design, development, and distribution of technologies are disconnected from the lives of their users: they value crowdsourced opinion over reflection, debate, discussion, and interpretation; they believe our lives can be governed through quantification and technical "fixes." As a result, they ignore diverse knowledge traditions around the world.

Morozov points out that corporate engineers and executives are not publicly elected. Most of them are barely publicly known, he says, yet they have amassed great power: "Google will be there forever. Democratic accountability will not be prevalent. You cannot file a public information request about Google. We are abandoning all the checks and balances we have built," he says, to keep our public officials in line. We are opting instead for "these cleaner, neater, more efficient technological solutions. Imperfection might be the price for democracy."[3]

What then about the legality of this situation? Tim Wu, an expert on monopolies and the technology industry, is the individual who coined the term "net neutrality" to describe an internet where any website or piece of content is equally accessible.

Two books by Wu have garnered significant public attention. The first, *The Master Switch*, describes what Wu calls the "cycle," a process by which technologies shift "from somebody's hobby to somebody's industry; from jury-rigged contraption to slick production marvel; from a freely accessible channel to one strictly controlled by a single corporation or cartel—from open to closed system."[4]

How did this come to be with the dominant technology companies of today? Wu's second book, *The Attention Merchants: The Epic Scramble to Get Inside Our Heads*, provides a one-word answer": attention.[5]

From selling advertisements to buying influence, attention is the commodity of the day that allows tech companies to gather so much of our personal data. Having power over our attention allows a company (or its technology) to direct where our minds and dollars flow—ultimately bottlenecking where we go for information, where we go to socialize, and where we buy and sell.

The control over what we pay attention to (and consume) now rests in the hands of the few. So how far along are we into the technological shift, from an open to a closed centralized system, that Wu describes in *The Master Switch*? The situation parallels the prestige enjoyed by massive companies that once occupied the throne in their respective industries, like General Motors did among automakers of the 1950s. When I spoke to him in early 2018 Wu explained to me that in the past, such companies were treated "as the best in the world. ... But by the 1970s it was like, 'Oh my god, what the hell was that?' It's like a drug thing where on the way up it's like 'Woo, this is a wild ride,' and then comes the hangover. That's the way it is with centralization, and with the tech companies. The hangover part is evidently on us right now."[6]

Wu then asks rhetorically, on behalf of the major players of the technology industry, "What is it all for?"

Going back to Wu's argument in *The Master Switch*, our conversation underscored a key point: Instead of building technological and commercial systems that make it as easy as possible to buy stuff or entertain ourselves, we need to recognize economically sustainable alternatives that consider users, societies, and their welfare in the design of platforms. We can design technologies that support "making human life meaningful" rather than reducing us "to conditions of servitude or idleness."[7]

Wu shares this position with George Soros, the billionaire founder of the Open Society Foundation and a long-time advocate for open democratic media. In a speech Soros gave at the World Economic Forum in January 2018 he echoed the critique that social media companies are engineering addiction, especially among adolescents. He likened tech companies to casinos that hook people to the point of gambling "away all their money, even money they don't have."

> Something very harmful and maybe irreversible is happening to human attention in our digital age. Not just distraction or addiction; social media companies are inducing people to give up their autonomy. The power to shape people's attention is increasingly concentrated in the hands of a few companies. It takes a real effort to assert and defend what John Stuart Mill called "the freedom of mind." There is a possibility that once lost, people who grow up in the digital age will have difficulty in regaining it.[8]

We cannot risk losing our diverse values as human beings by blindly giving in to today's technology. This is more than ever the case when we see how creators of new technology increasingly express doubt in the human being, pointing to our mortal fallibility. Such questioning sees death as a tragedy and our bodies as limitations. It fixates on escapes—other planetary colonies, post-apocalyptic bunkers, fusing our bodies with machines, or downloading our consciousness into machines—because it lacks an interest in the here and now and the central question: How do we support and heal our world as it is?[9]

Expressing doubt in human capability ignores that our species has always comprised diverse cultures, languages, ways of knowing, and values rooted in ecosystems and material relationships. Seeking to standardize, translate, and control human consciousness in a digital form would be an astounding and likely impossible technical feat. More importantly it disregards different human needs—indeed, the notion of the human itself—on the altar of technical transcendence. Some technologists write with enthusiasm about a "post-human world"—think of Ray Kurzweil and the Singularity—and they question humanism itself. But we must ask ourselves about the ethics of dragging billions of people and the communities to which they belong into a post-human world without their consent and against their stated values. Is this how we want to inaugurate our future?

Technology does not have to serve one master. The rich and roiling collection of the billions of people in our world have hopes and dreams worth listening to. We cannot afford tech that serves one pantheon: the gods of money, power, and individualism.

Diversity and Disconnection

I have painted a picture so far of disconnection between those that design and profit from the technologies that define our lives, and the world outside of their corporate offices and laboratories. I have not touched yet on *internal disconnection*, the way that the business and engineering divisions of powerful technology companies themselves fail to reflect the diverse worlds they attempt to reach.

I learned of this concern, not from a journalist reporting from the outside, or from a disenchanted former investor like McNamee, but from an engineer and insider to the industry: Tracy Chou.

Chou is Stanford-trained engineer who at age twenty-nine, in 2017, was featured on a cover of *MIT Technology Review*. The issue celebrated a "visionary" theme: "35 Innovators Under 35." The title of the article about Chou says it all: "Bringing Tech's Dismal Diversity Numbers Out into the Open."[10]

Taking a cue from Facebook COO Sheryl Sandberg, who had raised concerns about the numbers of women in the technology industry, Chou and several other women developed Project Include, dedicated to diversifying the workforces that populate Silicon Valley companies. Chou wrote a post on Medium titled "Where Are the Numbers?" requesting data from these companies around diversity.[11] As a result, she and her colleagues were able to gather statistics from over a hundred companies and identify that less than 20 percent of technology workers, and even fewer executives within the industry, were female.

I met Chou in the historic city of Guanajuato in central Mexico. We were both speakers at a UNESCO meeting concerned with technology, science, and the future of Latin America. In her speech, Chou discussed concerns that the technologies exported to billions by Western technology companies might reflect the biases within these organizations. The absence of women and non-Asian minorities in the most influential engineering and managerial positions within the tech companies, Chou argued, influence how technologies are designed, the data sets they learn from, and how they are applied across the world. As a result, misrepresentation, bias, and even discrimination become the norm. We cannot simply analyze a technology itself, she said, but we must instead consider its chain of production and its potential impact on billions of people.

Two factors convey why it's important to address Chou's concerns: First, the power of a technology produced in one location but scaled to billions is amplified because of the *network effect*, whereby a service becomes more valuable when the number of people using it increases. Second, technology sectors have emerged in cities such as Nairobi (Kenya), Cape Town (South Africa), Bangalore (India), Auckland (New Zealand), and more. The incredible economic success of Silicon Valley has inspired these new incubators, and they are likely to follow the lead, which may be to ignore or dismiss diversity issues.

Thus, the relatively simple issue of accommodating greater diversity within a technology company has more far-reaching and pronounced effects. As networked technologies further penetrate our lives, blurring our online and offline time, the places we go, and the things we interact with, we run the risk of "naturalizing" systems that are built upon biases.

My conversation with Chou recognizes the bias that sees excellence in science and technology, and the perceived skill sets needed to excel in these sectors, as being at odds with women. Instead of acknowledging the absence of women in the tech industry as a problem, we have witnessed increased justifications for the status quo.

Google faced a major PR crisis for an anti-diversity manifesto written by one of its employees that went internally viral and was eventually leaked to the press. This document, titled "Google's Ideological Echo Chamber," argued that the absence of women in the tech industry could be biologically rather than culturally or socially explainable.[12] It also stated that Google should not offer programs to recruit or support underrepresented minorities. It is likely, based on several articles in major news sources, that this opinion was shared within the company.[13] Indeed, Google has been in a dispute with the Department of Labor over its low rate of female employment (less than 20 percent) as well over wage disparity.[14]

The author of this Google manifesto, James Damore, was eventually fired from the company. But his views are not outliers. Lawrence Summers, the former Harvard University president, who is also a former Obama and Clinton administration cabinet member, used his platform at a 2005 economics conference to claim that women's underrepresentation in the sciences reflected their lack of aptitude for the subject matter as a function of "innate" differences between men and women.[15]

Much research has rebutted such claims, but we continue to hear arguments about nature versus nurture in discussions of gender representation in engineering and science. Missing here is the ethical question: Given that digital technologies are increasingly part of all our lives, why should approximately half the population in our world not be part of the process by which they are developed and designed?[16]

This issue takes on greater urgency because automated "intelligent" systems are being developed that will threaten many existing jobs and further widen the divide between women and men in the workplace. Indeed, a report authored by the World Economic Forum in January 2018 warned that women currently hold almost 60 percent of threatened jobs.[17] According to Chou these biases could be particularly insidious as we build intelligent AI systems and release them into our worlds: "We don't even fully understand...the design of the [technology] models themselves, the way systems are being used, the ways the models are being trained, and the bias of the data sets."[18]

Chou also tells me that when it comes to tech businesses we need to follow the money trail. Given that a significant amount of Silicon Valley funding comes from organizations led by white men, women and minorities are on the outside looking in. As a result, "All the power still stays within the same [groups]."[19] Who you are as a funder ends up affecting what businesses are supported and technologies are created, "how much (a tech entrepreneur) reminds investors of themselves...so that works against people who don't 'fit in' culturally or socially."[20]

As tech industries make decisions about who they hire and the technologies they create, they then shape other ecosystems: political systems, small- and medium-sized businesses, policing and surveillance systems, and more. As the systems they build become more "intelligent," they take on the biases of not only their engineers but the realities of an unequal world from which they learn.

Chou tells me that *passive inclusion* isn't the solution. It isn't good enough to simply enlist more black, brown, queer, or disabled persons to build products and services that serve the agendas of white male CEOs and investors. She challenges us to imagine a more distributed technology industry, where the needs and voices of customers and users in diverse parts of our world can drive the design and development of systems. Instead of seeing feminism, for example, as a quota-based approach to hiring more women, it can be envisioned as a new way of creating and designing technology, one guided by an appreciation of *difference*, seeing cultural, economic, gender, and generational diversities as strengths to build upon.

II Political Data Games: Targeted, Manipulated, and Motivated

6 Brave New Digital World

In 1958 the journalist Mike Wallace interviewed Aldous Huxley, the British author best known for writing *Brave New World*. This dystopian sci-fi novel, published in 1932, takes place in the fictional and future World State society, where human beings are produced in laboratories and assigned to different classes based on their intelligence and physical gifts.[1]

The previous chapters have shown how powerful tech companies operate like digital laboratories: They track, gather, and store data from the footprints we leave online. They use that data to create and update our individual profiles, which determine whether policing systems will view us with suspicion, for instance, or whether we'll receive medical benefits for preexisting conditions. But we have we have no ability to see the data they collect, let alone correct it.[2]

In the Wallace interview Huxley spoke of threats to democracy, with an eye on the present and the future. Rather than focus on political and social inequalities, he pointed the finger at technologies we use willingly every day—especially those that have the potential to distract us, like television. "We mustn't be caught by surprise by our own advancing technology," he said.

Huxley emphasized that to remain in power one must obtain the "consent of the ruled." And to get such consent, he said, requires "new techniques of propaganda" that have the ability to bypass the "rational side of man and appeal to his subconscious and his deeper emotions, and his physiology even, ... making him actually love his slavery."[3]

Today our ability to make rational choices, including political ones, is more compromised than ever. Most of us believe that tech companies aggregating hundreds of data points about us everyday do it to provide us with traditional, "personalized" advertising. Many of us are okay with that, and don't mind our data being collected to sell us products like soap or cars, or to recommend new music, movies, and TV shows.[4] Fewer of us are aware that companies use our data to sell us political agendas and politicians, often by spreading propaganda and *disinformation*, intentionally misleading material tailored to exploit our vulnerabilities and foibles. We never consented to having our democracies taken over by the same process that entices us to purchase one brand (Starbucks, let's say) over another (Dunkin' Donuts).

In this way we've become polarized by digital technologies that run our personal data through opaque algorithms, making billions in advertising revenue for the companies that produce them, while working against the basic values of people-centered technology. Polarization especially undermines our control of democratic functions. People, not machines or corporations, should have power over their own political lives.

If we can no longer trust what we see—if in fact we are even manipulated by what we see—we must learn to understand, confront, and overcome the forces that threaten our individual minds and hopes for democracy. Indeed, human beings, bots, and microtargeting algorithms all play a role today in closing us off from facts, multiple points of view, and the contexts behind the information we see. As Huxley said in 1958: "The whole democratic procedure … is based on conscious choice [made] on rational ground."[5]

Democracy also depends on public debate. It assumes that the playing field for such debate is level: we may have socioeconomic differences, but each of us should have equal opportunity to vote according to our experience, our values, and our best judgment. As citizens we may have different political views, but the events that inform our perspectives should be accessible and commonly understood.

Transparency, central to any democratic system, depends on having equal information—not information "suggested" for us based on surveillance and data tracking or gathering. But tech giants carry out these pervasive practices in hundreds of hidden ways. The usual suspects, Facebook and Google, are not the only guilty ones. Even online newspapers we rely on for unbiased analysis and reporting, including those like the *Guardian* that discuss problems with surveillance, host tracking software similar to what these very companies use.

Whatever passes for transparency today seems one-directional: tech companies know everything about us, but we know almost nothing about them. So how can we be sure that the information they feed us hasn't been skewed and framed based on funding models and private news industry politics? Data breaches may happen everyday, but how often do they involve citizens learning about government spying? Think of how rarely we learn of revelations like the ones the whistleblower Edward Snowden made in 2013.

Big technology companies, particularly Facebook and Google, have a huge role to play in the political lives of people across the world. Mark Zuckerberg could privately press a button to tell Donald Trump or Hillary Clinton supporters how to vote and, without anyone knowing, sway well over 300,000 votes—enough to tilt the balance in a divided election.[6] Google's impact may be even greater.

Although these companies profit from the public's use of their platforms, it remains unclear whether they acknowledge the *public responsibility* to be forthcoming about their political impact. Facebook, with over 2 billion users across the world, has but two-dozen employees working for its global governance team (according to several former employees I spoke with off the record). It's one thing to profit from hosting a place where we do public things like socialize or read the news, but it's quite another to be accountable to the public. Whether Facebook is accountable to its users and the principle of democracies seems questionable at best.

Politics and Media: A Virtual Reality?

At the time of his TV appearance in 1958, Huxley had just finished writing a series of essays titled "Enemies of Freedom." At the start of the interview Wallace asked him to clarify what or whom he referred to as an enemy. Huxley did not name any lone sinister individual but instead talked about "impersonal forces...pushing in the direction of less and less freedom." He suggested that "technological devices" could be used to accelerate the process of "imposing control." Today we often refer to these impersonal forces as "bad actors." I'll discuss some of them in chapter 7: the now-defunct Cambridge Analytica, for instance, and the Russian government, which has been accused of breaking legal and ethical boundaries to interfere in the 2016 US presidential election.

How have shifts in the production and use of media and technology come to impose control and thus threaten democracies? To get some answers, I interviewed David Axelrod, the chief strategist behind Barack Obama's successful outsider campaign for the US presidency in 2008. Axelrod is now a political analyst and host on CNN. He bluntly points out that politics in the United States and much of the world has turned into a *virtual reality*—that the internet and television, which have converged into one, now feed the viewer with perspectives and information that "affirm rather than inform."[7]

Just a few decades ago, and not only in the United States, Axelrod says, the few news networks that existed tended to focus on the middle ground, covering the news in ways that were far more similar than different to one another. This tendency in the United States carried over from the controversial and later abandoned Federal Communications Commission (FCC) fairness doctrine that originated in the late 1940s, when the mainstream TV networks were ABC, CBS, and NBC, as they remain today. The fairness doctrine required broadcast licensees to present controversial topics in ways that the FCC deemed honest, balanced, and objective.[8]

Today, however, news sources are *brands*. In the United States, Fox News speaks overwhelmingly to conservatives, including the current presidential administration, in an American version of what Axelrod describes as "state television." MSNBC positions itself as the resistance. Across the world, we see similar examples. Increasingly, viewers know what perspective they will receive without even listening to a single word uttered on any program. And with the internet, the situation is no different. We watch and view content that reinforces our positions, all aided by algorithms that recommend and personalize content.

How interesting, I thought, that Axelrod characterized CNN as a middle-of-the-road network when its reporting tends to lean toward a critical view of President Trump's transgressions. CNN, however, often excludes political positions that are significantly right or left of center. Does neutrality mean the absence of these voices, and is that erasure healthy for a democracy? Is the exclusion of these perspectives evidence that CNN is out of touch, given that the populist and supposed "anti-establishment" positions of politicians, such as Trump or Bernie Sanders, remain widely supported?

These questions bring other troubling ones to mind: Are we, as citizens, viewers, and users, really interested in unbiased, even objective, content? Is there even such a thing as impartiality? When Axelrod and I talked at some length about CNN's political analyst Chris Cuomo, he described Cuomo as a journalist who does not take sides. Then he questioned whether there is "a market for a robust middle ground, one that challenges political positions taken by conservatives and liberals alike."[9] Cuomo's *Prime Time* television show debuted in June 2018, airing at the same time as popular partisan shows hosted by MSNBC's Rachel Maddow and Fox News's Sean Hannity. Some viewed CNN's decision to put Cuomo in that slot as a "suicide mission." Nielsen ratings for January 2019 published in the *Washington Times* showed Cuomo, in his best month to date, still trailing with an average of 1.64 million nightly viewers to 3.25 million and 3.04 million (respectively) for Maddow and Hannity.

Axelrod and I next looked at the controversy surrounding Donald Trump and the media. His 2016 election was rife with questions and concerns regarding truth, information, and misinformation, which have continued to plague his early years in office. And yet Trump has continued to challenge and deride the mainstream media's "fake news" coverage of his administration—and both the left and right wings of the political spectrum continue to critique mainstream media. The president has recognized, according to Axelrod, that "if you are willing to light yourself on fire, or light someone else on fire, you can dominate any news cycle. People check the president's Twitter feed regularly for *a fix*, and I'm not just talking about people who are supporters."

Whether we speak of reporting on partisan television, online platforms, or even Axelrod's CNN, Trump is at the center of it. He continues to receive far more coverage on television and via social media platforms like Twitter than any other US politician. This has worked perfectly so far to solidify a political base, capture and maintain the attention of every voter, and disorient the opposition.[10]

How can such concentrated power emerge in a media and technology environment where so much information is available? Discussing Trump with me, Axelrod paraphrased the perspective of late US senator Daniel Patrick Moynihan: "You're entitled to your own opinion, but you're not entitled to your own facts." Describing today's political experience of media and technology as an "endless orgy of chocolate cake," Axelrod expressed concern about how the internet has sowed political divisions within and across societies.

Democracies operate slowly, and our societies and leaders need to make sense rather than simply exploit the disorientation that rapidly shifting technologies bring about. These technologies, according to Axelrod, "are coming faster and faster, creating a good deal of anxiety and a great deal of hyperactivity." It is up to us to intervene, to ensure that the systems we have embraced serve us all.[11]

Part of the challenge will be to look at what's behind the curtain, the *market logic* that drives our media and technology. Whether we speak of cable television or the internet, most news comes to us from private, for-profit corporations. These channels are created to make money. Their profits are transacted through our addiction and attention to what we view or click on. And they depend on our continued engagement with it. For example, CNN's constant framing of its programming as "breaking news" may mean that even if it is politically centrist, it leverages the frenzied sensationalism of a 24/7 news cycle. Axelrod acknowledged this point, agreeing that at the end of the day, if inscrutable private interests take our political systems hostage, then we must find an alternative. There is a role for profit-driven companies in our society, of course, but that role shouldn't endanger the values and interests of the 99 percent.

Do More Data and More Channels Mean More Democracy?

As my conversation with the man behind Obama illustrates, media environments across the world have changed as more channels, web pages, and apps compete for the attention of citizens. Daniel Kreiss, a scholar of politics and technology at the University of North Carolina, points out to me that the world looks completely different today, even in much of the global South, than it did in the 1960s.[12] In the United States, a few decades ago, one could run a political advertisement on television and reach over 90 percent of the electorate.[13] In Kreiss's recent book, *Prototype Politics*, however, he points out that we have witnessed a sea change: "Data has become more important because it helps candidates find the electorate, where they're paying attention, and get a message out in front of that."[14]

Most citizens across the world seem to accept personalization, even microtargeting, in their economic lives. Marketing analytics, like techniques that partition messaging and advertising to reach specific consumers (or demographics) have long existed.

What then has changed? First, we've experienced the exponential growth of data used not only to target us but also to organize the information we see. Combined with exponentially faster processing and cheaper storage, this rapid growth has made possible a type of fluid, even invisible, ordering of the world. Second, we've watched the use of analytics and personalization shift out of the commercial and economic spheres and into our political lives. That shift may compromise a major foundation of democracy, as the ability for the individual to participate politically without coercive manipulation is a central facet of democratic choice.

In the United States and a number of countries across the world, we see a weakening of traditional political parties and the rise of populism. Politicians such as Donald Trump, Rodrigo Duterte in the Philippines, and Recep Tayyip Erdoğan in Turkey express sympathy with aspects of authoritarianism (if not outright adoption of it). They have taken power, especially because many voters experience a sense of betrayal by establishment institutions and politicians. Such figures can rise to power as well through their ability to succeed within a technology and media climate where attention-grabbing content, turned into a spectacle on our Twitter or Facebook feeds, can be exploited for political gain.

Even in the pre-internet era, when television and other media networks were deregulated, we arrived upon a satellite-television world where hundreds of channels replaced a few.[15] Citizens today receive news from channels, networks, webpages, and applications whose numbers have expanded beyond imagination: we all experience the external political world through different "pipes of information." And yet, despite such a range of choices, different channels are often owned by the same holding company. Massive media corporations such as Disney and Time Warner control the television market, and engage even more viewers, offering more and different options and programs.[16] But, as the media scholar Robert McChesney cautioned in 1998: "The wealthier and more powerful media giants become, the poorer the prospects for participatory democracy."[17]

Aldous Huxley cautioned us in 1958 not to be surprised by our own advancing technology, its means of fueling propaganda, and its threat to democracy. In 2015, McChesney expressed cautious optimism about how technologies might support democracy: "We are in a position, in some respects for the first time, to make sense of the Internet experience and highlight the cutting-edge issues it poses for society, ... to better understand the decisions that society can make about that type of Internet we will have and, accordingly, what type of humans we will be and not be in the future."[18] To illustrate this point, let's next look at social media's role in shaping online political campaigns.

7   Cambridge Analytica and Global Disinformation

When I first became active on Facebook in 2009, I had no idea how massive the social network would become. And never did I imagine that the social media giant would be embroiled in a scandal involving the activities of Russian hackers and indirect "partners" like Cambridge Analytica, the firm that worked for Donald Trump in the 2016 US presidential election.

Cambridge Analytica was a political consulting firm that existed from 2013 to 2018. It earned a reputation for influencing electoral processes using a combination of data mining and analytics—by acquiring voter data, for instance, and developing psychometric techniques to hone strategic, personalized political messaging to these voters.[1] According to Analytica executives, the firm was involved in more than two hundred political campaigns around the world, including those in Kenya, Nigeria, Mexico, India, and Argentina.[2] (Company representatives have denied the firm's association to the Brexit campaign.[3]) Analytica worked early in the 2016 US primary campaign for the unsuccessful Republican candidate from Texas, Senator Ted Cruz. But it took the winning national campaign of Trump— and discoveries that unfolded in the aftermath—to earn the firm its current international notoriety.[4]

A quick overview of Analytica's relationship to the Jubilee Party in Kenya sheds light on the firm's practices and ethics. It first played a role in the country's 2013 presidential election, when the party took power under its current leader, Uhuru Kenyatta. During that campaign serious social and political rifts endemic to Kenya continued, and violence erupted throughout the country and region. Cambridge Analytica employed a strong ethnonationalist agenda in their PR platform, says the journalist Nanjala Nyabola, and embedded itself in the contentious and violent Kenyan election for profit.[5] In 2017, the party signed a three-month contract with Analytica and paid the firm $6 million to reelect Kenyatta.[6] Mark Turnbull, Analytica's managing director, has claimed that during those two campaigns the firm had "rebranded the entire party twice, written their manifesto, [and] done two rounds of 50,000 surveys" to determine Kenyan citizens' hopes and fears. "We'd write all the speeches and stage the whole thing, so [we had a hand in] just about every element of [the Jubilee Party's] campaign."[7]

Analytica's origins foreshadowed its future involvement with Trump. Robert Mercer, an American conservative billionaire who holds multiple patents in the field of computational linguistics, co-founded the firm with Steve Bannon, Trump's former chief strategist. Bannon, who was also the executive chairman of Breitbart News network, served as Analytica's vice president.[8]

Analytica suspended its operations in early May 2018 thanks to the data harvesting scandal I discuss in this chapter. Company representatives, however, claim innocence concerning charges of unethical behavior. They attribute the firm's closure to widespread negative publicity about its alleged purchase of the data that Facebook collected on tens of millions of US voters.[9]

In October 2016, just a month before the US election, Analytica's former CEO Alexander Nix, with whom I had some email contact regarding a planned interview, stated publicly that his company had acquired between four and five thousand data points about every adult in the United States (roughly 230 million people at that time). Carl Miller, the research director of the Centre for the Analysis of Social media, was troubled by the firm's ability to crunch that data into hyper-persuasive and hyper-targeted political messaging and even more concerned by Nix's offhand assurance that the firm did not see its methods as "intrusive."[10]

Psychometric Models and Data Mapping

Analytica collected data through a variety of means and sources. Those may include data brokers, the companies that buy information culled from a consumer's credit card history, or data-gathering sources such as supermarket rewards programs and movie-streaming habits, or automobile registries. This data would then be aggregated to create a "composite" profile of a citizen.

Analytica's recipe for success originated not from the data it gathered per se, but from mapping the data to complex *psychometric* models. One of these models is the OCEAN, which claims to understand human behavior based on a combination of five factors: openness to experience (O), conscientiousness (C), extraversion (E), agreeableness (A), and neuroticism (N).[11]

By the time Trump took office, news had spread about the "extremely individualistic targeting" Nix had boasted of before the election. On March 13, 2017, about two months after Trump's inauguration, I appeared on *Morning Joe*, MSNBC's flagship weekday news show, to talk about Analytica. During the interview Mika Brezinski and her panel asked me to explain psychometrics, data analytics, and how Facebook was used as a platform for Analytica's strategic messaging.[12] As I discussed the firm and the absence of personal data privacy legislation in the United States, most of the panel seemed shocked. One panelist—Rick Tyler, the former communications director for the Ted Cruz campaign—was not. Thanks to his campaign's experience with Analytica, he seemed quite aware of the potential uses of psychometric analyses to build targeted models of voters, which in turn could shape political campaigning online.

Here's why Facebook turned out to be the perfect psychometric laboratory for Analytica: The genius behind the Facebook platform lies not only its de facto role as the place to go for socialization online, but also in its ability to allow third parties to target it users with carefully framed ads and "news stories."

Even so, we can't assume that the mere ownership of data, combined with the use of psychometric modeling, directly impacts election results. Some have argued that Analytica's impact has been minimal.[13] Others admit that it remains unclear how (and at what level) strategic messaging affects our choices of who to vote for, or whether to vote at all. But we cannot overlook the pervasive and intrusive uses of data in political campaigns, especially these days, when almost any election is partly a story about data and the internet.

The Scandal Breaks

In March 2018, a former Analytica employee named Christopher Wylie revealed to *The Observer* that the British political consulting firm mined data from 50 million unknowing Facebook users in order to target them with digital political advertising. The firm was thrust into the middle of a major scandal that ultimately took it down. First to take center stage in the controversy was Aleksandr Kogan, a data scientist and lecturer at the University of Cambridge, who collaborated with Analytica.

Kogan admitted to harvesting the personal data of Facebook users with a survey app called thisisyourdigitallife (and subsequently transferring that data to Analytica). He developed the app independently when he was co-director of a company called Global Science Research. The hundreds of thousands of users who took the survey he administered were paid in the name of academic research. They consented to have their data collected by agreeing to the following terms:

> This app is part of a research program in the Department of Psychology at the University of Cambridge. We are using this app for research purposes—learning about how people's Facebook behaviour can be used to better understand their psychological traits, well-being, health, etc and overcome classic problems in social science. Users of the app will be presented with a description of the types of data we gather and the scientific purpose of the data. Users will be informed that the data will be carefully protected and never used for commercial purposes.[14]

The app collected not only the data of the consenting test subjects, but also the data of their unknowing friends on the Facebook network—and all of it eventually accumulated into an immense data pool. Some estimates claim it took data from as many as 87 million users, exceeding the initial estimate of 50 million.[15]

Adding to the confusion were the "terms of service," the conditions users sign on to when they create accounts with a given technology. On April 22, 2018, Kogan appeared on *60 Minutes*, one of the most widely watched weekly news shows in the United States. While talking about the terms users had to agree to in order to take the test, he stipulated that clicking OK gave the app's developer permission to "disseminate, transfer, or *sell* (user) data."[16] This was at odds with Facebook's developer policy, which prohibited third-party sales of app-gathered data. But, according to Kogan, neither he nor Facebook employees read this policy. By and large, Facebook users also fail to read these terms. Most of us would agree that they are "beyond comprehension," as the Republican US senator Lindsey Graham characterized them.[17]

It's clear from the Facebook perspective that *access to data is not a problem; it is a feature*. Indeed, Kogan described the prevailing assumption across

Silicon Valley: access to personal data is something about which "everybody knows [and] no one cares."[18]

As Cambridge Analytica put its immense pool of data into a complex "profiling system," the firm was able to extrapolate from it using psychological surveys blended with commercial data and voting histories. Then it assigned one of its thirty-two personality types to both surveyed and non-surveyed individuals. This allowed Analytica to customize campaign advertising and target specific individuals that fall under one or another personality type. Facebook's "dark post" feature, which allows specifically tailored posts to be only visible to individual users, made this technique possible.[19] As a result, it is nearly impossible to be aware of what another Facebook user may see, let alone to rebut or regulate such activity.

In the past, if an inflammatory piece of news surfaced, a user could respond with a counter-position because the story would be publicly accessible. The electorate could weigh different perspectives on that news story as they made their decision about who or what to support. Now, because campaigns can get such intimate demographic and psychographic data on voters, using digital footprints from their everyday lives, they can push persuasive messages *targeted to the individual voter and hidden from everyone else.*

This data-mining practice is relatively common. Facebook uses similar data analytics to categorize its users as very liberal, liberal, moderate, conservative, or very conservative, and it adjusts its advertising strategies accordingly. But the discovery of such data harvesting in tandem with psychometric models associated with Trump's controversial election caused widespread alarm, bringing the question of Facebook's role in the presidential election, and in politics more broadly, to the forefront of national discourse in the United States.[20]

Facebook Joins the Showdown

What followed was a series of accusations: Cambridge Analytica, Kogan, and Facebook each pointed the finger of blame at one another. Facebook's response amid the controversy was to ban both Cambridge Analytica and Kogan from the platform, stating that Kogan had violated its policy.[21] Analytica claimed that the company was unaware of Kogan's specific process for collecting the data. Kogan counterattacked, insisting that Analytica had assured him the process was legal, and even highly typical.[22]

Figure 7.1
Facebook's Mark Zuckerberg surrounded by media reps and members of the US Congress.

The scandal forced Zuckerberg and Facebook COO Sheryl Sandberg to answer questions. Both executives sat for media interviews, apologizing for their negligence around the supposedly unanticipated effects of how their systems were used.[23] Zuckerberg, over two days and ten-plus hours on April 10 and 11, 2018, answered questions from a bipartisan commission of the US Senate and House of Representatives (figure 7.1).[24]

Listening with rapt attention to both days of testimony, I noted how many questions revealed an absence of understanding and a lack of "digital literacy" by politicians about what is at stake with data, the internet, and our political lives.

Senator Orrin Hatch, for example, revealed his ignorance about how most social media platforms make a profit when he asked Zuckerberg: "How do you sustain a business model in which users don't pay for your service?" Zuckerberg responded, "Senator, we run ads."[25] Without the right language to articulate their questions, the group of politicians found it hard to confront Zuckerberg on more complex and urgent issues concerning the use and abuse of data and user privacy. When asked whether his company was a monopoly, Zuckerberg stated: "It certainly doesn't feel like that to me." The answer went unchallenged.

In his testimony Zuckerberg seemed to skirt tough issues by referencing the technology's humble beginnings back in his "dorm room." Did he mean to imply that because he never anticipated the problems the company faces today, he was somehow justified for not fixing them when they surfaced? Or was he saying there had never been strategic or corporate decisions that required deviating from the company's original game plan? These tough questions were never asked or answered.

Some Facebook critics have reported that the company's claim of naiveté and ignorance concerning fake news was disingenuous. They argued that Facebook intentionally spread disinformation to shift blame away from itself and toward other companies in an effort to fend off government regulation. For example, Facebook employed a Republican opposition research firm to discredit activist protesters and linked them to George Soros, the controversial billionaire who has given large sums of money to support progressive politicians and an open and neutral internet.[26]

Zuckerberg's pledge to better "police" Facebook was notable for its vagueness, specifically the promised solution of "better AI."[27] He told the US Senate: "It's not enough to just build tools. We need to make sure they're used for good. That means we need to now take a more active view in policing the ecosystem."[28] All this sounds good, but questions abound. How does Facebook, or for that matter any major data or technology company, characterize the "good" use of tools, "policing," or even what it means for Facebook to be "more active"? How such terms are defined and implemented makes all the difference. Should tech giants have the power to define these or should we?

Evasive Answers, Hidden Agendas

In Zuckerberg's testimony we hear simplistic and evasive requests: *Trust us: we will take care of any issues ourselves, and root out "bad actors" based on a set of principles that we refuse to clarify or disclose. Trust our technology independent of how it is designed and whom it serves.*[29] These appeals ignore a fundamental problem: the thousands of data points that are being collected, bought, and sold by politicians and corporations about every member of the electorate remain *hidden from public access or audit*. The issue affects everyone in our world with a mobile phone or internet access. And in some cases it has an impact on those who do not, but are still being monitored.

Zuckerberg, to his credit, did agree to four recommendations in his conversation with the Senate, though largely in abstract terms: (1) privacy legislation, (2) giving users control over who sees their content, (3) legislation concerning facial recognition, and (4) the Honest Ads Act, introduced on a bipartisan basis by three senators.[30,31] He also acknowledged that Facebook ultimately holds responsibility for the content its users see.

But the devil is in the details. At the same time as Zuckerberg's testimony, we learned that some Illinois legislators, with the support of lobbyists and with Facebook's active encouragement, have been pushing to gut a state law that would give the company "free rein to run facial recognition scans without users' consent."[32] (See figure 7.2.) This is consistent with reports from Europe. Despite impending changes to terms and conditions stipulated by General Data Protection Regulation (GDPR) legislation in the EU, Facebook managed to roll out its facial recognition technology by sending users a consent flow of manipulative "opt-in" notifications. These messages encouraged users to turn on the feature and, in effect, help Facebook fill in its "own facial recognition blanks." By opting in, European users consented to Facebook's "grabbing and using their biometric data."[33]

The Analytica scandal made public the collecting and monitoring of Facebook non-users, including those who have *never registered* for a

CityNews   FACEBOOK TO RECOGNIZE USERS IN PHOTOS   TUE 5°

■ IF YOU SEE NEWS IN ACTION OR HAVE A STORY IDEA, GET IN TOUCH WITH US. REACH OUT TO 5:27 PM

Figure 7.2
Facebook pushes for facial recognition technology.

Facebook account. Facebook claims it has done this "for [the user's] own good."[34] (We have to wonder why Facebook is so good at engineering connections between itself and our data, but so bad at anticipating or acting upon privacy issues.)

The scandal also forced many of us to reckon with a disturbing reality: how our data is mined and sold for commercial and political purposes. For many, this meant focusing on Kogan as a media figurehead, either a grossly unethical tech developer or an unaware scapegoat for Cambridge Analytica's malpractice. Or it meant propping up "whistleblower" Christopher Wylie, who went on to point out in April 2018 that a "lot of people" had access to this data, and that "it could be stored in various parts of the world, including Russia, given the fact that [Kogan] was going back and forth between the United Kingdom and Russia."[35]

It is too simple to label either Analytica or Kogan as a villain and see Wylie as a hero. (And it turns out that Wylie himself may also have had access to the data. His startup Eunoia Technologies "reportedly approached the Trump campaign in 2015 about elections-related work."[36]) Instead, let's consider the house of cards upon which our political lives rest in relation to personal data. Dirty technological tactics, such as hacking personal and other data, are commonplace in politics. The United States aims these tactics toward other nations, other nations toward one another, and most nations toward their citizens. Noam Chomsky described the Analytica situation to me, relative to US involvement in foreign elections and politics, as "a toothpick on a mountain."[37] Perhaps, but the larger question is one of control and power over data in relation to our political lives.

We must remember that it is not in Facebook's interest, nor would it be advantageous for other powerful corporations, to change its practices. Their ultimate obedience is to one figure: their bottom line. So then where to go? We can start by asking some imaginative questions, ones that avoid treating Facebook as a unique problem but instead consider what type of technology industry would work for all of us.

For example, how would a free service like Facebook function without monitoring its users and monetizing their activity through the sale of data? Could the social network giant be profitable without engaging in activities that lead to disastrous consequences?

The issues I have raised are global, not just relevant in the United States, or even the Western world. If anything, they are even more potent in parts of the world where independent media are weaker, including in countries under authoritarian rule. The examples are piling up to reveal how platforms like Facebook have promoted conspiracy theories and blatantly false news and rumors, fueling violence and discord.

The two incidents I next describe expose these problems in action, the first from Myanmar and the second from across the world involving "cyber troops." They show how Facebook and the popular applications it owns, such as Instagram or WhatsApp, can be used to instigate and spread rumors and hate speech, and become the mass media Facebook users view.[38]

Rohingya Silencing

In 2017, international concern heightened over accusations that Myanmar's military was engaging in ethnic cleansing of the Rohingya Muslim minority. Myanmar is a mostly Buddhist country, and the Rohingya Muslims living there are widely considered to be among the most repressed and persecuted minorities in the modern world. The Myanmar government essentially treats them as undocumented immigrants. In late August 2017 the government declared the Rohingya a terrorist organization after insurgents attacked police posts in the state of Rakhine and killed twelve members of the security force. It then instituted a program carried out by Burmese troops and backed by Buddhist mobs to torch villages, gun down civilians, and rape or otherwise torture the Rohingya. That forced more than 400,000 Rohingya to flee to Bangladesh.[39]

Attacks by a dominant population targeting its own ethnic and religious minorities have occurred throughout history. In the case of the Rohingya, manipulated media reports surfaced on Facebook that justified and even perpetuated the genocide. The situation is eerily similar to events in the African nation of Rwanda between April and July of 1994. At that time, when the most popular medium in the nation was radio, false narratives and sensationalized fearmongering were used to legitimate attacks that broke out.[40] Radio broadcasters working for the Hutus, for example, would refer to the Tutsis as "cockroaches."[41] The national station Radio Muhabura, which promoted unity rather than divisiveness, was less accessible and therefore less listened to across the nation.[42]

Unfortunately for the Rohingya, Facebook turned out to be just as influential in the Burmese context as dominant radio stations had been in

Rwanda. Facebook designated the Arakan Rohingya Salvation Army (ARSA), a Rohingya insurgent group, as a "dangerous organization" for its supposed engagement in "terrorism."[43] The company ordered its moderators to delete any content "by" the group or "praising it." Facebook claimed the decision was made internally, with no direct request from (and in no direct response to) the government of Myanmar to follow its own recent declaration of ARSA as a terrorist group. Rather, the company said it based its assessment on allegations of the group's "violent activity" and not on its "political aims."[44]

At the same time, the Burmese military responsible for the violence had a verified Facebook page with 2.6 million followers, even though the top human rights official in the United Nations had accused the military of disproportionate violence in a "textbook example of ethnic cleansing."[45] In addition, Myanmar's leader, Aung San Suu Kyi, was widely understood to be stoking anti-Rohingya fervor with her social media posts. Her spokesman, Zaw Htay, re-posted Facebook's announcement that it would censor all content in support of ARSA. That post was shared nearly 7,000 times.[46]

Criticism of Facebook's decision to censor reports of human rights violations of a persecuted religious minority intensified given the company's professed policy to allow graphic content that was "newsworthy, significant, or important to the public interest."[47] Rohingya activists dedicated to stopping the genocide complained that Facebook was censoring posts documenting the brutal campaign against the Rohingya.[48]

The issue brings up many serious questions: How did Facebook decide that images and videos of the plight of hundreds of thousands of Rohingya refugees were not "of public interest"? How did it occur in the first place that Facebook could ban an organization representing a persecuted Muslim minority, rather than the military considered to be partaking in ethnic cleansing as defined by international consensus? Does the company actually know much about the situation in Myanmar and its brutal treatment of minorities throughout history?

Although Facebook continues to insist it is neutral, there are political stakes to the ways the company moderates content. Its belief that it can respond to and censor "violence" on its own terms—rather than through deep political or cultural analysis, or by relying on actual experts—speaks to the company's unwillingness to be transparent about its political power. And indeed, this is not merely a story about Myanmar. For example,

Facebook has recently been exposed for its role in fueling anti-Muslim violence based on false information in Sri Lanka.[49]

In an interview, Zuckerberg initially used the Myanmar incident as an example of Facebook's effectiveness in "tackling hate speech," claiming that its systems had effectively detected and removed these messages.[50] The pride with which Zuckerberg articulated this statement indicates that Facebook actually desires to act, not neutrally, but in a positive, and programmatic way—that is, to convert its approach to censoring hate speech into publicity for the company. In attempting to "tackle" hate speech, it's necessary to determine what constitutes hate speech to begin with—a task normally left to national governments and determined by civil and criminal courts rather than a bunch of executives and engineers sitting thousands of miles away.

Of course, private companies have the freedom to allow or disallow, with some legal restraint, what kind of speech they want to represent or host within their private enterprises. But these decisions must always be seen relative to political urgency, and especially when it comes to events unfolding around the world. Just as it became hard to ignore Facebook's role in incubating alt-right and white nationalist fervor during and after the 2016 US election, the case of Myanmar reveals that Facebook's so-called hands-off approach to political arbitration could in fact legitimize violence and hate.

Knowing this, we should look with a wary eye toward Zuckerberg's statement that "[Facebook] is building artificial intelligence to help better identify abusive, hateful or false content" even before this content is flagged by users. This allows Facebook to escape the responsibility it should have regarding hate speech and how we want to deal with it in our communities.[51] This responsibility requires a human touch, but it's also unclear whether the appropriate humans for this job should be those employed by Facebook. Why can't social media platform support the autonomy and power of user communities on the ground, for instance by enabling conflict-resolution or reconciliation processes, rather than continuing to direct power to the one place it always does: itself?

Cyber Troops across the World

Tech companies continue to wield their massive power and threaten political democracy by deploying "cyber troops." A University of Oxford study that defines the term as "government, military, or political party teams committed to manipulating public opinion over social media" notes that cyber troops are now being used in at least twenty-eight countries around the world.[52]

Many countries are experts at managing and manipulating public opinion online. The work of cyber troops is actively carried out "on the ground" when governments and political entities pose as everyday social media users, sometimes hiring people or using bots to do so. As with traditional propaganda, the goal is to make a government or party look good, secure public support for its goals and policies, and squelch critics. They can use misinformation, lies, and the manipulation of a population's basest prejudices in the pursuit of political ends.

President Rodrigo Duterte of the Philippines—a controversial figure who has attracted international attention for a bloody "war on drugs" that has killed more than three thousand of the nation's citizens—has admitted to relying on cyber troops to get elected. But unlike Cambridge Analytica and other firms that use individual targeting methods for campaign advertisement *prior* to elections, Duterte called in the cyber troops following his election in the name of legitimizing his violent regime, spreading propaganda, and amplifying messages that support his policies.

These digital foot soldiers in the Philippines consist of volunteers and private contractors who use different strategies to achieve the same overall goal. They generate positive social media interactions to reinforce the government or its political ideology. Or they instigate negative interactions that involve verbal abuse and online harassment, called trolling, to target government dissenters. The individual targeting methods of cyber troops silence political dissent online in ways that pose real-life danger like that witnessed in Myanmar. But rather than masking the plight of hundreds of thousands of religious minorities through mass censorship, negative cyber attacks call attention to specific online users, making them vulnerable and visible to Duterte's "war on drugs."

Governments use cyber troops in blogs, iPhone apps, and other online resources by soliciting volunteers to help share and endorse government content, and sometimes to galvanize the government's existing supporters against their critics. In Ecuador, for example, the government-hosted website Somos investigates and identifies government critics on the web. Then it sends updates to a list of subscribers, and encourages pro-government users to respond to and even attack those users with opposing or critical views.

In Sudan, the government maintains cyber troops that actually infiltrate Facebook, WhatsApp, and other services in order to spread its leaders' messages to users throughout the nation. In India, the world's largest democracy, WhatsApp has become a cyber troop platform by which fake news and religious hatred are directed toward the Muslim minority. In May 2018, in the southern Indian state of Karnataka, the nation's two major political parties claimed to have access to more than twenty thousand WhatsApp groups each, allowing them to rapidly mobilize supporters. Because a WhatsApp group is a closed conversation involving only its members, it is difficult to access or rebut from outside. Thus it's become an effective tool in spreading misinformation and radicalizing political sentiment. The nation's prime minister, Narendra Modi, is well known for his use of "WhatsApp warriors" who run grassroots groups known to post content warning voters about Muslims. A spokesperson from Modi's Bharatiya Janata Party in Karnataka describes the effectiveness of this tool: "Promoting one's manifesto is easier than ever before. … We get to know the ground reality in moments."[53] And in Brazil, with the election of Jair Bolsonaro as president, cyber troops were active in spreading misinformation and rumors via WhatsApp.[54]

Bots used by government actors step up cyber troop activity throughout the world, especially in places like South Korea, Syria, Turkey, and Saudi Arabia, where the unpopular voices and ideas of the government are sometimes amplified and inflated with fake likes, shares, and retweets.

The cyber troop example complements the troubling censorship and content moderation practices we see with many tech companies. As long as we rely on algorithmically determined content, cyber troops—whether they consist of real individuals, bots, or both—can sensationalize ideological messages throughout the web, making users vulnerable and susceptible to political manipulation. And, as in the case of the Philippines, government-imposed threats can put users at risk.

The remedy to political manipulation is not simply a question of making data visible or accessible but rather about infiltrating technological systems from the ground level, entering into them as unregulated users, and taking advantage of the rules already in place to manipulate and influence them. The threat posed by this kind of ground-level infiltration, on the user-end, paired with the lack of regulation and transparency on the other hand, makes for a very uneasy and misleading political terrain.

Democracy relies on access to information and the free sharing and debate of ideas. But the same democratic principles that so many seem to love about the internet, and social media specifically—its supposed support of free speech, marginalized voices, and political and social movements—have a dark side. They can also threaten independent media, open conversation, lead to violence and destruction.

Except for an occasional Tweet, the former US president Barack Obama kept a low profile his first year after leaving office. Then, in January 2018, he appeared to Netflix viewers in the premiere episode of *My Next Guest Needs No Introduction, with David Letterman*. Obama and the former late night talk show host, himself a recent retiree, covered a range of topics, from the humorous to the politically substantive.[1]

Strikingly, Obama spoke of technology and media today as a *threat to democracy*, saying that those who watch the politically conservative Fox News channel in the United States are "living on a different planet" than those who listen to National Public Radio.[2]

Obama's point might be even more significant when we apply it to the *online experiences* that politically polarize us. He told Letterman:

> One of the biggest challenges we have to democracy is the degree to which we don't share a common baseline of facts.... [Citizens across the world] are operating in completely different information universes.... At a certain point you just live in a bubble. And that's part of why our politics is so polarized right now.[3]

Obama himself exploited social media and citizen data to engineer two successful political campaigns, especially via his insurgent 2008 election that stunned the world.[4] Compared to all the hue and cry about Russian interference in the 2016 US election or the Cambridge Analytica scandal, Obama's point is more fundamental—our televisual, radio, and online media isolate us. This is the challenge he believes we face today: "How do we maintain that sense of common purpose, [of being] 'in it together' as opposed to splintering and dividing?"[5]

Today we either select the channels we watch based on existing political perspectives, or we go online to see the perspectives of our trusted friends (who often share our political perspectives). We think that what we see, via our search results and online feeds, is part of the "open" information universe. We are only beginning to learn how that universe is skewed, based not only on our choices of whom to follow or "friend," but also the content that algorithms show and hide.

The issue becomes more troubling because many technologies are designed to get attention by feeding us information that reinforces and sharpens our existing attitudes, even those based on biases we have and may not be conscious of ourselves.

A great example of this, illustrated by the technology scholar Zeynep Tufekci, is the Google-owned YouTube recommendation system. Calling it the "great radicalizer," she points out that the video site's recommendation algorithm "seems to have concluded that people are drawn to content that is more extreme than what they started [their search] with—or to incendiary content in general."[6]

While watching Donald Trump rallies on YouTube, for instance, Tufekci noticed that the videos next up to "autoplay" included rants from neo-Nazis, white supremacists, and other extremist right-wing fringe groups. The suggested videos connected to Hillary Clinton or Bernie Sanders rallies took a conspiratorial left-wing approach, with claims that the US government planned the 9/11 attacks, or that the government itself was a secret cabal. "It's as if you're never hardcore enough," she tells us. Fair enough, but is that the right design approach for a technology that has well over a billion users and is the largest consumer of bandwidth on the internet?

The effects of these autoplay suggestions on us, as users, deserve scrutiny. Jonathan Albright, from the Tow Center for Digital Journalism, points out that the video-sharing platform's algorithms have spread "crisis-actor content." The term applies to videos in which people pretend to be survivors of a traumatic incident (let's say a mass shooting) to further a particular agenda (let's say to push gun control or conspiracy theories). YouTube's role in hosting and spreading such videos, says Albright, "is akin to a parasitic relationship with the public." This genre of video is troublesome, since it not only targets an individual psychology but it also has the potential to trigger *mass public reactions*.[7]

Crisis-actor and algorithm-driven content are not just a YouTube problem, warns Tufekci. They are pervasive across other major technology platforms that host billions of users.

Ethan Zuckerman calls the effect of this phenomenon "normalizing the abnormal." Building on the work of the media scholar Dan Hallin, Zuckerman points to three common spheres of media discourse: consensus, legitimate controversy, and deviance. In democracies, most conversations and debates should fall within the category of "legitimate controversy"—where we embrace different perspectives in the spirit of tolerance and reason. But now, our online world and media ecosystem have made deviance the new normal. And we experience this deviance individually: the internet user sitting right next to use may be seeing entirely different content despite searching for the same subject.[8]

Tufekci, alarmed by this new normal, states: "We are building a dystopian society just to make people click on ads."[9] She and many other journalists and scholars warn that our media systems and technologies are further closing us off from one another, placing us in black boxes, echo chambers, and filter bubbles.[10]

In the Letterman interview Obama put filter bubbles into context when he described the results of an experiment done in 2011 during the height of the Arab Spring and Egyptian revolution. When googling "Egypt," US-based conservatives saw results related to the Islamist (though not radical) Muslim Brotherhood. For liberals, information about "Tahrir Square" came up. And for moderates, "Vacation Spots on the Nile" surfaced.[11] This example shows how online searches for news about one of the most important political events in recent memory obscured the complete picture. These filters close off our political awareness and consciousness. And they close us off from one another.

We are much more complex, however, than our Democratic or Republican, or left- or right-wing selves suggest. Many of us know the opinions of those with whom we disagree, though we may struggle to humanize their perspectives and reasons for believing what they do. According to the political scientist Daniel Kreiss, filter bubbles may contribute to a hardening of political identity, introducing a social world online that makes our existing opinions seem normal and natural.[12] As polarization increases, the need to find common purpose, to see our lives as interwoven, becomes more challenging. It becomes difficult to imagine how a positive outcome for someone else also may benefit us. It is far easier to see life as a zero-sum game.

In their book *Democracy for Realists*, Christopher Achens and Larry Bartels come to a conclusion based on their analysis of findings from social science research: our identity, including our partisan loyalty, is what ultimately matters in voting behavior, not our education and political literacy.[13] When the "social world" made visible to us by Facebook, YouTube, or Google influences us to quickly dismiss or accept a policy position based on our sense of identity, rather than on the issue itself, it threatens a central premise of democracy: a public that can rationally interpret, debate, and reflect on different points of view.

Search Engines: Political and Racial Manipulation

What about the internet's most visited website, Google.com, the de facto search engine of the world, dominating 86 percent of the global search market?[14] It too has massive power over voting behavior and perceptions of political reality.[15]

The psychologist Robert Epstein and his colleagues have recruited control groups to study and analyze the different aspects of the participants' interactions with the Google search engine: the way it *orders* the results, the information it *filters* out, and its use of the *auto-suggestion* feature. While these studies look at Google's impact, they can be applied to consider any technology platform that consumes our attention and filters what we see. And they show that underlying the many conversations and different political viewpoints we may have, the ordering and filtering of search results powerfully shapes what we see and think.

Let's look at the order of results: Today we commonly hear of organizations and businesses that justify spending many millions of dollars on a process called *search engine optimization* (SEO) to make their identities visible online. Why is this so important? An answer lies in the study of our online behavior: when presented with search results, over 50 percent of us click on one of the top two items on the list, and more than 90 percent of our clicks go the first ten items on the first page of results. If a perspective or source of information is not in this list, it essentially fails to exist for us. So it makes sense, says Epstein to try and "trick Google's search algorithm—the computer program that does the selecting and ranking—into boosting [a business or organization] up a notch or two."[16]

Epstein and his team devised a way to set up a mock search engine (using real web pages for authenticity) to compare how three different groups made choices about political candidates in foreign elections (to help ensure impartial results). They observed an effect that they call the search engine manipulation effect (SEME).[17]

Their data show that for undecided voters, the manipulation of search results through suggestions can transform an election predicted to have a 50–50 outcome at the start into a win with a 90 percent majority.[18] Are users aware of this bias at play? Epstein found the answer to be an overwhelming no. This held true even when the results were *masked*—in other words, when the order of results shifts, for instance, by placing a less favored candidate in the third or fourth position of the search outcome.

The research team's most disturbing observation resulted from a study it conducted in 2014 using information about candidates in a three- way race for prime minister in India: a whopping 99.5 percent of the participants had no clue that "they were viewing biased search ranking." Even in cases where users were very familiar with the candidates, the shift was significant. SEME, explains Epstein, "is a force you can't see; but unlike subliminal stimuli, it has an enormous impact—like Casper the ghost pushing you down a flight of stairs."[19]

Consider that in the 2016 US election, Hillary Clinton won the popular vote by about 2 percent, and that the electoral victor, Donald Trump, was largely successful thanks to his wins by less than 1 percent in Pennsylvania, Ohio, and Michigan.[20]

Contrast this with the conclusion made by Epstein and his colleague Ronald Robertson: Google can shift up to 25 percent of the national elections in the world.[21] With undecided voters and certain demographics the numbers are much higher. The message behind this research is clear: the hidden, invisible ways in which search engines function perceptively transform elections and our political lives.

This is not merely an issue related to search. It is also about *what we see online*, and the impact this has on our political landscape. For example, an article for the journal *Nature* by the researcher Robert Bond described a 2010 Facebook experiment in which the company sent "go out and vote" reminders to 61 million of its users.[22] The study revealed that 340,000 of those people voted who otherwise would not have. This raises many important questions including: How did the company determine which users would receive the messages, and did it target them based on a particular political identity? What potential for misuse lurks behind Facebook's agenda here?

What about the impact of these biases on vulnerable populations? Let's look at one of the most marginalized groups in the United States, the African American population. Safiya Noble, a scholar of technology and race, points out in her 2018 book *Algorithms of Oppression* that data discrimination exacerbates existing prejudices because algorithms reflect and normalize the biases of algorithm designers (or company executives), who tend to be mostly white and male.[23]

In 2018 *Time* magazine published an excerpt from Noble's book that recounts her search parameters and results, dating back to 2009, for the term "black girls."[24] Despite not including the words *porn* or *sex* in the search query, her first page of results included words such as *pussy*, *sugary*, *hairy*, *booty/ass*, and *porn star*. This was consistent with the autosuggestions she received for a number of different queries made on the site. Noble claims that the hidden engines shaping visibility and prioritization via Google reflect "women's historical and contemporary lack of status in society—a direct mapping of old media traditions into new media architecture."[25]

One of the major political issues within the United States relates to police shootings of African Americans. The first tragedy to polarize public response in this way occurred in Sanford, Florida, in February 2012, when George Zimmerman, a vigilante who was the "neighborhood watch" coordinator in community, shot seventeen-year-old Trayvon Martin, who was unarmed. The incident launched a firestorm that only intensified after a jury acquitted Zimmerman in July 2013. The Black Lives Matter movement, started by three African American women, was born out of Martin's shooting and picked up steam as a rash of other police shootings of young black men and women became staples of the political news cycle.[26]

How did Google represent this crisis? The search suggestions cast Martin as a "thug" and Zimmerman as a "hero" (see figure 8.1). In this manner the event is framed to attack the legacy and identity of an unarmed black teen victim.

The control of the search engine market by Google allows it to *define political events*. This practice threatens a democratic political environment built upon values of equality, diversity, and justice. The myth that our world has entered into a "post-racial" society, she says, combined with the dominance of corporations such as Google, means that we now accept the unacceptable: At Google, Facebook, and other large tech companies,

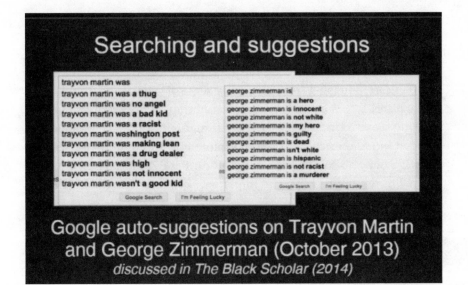

Figure 8.1
Google search suggestions assign blame to the victim.

for instance, we see massive underemployment of minorities and women, particularly in positions of management and engineering. And yet "intelligent" technologies seem to contribute to some of the biases that may have led to their underemployment to begin with.

Noble calls for public custodianship and regulation around how we find and navigate information online. She points out that public libraries could better develop search technologies that overcome racial and political biases, given their research into alternative methods of classifying information and their accountability to the public. The stakes are high—the disconnected commercial interests that drive the most popular platforms of the internet have the potential to affect electoral, racial, and even body politics.

These concerns parallel a disturbing study recently published in *Science* in March 2017. The study findings revealed that AI technologies might impose cultural biases and stereotypes on the texts they analyze. For instance, the algorithmic AI tool called "word embedding," when applied to the human resources task of scanning CVs, showed a 50 percent bias toward names that were European American versus African American, even when the CVs documented nearly identical education and experience.[27] Similarly, pleasant words were far more often associated with a list of European American names, whereas African American names were more easily associated with negative words.

Despite these troubling findings, the study co-author Arvind Narayanan explains that word embedding has been "spectacularly successful in the last few years in helping computers make sense of language."[28] To the credit of Narayan and his co-authors, however, the study does not assume that "making sense" of language is a neutral engineering task. They argue that we need more research not only to explore how machines make sense of language but also to analyze their political implications. This study and others I have described share an important perspective: AI not only learns our prejudices, it mimics them.

Algorithms and Data: Opening or Closing Our Worlds?

It is entirely possible to take a different approach toward these issues—to claim that overall, the systems so many of us have turned to are not just more efficient, but actually deliver and support "human values." This is the perspective taken by Michal Kosinski, a data scientist and psychologist most well known for his provocative and fascinating work in studying humans through *digital footprints*, or the data users give up in their interactions with digital devices and systems.

Now a professor at Stanford University's Graduate School of Business, Kosinski was a doctoral student at Cambridge University when he pioneered research showing that with these footprints he could make predictions of a user's gender, psychology, sexuality, religion, and more. Kosinski has been linked to the techniques used by Cambridge Analytica, though he has nothing to do with the political firm, and he describes it and its methods as unsophisticated and not worth our concern (see figure 8.2, which shows a headline conveying a conflicting opinion). Nonetheless, Kosinski has been portrayed as a protagonist in the Analytica discussion, particularly via an attention-grabbing piece titled, "The Data That Turned the World Upside Down." In that article the authors delve deeply into his psychometric research work and apparently his mind, as they explain his "bad feeling" when Aleksandr Kogan contacted him in 2014 for permission to access Kosinski's database in preparation for a secret project that Kogan wasn't at liberty to disclose.[29]

In chapter 4, I explained how our digital footprints reveal more about us and, analyzed together, may know us better than our acquaintances, friends,

Figure 8.2
A sensational headline connecting Cambridge Analytica to Facebook and Trump (source: Motherboard).

or close relations do. What does Kosinski have to say about the ethics of collecting this data? Surprisingly, he argues that data-tracking and algorithmic technologies open up the world to us, rather than close it down. He notes that in the past, human beings lived in a more limited and controlled information universe. What we learned, how we thought, what we believed were all largely controlled by the societies of which we were part. The priest, teacher, politician, and parent—all of these individuals not only nurtured us (hopefully) but also socialized us and influenced how and what we think. They controlled us far more than the internet does today, he argues.[30]

Kosinski sees the role of the internet as similar to how the printing press allowed the written word to reach new audiences, fostering and inspiring new communities to reflect and create their own documents.[31]

How do we reconcile this optimistic view with the concerns raised in this book? What about the "new connections" we experience online, controlled by forces over which we have no control? What about the fact that what we see online tends to reinforce existing political biases? Kosinski disagrees with the negative premise behind these questions.[32] He argues that algo-rithmically controlled recommendation systems do not merely echo our views, or wall off our perspectives. Instead, he believes they actually *expand* our worldviews.

Kosinski gave me the simple example of someone who is pro-gun own-ership, yet also likes action movies and is a fan of Britney Spears. Imagine an algorithm suggesting content for this person based on "liking" Spears. Would the content necessarily confirm the individual's pro-gun bias, or perhaps select dissenting content based on the fact that Spears has spoken out for gun control?[33] The latter is entirely possible, and in Kosinski's esti-mation, more common than not online.

What, then, can we say about the impact of Cambridge Analytica? Kosinski described the political agency as "totally insignificant" and far less adept at using data for psychometric political predictions compared to a number of other organizations.[34] When I asked him *which* others, his answer left me dumbfounded: *every big public-relations company.* (When I pressed for specifics, he mentioned only one: Quantcast.) Regardless of details, Kosin-ski kept going back to the same points: There is nothing exceptional about Analytica except its success in acquiring so much data, thanks to Facebook's negligence. There is nothing they did with this data, in his view, that was relevant or effective.

Can the type of targeting that Analytica undertook be problematic in Kosinski's mind? Potentially, but not compared to the incredible benefits he believes algorithmic personalization offers. *Personalization,* the process by which a user adapts an interface or system to suit their preferences, may be far from transparent, but for this Stanford data scientist and psycholo-gist, it actually is the *gateway to transparency.*[35] Algorithms give us paths to experience new information, he insists. They open up our worlds in ways that are digestible and efficient, rather than place us in filter bubbles and close us off. Letting privacy go, he argues, is a bit like fire—it can warm one's home, but at certain points may need to be controlled.[36]

I was both troubled and fascinated by the perspectives of Kosinski. But our conversation left me with one primary question: Is the choice between privacy or transparency a false one? Is it misleading to assume that we can only have one if we're willing to do without the other? And on whose terms are these values defined and implemented?

9 Bernie Is Born

The 2020 campaign of Senator Bernie Sanders for the US presidency is alive and well I as write this chapter. But in 2016, partly thanks to his team's creative use of technology, his campaign shook the country's political establishment. Sanders represents Vermont, a state with a population of less than 1 million, which is less than 0.3 percent of the overall national population.[1] In the month before launching his campaign in 2015, Sanders was polling around 3 percent without a single declared supporter from the US Congress.[2]

And yet, within twenty-four hours of declaring his entry into the race on the Democratic ticket, Sanders had raised $1.5 million, more than 100,000 people had signed up to join his campaign, and 35,000 of those had made donations.[3] Within six months Bernie had forced a fierce contest with Hillary Clinton and piqued the attention of the Republican-primary frontrunner Trump, who leveled criticism at Sanders by calling him a maniac and misrepresenting him as a communist.[4]

By the end of his campaign Sanders had set a record—receiving more distinct contributions in a presidential primary than any candidate ever, far more than Obama had received in his groundbreaking 2008 campaign.[5] According to several polls conducted in October 2018, Bernie was still the most popular politician in the United States.[6]

Sanders refused to take any corporate or Super Political Action Committee (Super PAC) funding, and he managed to keep to this promise throughout his primary candidacy. All in all, the campaign for his 2016 run raised $218 million online, almost completely from small-dollar donors. The average donor donation total became a rallying cry for supporters to chant before Sanders appeared in public: 27 dollars![7]

As a longtime populist progressive outsider, Sanders's pleas for greater economic equality, and for support of the shrinking middle and working classes, struck a chord with many in the nation, including those who identified as political independents. His strident criticism of the nation's largest banks and his calls for their regulation seemed to continue where the Occupy Wall Street movement of 2011 left off. And despite being one of the older politicians in the United States, he achieved rock-star status with young voters.[8]

Sanders, like Trump, emerged when a significant percentage of the national electorate were already disenchanted with the mainstream of the country's two major political parties. Many saw both parties as corrupted by corporate and lobbyist money.[9] The time seemed propitious for Sanders, who argued for renewed investment in infrastructure and renewable energy–powered industries. This outsider approach was similar to Trump's "Make America Great Again" campaign in its criticisms of free trade policies. Yet unlike Trump, Sanders's solution was not merely to bring jobs back but to radically transform the US economy in a pro-worker and middle-class direction.

Fair enough, Sanders's voice and positions came at the right moment. And his team was able to use the internet as a megaphone. As an example, consider one surprising day in Portland, Oregon, in late March 2016. As Sanders gave a rousing speech to a crowd of 11,000 just fifty-two days before the Oregon primary, a bird fluttered nearby.[10] Rather than treating it as a distraction, Sanders welcomed its presence. As he did, the bird perched on the podium, eliciting widespread cheers from the audience (see figure 9.1).[11]

The event lived on with the internet's help. Bernie Sanders gave birth to "Birdie Sanders." Politico immediately posted a video of the event on Twitter, which quickly went viral, gathering nearly 50,000 "likes" and 36,000 retweets within three days.[12] The hashtag #BirdieSanders became one of Twitter's trending topics as news of Birdie penetrated other media channels and sources, including major news outlets and the blogosphere.[13]

More than a decade earlier, Howard Dean's primary campaign for the 2004 Democratic Party nomination had become the first of its kind to do significant online organizing. Joe Trippi, a strategist who ran the campaign, attributed the internet's expansion since that time as a factor in how Birdie (and Bernie) Sanders became so popular: "The bird thing … if that had happened in 2004," he said, alluding to the technology available at the time, "there is

Figure 9.1
The Birdie Sanders moment.

not a damn thing we could have done. Now, you have the prowess to take advantage of [the Birdie moment] and the network that they built on."[14]

Despite valiant efforts and the unpredictable, fortuitous appearance of the bird, the Sanders campaign ultimately fell short. Bernie conceded to (and endorsed) the insider Hillary Clinton on July 12, 2016, even amid accusations of Clinton's collusion with the Democratic National Committee.[15] The story of the Sanders campaign is insightful, however, considering its use of digital strategies that shape populist and grassroots politics.

How did Birdie Sanders go viral and become one of several memes that not only supported this campaign, but also turned Bernie into a household name? How did one of the nation's oldest politicians become so popular with first-time, twenty-something voters? How did Sanders avoid corporate money to the tune of hundreds of millions of dollars—the typical bedrock of political campaigns—and remain viable with an average donation of twenty-seven bucks from millions of Americans? What did this campaign reveal about the internet's potential to be a great leveler, to open up opportunities for citizen participation and people's power over politics?

I give examples throughout this book about how technologies are designed and deployed in relation to the purposes, values, and ideals of those they serve. In a campaign like Sanders's, the technological challenge would be completely different than the campaigns in which Cambridge Analytica had a hand, for instance, with candidates who were either extremely wealthy (Trump) or supported by big-moneyed interests (like Ted Cruz) or involved in other power plays across the world (like Uhuru Kenyatta).

What was the difference? Instead of seeing technology primarily as a way to collect, monetize, and target voters based on data, the Sanders team realized it could also wield the power of the internet to *bring people together*.

Revolution Messaging and the Online Campaign

I spoke several times in 2018 with the team at Revolution Messaging, the politically progressive digital agency that led the Sanders online campaign. The organization had already observed the potential of the internet to galvanize the masses during the campaigns of Howard Dean in 2004 and Barack Obama in 2008. Hoping to further move the American public toward solidarity and action for Sanders, Revolution spread the word via emails and social media posts while also using the internet to fundraise with great success. The campaign received a record 43 percent of its donations from mobile devices.

Keegan Goudiss is Revolution Messaging's director of digital advertising. He told me that instead of microtargeting voters (the tactic Analytica had used), Revolution focused on reaching millions of US voters who were looking for an alternative to traditional establishment politics.[16] "It's very hard to run a successful, intricate microtargeting campaign," Goudiss explained. "For Bernie our intention was...to start with a wider net...to reach more donors" rather than rely on 2,000 rich citizens.[17]

Sanders's online success allowed him to stay true to his word: not only did he avoid corporate or lobbyist (PAC) money, but he also eschewed the traditional donor events usually restricted to extremely wealthy campaign contributors.[18] In short, putting the weight of the campaign into online outreach allowed Sanders to maintain his populism and perceived integrity within the voting public.

Although Revolution Messaging pioneered and implemented the strategies it used for the Sanders campaign, the agency has used similar techniques to support other grassroots causes.[19] (See, for example, my profile of the Daily Action campaign further into the chapter.) Each of these movements has produced significant successes in impacting awareness, putting pressure on policymakers and corporations, and gathering evidence to show that citizen engagement can make a difference.[20] But despite the organization's passionate support for grassroots causes, Revolution remains agnostic about technology. The agency does not buy into any platform or brand hype but focuses instead on the digital strategies that might best support a particular outcome. Although there is a use for nearly every technological platform, according to the Revolution perspective, the organization maintains a healthy skepticism about any single tool or system, particularly corporate black boxes such as those owned by Google, Facebook, or Twitter.

Content Is King

During my hour-long interview with Arun Chaudhary, a senior strategist at Revolution Messaging, he repeatedly surprised me with lively statements about successful digital campaigns.[21] Comparing methods at Cambridge Analytica and Revolution, for instance, he pointed to Analytica's lack of transparency: "They are selling snake oil—they are selling the sizzle. The difference is that if you ask [Analytica] what their secret sauce is, they won't tell you, and if you ask us we will say there is no secret sauce."[22] For Chaudhary the willingness to be forthcoming about a campaign's methods in reaching voters is key.

Instead of playing the data-surveillance game or hiding behind hysteria-producing "snake oil" algorithms, Chaudhary said, Revolution focuses on the internet's potential to inspire voters to pool resources and ideas, thereby empowering their *agency* in fighting for causes in which they believe.[23] And part of that potential depends on the voters' engagement with the candidate and the campaign message.

That's why, as Chaudhary emphasized once more, *content is key*. Despite all our speculations about invisibly collected granular data and its value, we must also remember that what makes the internet go around is *video* content. At the end of the day people want good stories.

How does that inform what Revolution does? Birdie Sanders did not magically become a meme without the story. The process involved a combination of diligent observation, preparedness, and quick decision-making by a few key Revolution staffers. As Chaudhary explained, it required having a photographer on hand, ready to shoot the exact moment when the bird just happened to land, as well as the ability to spread the story instantly by email.[24]

Throughout my interviews with staffers at Revolution, I sensed a trust in the voting public and a faith in democracy. Even in an era where many are critical and cynical about the state of US politics, the agency believes that voters want to participate in political change. As Chaudhary told me, "[Our goal] is to give people something to give a shit about, that they can then look up themselves without us spoon feeding it to them."[25]

For example, consider Sanders's incredible success, especially as a senior citizen, in attracting young voters, whether they were eligible to vote in a national election for the first time in 2016 or were part of the broader millennial age range.[26] Revolution staff explained their strategy to *go where these voters are*—to platforms like Snapchat and Instagram. But again, instead of using these platforms to disorient or target users with specific content, the goal was to present Bernie "as is," someone with the same desire for change as so many young people. The question isn't "How do you talk to kids, but how you invite them into the same conversation as the one adults are having." That is what younger voters want so they see, that Bernie "is their guy, he's not mysteriously forced onto them."[27]

People crave transparency and great stories. In the work the agency performs for many political causes, Revolution continues to use the strategies of emotional connection and direct communication that proved so effective with the Sanders campaign.

Crowd Power

Revolution's Keegan Goudiss shared a company view with me that echoes the voices of activists during the Occupy movement: "For too long in this country, you had to be a millionaire or have millionaire friends in order to run a national or statewide campaign. Candidates spent their time dialing for dollars and hobnobbing with wealthy donors, and so it's little wonder we ended up with an economy that favored the 1 percent. We set out to change that dynamic."[28]

Goudiss's message carries the Occupy Wall St. analogy a step further: technologies can and should be designed to support the billions of people who use it—call them the 99 percent—not solely the economic interests of big tech companies. The Revolution Messaging approach applies not only to technology design but also to its strategic *use* of technology—as a tool to amplify the power that people have in numbers.

The most prominent Revolution design to date is the Revere Technology Suite, a mobile system used for politically progressive causes. Revere integrates three functions in tandem—texting, calling, and data collecting—to allow organizers to gather lists of supporters and plan campaigns. With the option to choose from a range of activities, users can focus their efforts to make the greatest possible impact on a given cause. The technology also allows organizers to assess user response to their messages.

Because Revere's design focuses on activating and coordinating user voices rather than invisibly monitoring us, it can leverage the scale of the masses to push forward candidates and policies from within their ranks. My interviews with the Revolution team reveal a vision of technology as an instrument through which to involve citizens who share their stated progressive values, independent of where they are, how much money they make, what demographic they represent, or what issues they are passionate about. Instead of being seen as a technology or data company, Revolution wants to be identified with one word that defines its expertise: *organizing*.

This perspective can allow us to rethink how technologies are used, and to focus on the potential of the *crowd* to fund, publicize, and power a campaign. Dae Levine, Revolution's senior vice president, told me: "[We want] everyone to participate, wherever they are and in whatever way they can. It's not about donors who give money; it's not about people who give time on the street; it's not about people who volunteer in the office. It is about everybody doing everything. [We] develop our strategies and tools to allow everyone to participate in whatever way they can."[29]

With the Sanders campaign we saw this approach and its logic in action: If more people can be reached, even if individually they contribute less, great successes can still be obtained. By activating the masses, $27 average donations came together to set fundraising records.

Linking Online and Offline Efforts

In conversations with the Revolution staff, I learned of another powerful digital strategy: thinking across technologies and beyond the contrived divisions of online and offline. Building and supporting movements, particularly when billionaires don't fund then, requires working with whatever methods bring people together to make change possible.

Revere technology can be been applied to a range of tasks related to out-front or behind-the-scenes political organizing, including text messaging campaigns, poster designs, and in-person protests. Its use in the Daily Action campaign was remarkable.

This is how it works: Daily Action sends more than 250,000 subscribers a daily text with a brief descriptive message about a single action or cause. It then offers a quick and easy way to make calls in support of that action: users can contact members of Congress, government agencies, or a business—just by touching the corresponding phone numbers that appear on the screen.[30] By constantly monitoring the news cycle, team members can handpick causes and campaigns that need timely attention and support.

Carla Aronsohn, who led Revolution's work with Daily Action at the time of my interview, explained the team's thinking at this current political moment in the United States: "It is frankly exhausting to watch the news every day. This is part of why Daily Action feels so accessible—it [provides] 160 characters a day about one issue that you can take real action on."[31] The thousands involved with this campaign engage in one manageable daily task that allows them to contribute to change. The choice to participate remedies what Aronsohn sees as a general sense of helplessness felt by many US citizens today.

Daily Action has achieved significant success using Revere technology to mobilize opposition to the Trump administration. It has also sparked new campaigns and movements. After the February 2018 mass shooting in Parkland, Florida, claimed seventeen victims at Marjory Stoneman Douglas High School, students from the school and other youth across the nation pressured the US Congress to ban assault weapons.[32] In the weeks following the shooting Daily Action's efforts resulted in 10,000 calls made to specific legislators and companies that supported the National Rifle Association (NRA).[33] Some organizations targeted by the campaign, such as the First National Bank of Omaha, have decided to divest their support from the NRA.[34]

As a follow-up to Daily Action's use of text messaging and targeted phone calls, the third Revere technology—collecting data from those calls—is especially important. It arms organizers with evidence about citizen commitment to the causes they advocate and support. During the Trump candidacy and the first days of his administration, for instance, Trump downplayed the Standing Rock protests against the Dakota Access Pipeline as insignificant and deemed the issue itself as unimportant. Revolution's Revere technology presented statistics to the contrary, tracking what calls were in support of the protests, to whom they were made, and for what period of time. And to compensate for the lack of internet access at the site of the protest, the agency provided activists with free texting platforms to assist their on-the-ground coordination.

Perhaps because of the pressure brought on by more than 10,000 phone calls in the first few weeks of Trump's presidency, the White House closed its direct comment line. In response Revolution created a system dubbed "White House Inc." to reroute calls to Trump-owned businesses and properties, as well as to selected members of Congress.[35] Compared to initiating email campaigns and relying on social media posts, phone calling has been shown to be a far more effective means of placing pressure on politicians and businesses, and of making democratically necessary two-way conversations possible.[36]

Revolution's strategy here merges different approaches toward organizing: use new technologies (a mobile phone app and a database of enrolled supporters) in combination with older technology (phone calling) to contact selected public servants and organizations. And then bring the crowd together as much as possible through in-person protests. In this way, Revolution has been able to make waves around specific policy issues. Some of those involve clean water, climate change, and the protection of special counsel Robert Mueller's investigation of potential Trump administration collusion with the Russian government.

Through their campaigns, the Revolution team learned that text messages get far greater real-time engagement than emails or posts on a website. With the Sanders campaign, SMS (short message service) texting technology provided supporters with locations for phone banking and debate-watching parties, thus bringing people physically together to organize as they saw fit. Voters also received text messages from Sanders himself. Aronsohn told me how Bernie's texts created "a very personal experience" that the affirms the "power of people's responses to texting and taking action."[37]

Revolution has also employed similar organizing strategies for economic justice causes. I learned of their efforts with the Independent Drivers Guild, based in New York City, which brings Uber, Lyft, Juno, and Via taxi drivers together to fight for better working conditions, earnings, and benefits (see chapter 11). The approach here again is to use online technology as a means of activating *offline* organizing.

Opening a Pipeline for Change

We've seen in previous chapters how big tech companies tend to thrive on the amplification of spectacle and misinformation, and on the data and attention they can get from their users. But this is not the only destiny for technology. For organizations like Revolution Messaging, the goal could not be more different: use technology to connect people who together build a pipeline for change.

The Bernie Sanders story shows how causes that begin as grassroots campaigns can transform a nation's politics. No matter which side of the political spectrum we identify with, it shows technology's potential to bring power to people, the ultimate guiding light of democracy.

After much speculation, from the media and his 2016 supporters, Bernie Sanders announced his 2020 presidential bid on February 19, 2019. The Democratic field to that date (numbering 12) already offered a bigger choice for progressive voters than in 2016, including minority and female candidates. Many candidates have adopted Bernie's ideas, ones that just a few years ago may have been seen as on the fringe and unrealistic. But since ceding the 2016 race to Hillary Clinton, Sanders has continued to build "a digital media empire" by effectively using and producing viral video content that champions progressive causes, harnesses social media, and understands its ripple effect in ways that challenge Trump's already enormous social media presence. And there's no doubt Bernie is the still the "king of grassroots fundraising."[38] The Sanders campaign's ability to raise a record-breaking $5.9 million, from more than 225,000 donors within twenty-four hours of his entry into the race, shows that he will be a force to be reckoned with for 2020.

This chapter has shown how digital strategies that privilege powerful storytelling and content, mobilize the crowd, and work across media online and off, are likely to have an impact in many elections—domestic or international—into the future. Scott Goodstein, Revolution's founder, summed up what he feels the Sanders 2016 campaign has already given us: "a little hope for the future that if you have the right message and a little money to try to experiment, that message can be heard by a massive amount of people."[39]

Every day, it seems, the news cycle offers us a story about technologies used for political manipulation—not just within but also among nations. Our anxiety looms large when we think about where digital technology and data are headed—and it looms even larger to think we have no control over their direction. Our intimate technologies feel hijacked by forces we barely understand.

Perhaps this explains why Cambridge Analytica dominated the news for weeks, why Russian manipulation in the 2016 US election is still so unsettling, or why a potential trade war between China and the United States, which has included a great deal of wrangling about hacking and trade secrets, centers on the technology industries in both countries. We can't stop talking about these sketchy digital acts even though their full impact remains unclear.

When I spoke to the political activist Noam Chomsky in March 2018, he downplayed the ultimate impact of Russian hacking and Cambridge Analytica (see chapters 1 and 7). He pointed out to me that the US use of technology to manipulate the politics of foreign nations is far worse.[1]

I had a chance to discuss these issues with Laura Rosenberger, an international affairs strategist and term member of the Council on Foreign Relations who has worked for multiple US administrations; during the Obama years she directly advised the former secretary of state Hillary Clinton on national security priorities. In areas of technology, Rosenberger is one of the most experienced individuals connected to the US government—an expert on Chinese firewalls and cybersecurity used by China, the National Security Agency's PRISM spying program, as well as cyber troops and hacking efforts across the world.

I asked Rosenberger about the international use of technology, whether it be to manipulate a state's own citizens or those of other nations. In response, she immediately turned the discussion to China.

When the internet arrived in China, so did the expectation that it would open up the world's most populous nation to the rest of the globe paving the way for democracy. That hope has only been partly fulfilled, as Rosenberger explained, because China's powerful technology companies are in lockstep with the government: they ensure that internet content supports the state's surveillance machine and its power over popular online services—buying and selling, communicating, and searching for information.[2] TenCent, Baidu, and Alibaba dominate the market within the country and affect billions across the world in the social media, search, and e-commerce markets respectively.

To take further advantage of this control, the government enlists hundreds of thousands of people to serve in its so-called 50-cent army (figure 10.1). The mission for these troops is to "praise and distract," either by responding online to negative comments about the Communist Party, or by writing "posts that cheerlead for the government." The 50-cent army, comprising actual government employees, has provided approximately 448 million online comments every year.[3]

Figure 10.1
China's 50-cent army (source: Sohu).

Perhaps more ominous is China's use of citizen-collected data to score, classify, and organize the populace. In 2014 the government announced a "social credit system," using data gathered from approximately 1.4 billion Chinese. An official government document describing the system sends a clear message: "Keeping trust is glorious and breaking trust is disgraceful."[4]

The social credit system rates citizens based on several factors including: what they post or how they portray themselves on social media; their personal habits (smoking cigarettes, playing video games); what they buy online; and whether they've refused military service. After assessing these behaviors, the government "awards" citizens a rating intended to indicate their "trustworthiness," which in turn affects their ability to get jobs, send their kids to the best schools, and board an airplane.[5]

It's terrifying that Chinese schools have become sites for intense surveillance and biometric identification. Cameras now record individual facial expressions to create running scores for students based on the attitudes or emotions their faces reveal; these can be combined to "read" the mood an of entire class. Students and professors alike have expressed concern that instead of supporting the open practice of learning and experimentation as a process, such "black technology" could filter and compress students' educational and life experiences, or be used to invoke a form of social control.[6]

The problem becomes more extensive when facial recognition and surveillance stitch together profiles of citizens across the whole of Chinese society without boundaries or transparency. According to an article published in the *Atlantic* in February 2018, the Chinese have set up an infrastructure whereby "a vast accompanying network of surveillance cameras will constantly monitor citizens' movements, purportedly to reduce crime and terrorism."[7] As the *Atlantic* co-authors argue, "The expanding Orwellian eye may improve 'public safety,' [but] it poses a chilling new threat to civil liberties in a country that already has one of the most oppressive and controlling governments in the world."[8]

The Chinese internet functions more like a nationwide *intranet*, since most digital communications are controlled, manipulated, and stay within the country. But China has other ways to use technology to support its interests. As Laura Rosenberger explained to me, the acquisition of Grindr by the Chinese technology firm Kunlun Group allows the state to access data of foreign citizens, especially citizens of global adversaries. Because

Grindr is a dating app for bi, trans, and queer users who mostly come from the United States and other Western nations, the personal data available to the Kunlun Group (and potentially the government) could reveal sexual preferences.[9] A similar threat comes from Facebook; the company disclosed in June 2018 that it has been sharing data with several Chinese firms, including the popular device maker Huawei.[10]

Even Apple, whose iPhones are very popular within China, must play by the rules of the Chinese game if it intends to stay in the country. One troubling example of this is the requirement that Chinese users' iCloud accounts be housed within a new Chinese data center, thus giving authorities easy access to the data of their citizens.[11] This is a major change from Apple's stance on user data and the company's claim that more than any other large Western technology company, Apple protects privacy. In the past, the cryptographic keys needed to access an iCloud account lived within the United States. For Chinese users, that will no longer be the case.

Each scenario I've mentioned in this chapter illustrates a point of leverage in the digital war games played out within or between nations. In the case of China, leverage comes from the ways data can be acquired and then, as a result of the kinship between the business world and government, be used to shape *political and economic* outcomes. These activities are legal and increasingly widespread, because of one key factor: the money made in the transaction.

It is also well known that the Chinese government has conducted many extraterritorial hacking attacks, for example of the US Office of Personnel Management, which allowed it to access personal details of up to 21.5 million Americans.[12] This included Social Security numbers, names, dates and places of birth, and addresses. The hack was conducted so adeptly that James Clapper, the former US director of national intelligence, stated, "You have to kind of salute the Chinese for what they did."[13]

Rosenberger described activities like these, both legal and illegal, as *information operations* (info-ops). She believes info-ops might allow China to be behind the next iteration of political meddling in foreign elections, along the same lines as Russia's recent activities.[14] This might be accomplished using *deep fake* technology, which allows video and audio content to be doctored, edited, and manipulated in such sophisticated fashion that it is often convincing to the human eye.[15]

The United States, too, hacks its adversaries, and even its citizens, in spying programs like PRISM, the National Security Agency program that Edward Snowden exposed.[16] For example, working with Israel, PRISM has been alleged to be behind the Stuxnet computer worm attack on Iran's nuclear program.[17]

Technologies should benefit citizens and users, rather than merely the self-interests of their state or corporate creators. This is an issue not just in the West but worldwide. Supporting political positions and candidates outside of the corporate lobbyist-funded establishment is one mechanism of checks and balances some of us can use to support democracy and resist programs of surveillance and social control. We can also vote with our feet and demand privacy-oriented services and platforms that refuse to hand user data over to governments or other shady third parties. What direction we take in the digital world will determine whether technology brings us together in the spirit of democracy, or whether it divides and manipulates us.

III Gig Economy Blues: Corporate Windfalls or Living Wages?

11 Disrupting Jobs and Lives

On February 5, 2018, Doug Schifter, a registered taxi driver from New York City, committed suicide in front of city hall.[1] Just a few hours earlier, he had commented on Facebook to explain his struggle for survival in the digital gig economy, stating: "Companies … count their money and we are driven down into the streets we drive becoming homeless and hungry. I will not be a slave working for chump change. I would rather be dead."[2]

Disruption, a buzzword in tech circles that refers to groundbreaking innovation, often has a negative impact on the lives of people whose jobs are subject to shifts in technology (see Schifter's fellow drivers in figure 11.1). Schifter used to clock forty hours per week when he started driving in the 1980s, but by 2018 he worked more than a hundred because agreements made by city officials had opened the streets to Uber and Lyft vehicles. Unfortunately the extra hours didn't help him climb out of the financial hole these tech platforms created.

In the five months that followed Schifter's death, despair over work-related struggles contributed to five other suicides by taxi or livery drivers in New York City.[3] These stories are reminders that spates of suicides triggered by new technologies are neither new nor rare. Over the past two decades, for instance, more than 300,000 farmers in India committed suicide thanks to to another technological disruption, when Monsanto commodified and monopolized the cotton-seed market by introducing its genetically modified seeds.[4]

In 2016 some 13,000 taxis cruised the streets of New York. But the number of for-hire vehicles is now more than 100,000, thanks to the emergence of companies like Uber and Lyft.[5] The city has traditionally attempted to balance cab supply and passenger demand by controlling the number of

Figure 11.1
Taxi drivers, mourning Doug Schifter in 2018, organize to protest unfair working conditions (source: Black Car News).

taxi medallions at a given time. (Buying a medallion allows an individual driver to own and operate a cab, and to hire additional licensed drivers if they chose.[6]) With Uber and Lyft, however, almost anybody can drive. As new unregistered drivers flood in, they increase competition, threatening wages and the availability of work. As a result, many registered and unionized taxi drivers have seen their wages drop and their hours increase.

Uber and Lyft, though often thought of as transportation companies, are first and foremost technology companies backed by Wall Street.[7] They promise efficiency for users and charge less expensive fares than registered and licensed city taxis. That's great for consumers, but not so great for workers. Traditional taxi drivers like Schifter are in trouble. For example, one of the drivers who committed suicide in May 2018 had borrowed $700,000 three years earlier to purchase his medallion, and could no longer meet the payments. But also in danger are the drivers for Uber, Lyft, and other rideshare companies. Their wages are shockingly low, they receive no benefits, and they have no ability to collectively bargain.

Meanwhile, we see a windfall for the tech companies. They "disrupt" transportation systems yet offer little more to workers than basic wages in an insecure digital gig economy. As a result their corporate value increases as the public becomes increasingly addicted to their services.

None of this would be possible without the compliance of national, state, and city officials, hurting workers and endangering the sustainability of the public transit systems they are supposed to oversee and innovate.

Without a doubt Uber and Lyft have proven that they can provide a service consumers want at a price that makes a traditional taxi service nearly obsolete. And they have done so brilliantly by enlisting our mobile phones to facilitate a flexible transportation infrastructure. But questions remain. Why and how are their prices so low? Most economists and analysts have come up with few answers. But some research does show that the companies themselves subsidize the low prices users pay.[8] Why? The strategy may not even be aimed to make a profit in their current form as taxi services, but instead to get intimate details of travel and transportation from millions of users, which in turn can inform a new fleet of automated, driverless vehicles.[9]

Remember, *data* is the oil of the digital economy.[10] Yet the outcome for the worker may be disaster and death. As it stands now, drivers and users provide the data and the labor power that make the business run while Uber and Lyft take home the profits, control the data, and own the technology used to gather it.

We cannot remain naive about the potential harms digital platforms may have on the economic security, health, and working conditions of laborers. Uber drivers, for example, are not insured as they drive to pick passengers up, only when they actually have passengers in the car.[11] And the actual wages they receive are stunningly low. According to research conducted by MIT's Center for Energy and Environmental Policy, the median profit earned from driving is less than $4 per hour for Uber and Lyft drivers.[12] This means that 74 percent of drivers are making less than their state's minimum wage.

Uber's chief economist has disputed MIT's non-partisan, academic study of his company's wages. But the power of the gig economy to shift responsibility, stress, and pain to workers—and away from the companies they work for—is nevertheless clear. Tech companies talk about "disruption," "innovation," and "sharing." But they remain oddly silent on basic principles regarding the dignity of work, rules about which have been respected for the most part in wealthier nations. These include establishing floors for hourly rates and number of hours worked, collective bargaining and unionization, and health and other benefits.

According to a 2018 UCLA Labor Center report, 44 percent of drivers in Los Angeles have difficulty paying for work expenses such as gas, insurance, and maintenance.[13] More and more drivers are investing financially—for instance purchasing a car in order to drive for Uber or Lyft—and thereby becoming locked into the work, which they must then continue to do in order to pay off their debt. According to the report "a lack of transparency and vague and shifting conditions of employment" have caused almost 50 percent of drivers to feel as if they are not receiving the income they have earned.[14]

How can we be okay with a digital world that makes astronomical amounts of money for a few executives or investors at the cost of small businesses and, ultimately, middle- and working-class people? It doesn't have to be this way.

Driving Toward Economic Justice

Today we have the opportunity to innovate with technology, and to do so in economically fair and balanced ways. When it comes to transportation systems, we can begin by examining the relationship between drivers and the tech companies that have stepped in to manage their access to work. Establishing wage floors and secure benefits for taxi drivers, as well as protecting passenger data, might be the first steps in finding a more equitable way to do business.

Digital technologies are transforming economies over the world. Most forecasts about the future of work tell us that a huge wave of automation is coming, and that it could affect 47 percent of jobs in the US labor market over the next twenty years.[15] Some 400 to 800 million people throughout the world could lose their jobs to automation, another source estimates, while 75 to 375 million may be forced to change occupations.[16] It doesn't matter whether you work in nursing, teaching, sales, accounting, farming, plumbing, truck driving, or the medical field—the robots are coming. This makes it essential that all of us, not just the .001 percent of us who build automated and gig technology, come to the table to discuss what kind of technological future we want. We can embrace automation with our eyes open by identifying types of jobs that might be created through this transition, and by ensuring that the people of our world are economically secure.

It is common today to think of the main instrument of worker organizing, our unions, as relics of the past. Union membership in the United States and abroad has declined steadily since the 1970s. But this is not the whole story—indeed, unions are still alive today and are re-imagining their role in relation to the digital economy.[17] They, like workers' councils, universal basic income programs, and other examples I discuss, represent just a few of the ways we can protect fair wages, benefits, pay standards across industries (rather than just within one), workplace safety, and pay equity for women and racial minorities.

Why is economic justice and balance so important today? Because the world faces staggering inequality—its eight wealthiest people have as much wealth as its bottom 50 percent, a total of more than 3.6 billion people.[18] In the United States, the top one-tenth of 1 percent of the population earns nearly as much as the bottom 90 percent of the population combined; its three richest people hold more wealth than the bottom 50 percent of the country.[19]

As of early 2019 unemployment was at an all-time low in the United States, fueling rhetoric that economic and stock market growth is helping everyone. But that claim doesn't stand up to scrutiny. In the United States, even if paid a $15 per hour wage, the average American can't afford the rent for a modest one-bedroom apartment. With the current national minimum wage (of $7.25), one would have to work 2.5 full-time jobs to afford a one-bedroom apartment in most of the country.[20] What is work after all if it doesn't provide the worker with housing, education, healthcare, and basics sustenance? Ensuring that workers have secure, well-paying, and dignified work is an objective that we can still pursue. But it won't be easy to achieve.

When I was in the East African country of Tanzania in summer 2018, I was surprised to see a vehicle advertising Uber services "now in Slipway," a hotel in Dar es Salaam (figure 11.2). The vehicle was not a taxicab or an automobile, but a three-wheeler rickshaw that Tanzanians refer to as "Bajaj," after the Indian company that makes it.[21] Similar to the ways in which Google and Facebook capture, monitor, and absorb our expression and communication online, or the ways Amazon absorbs our processes of

Figure 11.2
An Uber three-wheeler in Dar es Salaam.

buying, selling, and storing data via its cloud platforms, Uber too is in the business of capturing and co-opting. Having made headway in developed urban settings, the company has now set its sights on the transportation industries of the developing world.

The rickshaw has long been a transportation staple of the "informal economy." Ride prices are traditionally negotiated up front and payment is made in cash. But now, in East Africa and other developing countries, the relationship between a customer and driver on the street can be controlled through an app produced by a company located thousands of miles away. This disruption in the informal sector threatens the traditions of economic exchange across cultures and regions.[22] By positioning themselves as the go-between, large technology companies from Silicon Valley to Shanghai create dependencies in the people and places where they expand, all the while extracting value from them.

Such intervention runs the risk of being exploitative and intrusive. Even if we take into account the ability of people across the world to develop their own alternatives—and to appropriate, work around, and subvert the systems created by large technology companies—we have to recognize that the playing field is not level. The strategy of baiting customers with artificially low prices, thereby creating dependence and knocking out the competition, is working for Uber. But few small companies can withstand such a strategy for long. As of February 2019, Uber's valuation, in anticipation of its planned initial public offering (IPO) later in the year, is slightly less than $90 billion, up from $72 billion in 2018.[23] Even though that figure had earlier been projected as high as $120 billion, investors remain confident that Uber can dominate the domestic and global transportation markets, and perhaps even use the data it gathers to lead the way in the automated car industry.[24]

How Guilds and Alliances Push for Change

Many rideshare workers, and gig economy workers in general, have another job. But this *part-time-ization* of work, to quote the driver organizer Biju Matthew, is dangerous. It means that in places without strong worker protections or corporate regulation, employers have the ability to exclude gig workers from the benefits associated with a full-time job, no matter how many hours they work. This is because they are often treated as *independent contractors*. In addition, workers in the gig economy have no guaranteed minimum wage when they are paid by the task rather than the hours it took to perform it. So the technologies that promise flexibility and access to work may also bring with them a hefty dose of insecurity for the laborer.

Bhairavi Desai, the executive director of the New York Taxi Workers Alliance, has been a labor leader since 1997. Nearly a year before Doug Schifter's suicide, she spoke publicly about the extreme economic and psychological pressures on her drivers, and worried about their long-term ability to cope: "Half my heart is just crushed," she said, "and the other half is on fire."[25] Desai has begun to work with the Independent Drivers Guild (IDG), representing approximately 70,000 drivers in New York City.[26] The IDG, having won several legislative victories for its members in the past two years, is showing that workers can push back. For example, in August 2018 the New York City Council agreed to mandate a minimum pay rate for app-based drivers and to limit licenses for new rideshare vehicles.[27]

How did the guild do this? It used the Revere text messaging communication technology (discussed in chapter 9) to bring drivers together, an otherwise impossible task given that so many are working such long hours driving in different locations. Uber had not allowed customer tipping until July 2017, when drivers and advocates used phone calls, text messages, and social media channels to push back.[28] As of June 2018, the IDG was able to win a benefits package for an estimated 43,000 drivers across the state thanks to a "black car fund," notably not paid for by Uber or Lyft, but by riders via a 2.5 percent surcharge.[29]

The guild has now turned its efforts to pushing for a minimum pay rate for all drivers, which would increase wages by an estimated 37 percent.[30] The petition it circulated presents a number of astonishing statistics in relation to rideshare drivers in the city. It not only confirms the under-minimum wage payment drivers currently receive, but also notes that workers are averaging eleven-hour working days. Despite this, workers report that their financial well-being has suffered. Most come from households with an overall income of less than $50,000 per year and more than half of them struggle to balance this meager income with caring for a dependent under the age of eighteen.

Agencies, Advocacy, Legislation, and Courts

In 2018 the National Employment Law Project (NELP) released a report to consider New York City's "first-in-the-nation proposal regarding the wage standard for Uber and Lyft drivers." Rebecca Smith, the director of the Work Structures program at NELP, concludes: "Companies like Uber and Lyft have used the hardball tactics of the gun and tobacco industries to buy political influence and preempt or override local policies intended to protect consumers and drivers."[31] From the report's findings Smith determined that "Uber alone would be the largest for-profit private employer in New York City—if Uber drivers were classified as employees rather than independent contractors. Uber's mark-up in New York City is a whopping 600 percent of its operating costs. These companies clearly can afford to pay a living wage, but simply choose not to do so."[32]

Other states have responded to concerns that the expansion of digital gig economy jobs will result in larger numbers of workers not covered by minimum wage laws. In April 2018, the California Supreme Court issued a ruling that would make it far more difficult for tech companies to claim exemption from state wage laws because their independent contractors are not employees.[33] But the companies are pushing back. Lyft, Uber, Handy, TaskRabbit, Postmates, DoorDash, and others have appealed to state politicians to warn that the court's ruling will "stifle innovation."[34]

What about outside the United States? In the past few years, some nations have begun to challenge rideshare and other gig economy apps, imposing increasing regulations.[35] In one notable case the Transport for London (TFL) agency decided not to renew Uber's license to operate in the city because the company was "not a 'fit and proper' private car-hire operator." TFL "cited four areas of concern, including [Uber's] approach to reporting criminal offences and carrying out background checks on drivers."[36] In addition, Singapore, which held no restrictions on Uber prior to 2014, implemented new laws requiring Uber drivers to register with the Land Transport Authority and acquire a Taxi Driver Vocational license.[37] The new laws also stipulated that rideshare apps should include customer service, price, and fee info, and provide passengers with the option to not disclose their destination info before booking.

In a landmark win for taxi associations, the European Union's top court ruled in 2017 that Uber was a transportation service, making the app subject to the same regulation as taxi services.[38] This was a significant setback for Uber because the company insisted on classifying itself as an "information society service," a designation certain to lighten the load of applicable regulations.[39]

Many app-based labor platforms brand themselves as "technology" companies rather than identify with traditional industries, even though they are dramatically reshaping the way people work in those industries. Gig economy tech companies often promote themselves as providers of neutral and efficient resources, insisting that the service they provide is merely to increase efficiency by connecting the housing, driving, and labor we offer to one another. That would be true if the output of such innovation benefited the environment, the workers, and their livelihoods, rather than managers, investors, and—yes—consumers. But when we look more deeply, as the EU court did at Uber, we often see that these promises of efficiency don't pan out. A recent New York City–based study has shown that, rather than reducing car ownership and making transportation more efficient, Uber and Lyft's carpool services add to traffic congestion and take riders away from public transit.[40]

The above examples reveal that political will and hard work can change the disturbing trajectory of a digital economy in which middle- and working-class people spend more time on the job for less money to try to compensate for the precarity of the gig economy. In Silicon Valley itself, thousands of security officers subcontracted by tech companies have agreed to a union contract to raise their wages, thanks to the efforts of labor advocacy groups like Silicon Valley Rising.[41] Ironically, some of these officers were homeless while they performed their duties, which included preventing other homeless persons from settling close to corporate offices. Perhaps the power of organizing will help them achieve greater economic security.

Regulating the App-Based Economy

Beyond rideshare platforms, a host of other app services have received pushback in tourist destinations like Paris, Berlin, and especially Barcelona, where it's illegal to rent an entire property without the landlord's presence unless it's registered as a hotel.[42] Airbnb has been at the center of significant controversy around the world, particularly in cities with rising rent prices, where landlords can make more money renting apartments short-term to tourists rather than to full-time residents.[43] Landlords in Paris faced fines up to 25,000 euros after a government crackdown found that many of them were renting apartments far longer than the legal limit of 120 days per year; across France, 44 percent of flats listed on Airbnb were available for rent full-time. [44] And in Berlin a law passed in 2016 allowed landlords to evict tenants they found subletting their apartments through Airbnb. The legislation banned short-term rentals unless approved by the Berlin Senate.[45]

Some cities, states, and nations have attempted to balance the effects of private corporate power with measures aimed to benefit the public. One of the most controversial has been the "head tax." Consider what happened in Seattle when, in May 2018, the city council passed the so-called Amazon Tax. The tax required large employers headquartered in the city (the retail tech giant Amazon being by far the largest) to pay $275 per year per full-time employee; those tax dollars would be allocated to fund affordable housing and services for the many homeless people the companies had displaced and disrupted. (Amazon alone has turned more than 8 million square feet of real estate into office space.) But the city council voted to repeal the tax a month later. Amazon threatened to halt expansion that would bring even more jobs to the city, and some council members succumbed to fears of the long-terms effects if the tax alienated other businesses looking to locate in the city.[46] This legislation should prompt governments and regulatory bodies to ask pertinent questions, such as: How much weight should they give to the economic growth and jobs that tech companies bring given the potential for massive disruptions to the city's existing population? And how do they determine an appropriate target to help shoulder such disruptions?

The examples above show how Uber, Airbnb, and many other app-based "work" platforms are transforming the fabric of our cities—the ways we work, live, and exchange goods and services. These companies consider themselves mediators, bringing any of us with an extra bed or access to a car, for instance, into contact with someone who wants it. They rarely own anything and use automated technologies to hire almost no one, and thereby minimize costs and overhead as much as possible.

It is likely that, as the effects of technology on our economies become increasingly palpable, there will continue to be a push and pull relationship between activists, governments, and tech companies. Most analyses conclude that within Europe and North America about 25 to 40 percent of work currently has a connection to the gig economy, though that is complicated by the fact that there are traditional gig economy workers who are not accounted for in these studies.[47] They can include, for example, workers hired on the street or through referrals, without the use digital platforms or apps. Indeed, the growing importance of the gig economy has influenced robust conversation in Europe around *flexicurity*, a set of policies to recognize the shifting nature of work while guaranteeing workers a universal safety net in terms of income and/or benefits.[48]

In non-Western nations, it is likely that the gig economy is even larger, with significant percentages of the population falling into the "informal" or unregistered labor market. As the gig economy turns digital, governments have an opportunity to threaten and monitor the informal economy by tracking and surveilling these populations through biometric technologies, digital banking accounts, and social security–type numbers.

Predictions conclude that, moving forward, it is likely that new technologies will push work into the independent, flexible, and "contingent" category, loosely and broadly defined by job instability, temporary work, and a lack of regular schedules.[49] Yet as this occurs, there seem to be few guarantees about what this means for worker security, or the overall economic security of citizens across the world.

In the United States, only a few members of Congress have directly discussed the future of work and economic security as a central part of their political platforms. One push, however, comes from the populist right wing, in the form of criticisms of immigrants and the outsourcing of jobs. Donald Trump's "Make America Great Again" 2016 election promise to revive America's past glory is a well-known example. What was the "great" era, according to this campaign and the president it brought to office? Supposedly any time before globalization facilitated migration, outsourcing, and a globalized supply chain. But according to the MIT economist Frank Levy, the globalization of labor barely holds a candle to what automation may make possible. Levy predicts that in the United States by 2024, 76,000 truck-driving jobs, 210,000 assembly jobs, and 260,000 customer service jobs are likely to be wiped out.[50]

As I stressed at the beginning of this chapter, today we have the opportunity to innovate with technology, and to do so in economically fair and balanced ways. In the next chapter I explore additional ways to ensure balance and equity in work (and for workers) of the future.

12   Protecting Work and Workers

Ever since the Industrial Revolution, the tensions between capital and labor have been fueled by the relationship between human beings and machines. Let's look at this in the simplest of terms: From a business perspective, paying less for labor represents a savings in capital. From a human perspective, however, getting paid less can cost workers more than lost wages. This economic challenge is clear and present within the digital gig economy as a struggle between two values: (1) accumulating as much wealth as possible, and (2) striving for a more economically equal society.

What are the costs to workers as we shift to a digital gig economy? What mechanisms can we put in place that will really do the job of protecting workers? Must making money and protecting workers be mutually exclusive objectives?

In this chapter I explore some answers to those questions, first by examining the changing dynamics of labor unions—the main instrument through which workers organize—and then by looking at the rights of contract workers and the effects of automation on businesses and laborers as well.

The Big Picture for Unions

In recent years the power of workers has declined as unions have weakened. This is especially true in the United States, and to a lesser extent in countries across the world. No matter where we look, we can see how the declining power of labor, combined with dramatic shifts in technology, presents a profound threat to workers. The trend is clear: we're heading toward a digital economy in which the vast majority of the population has little power.

In the United States, politicians and courts have stripped the nation's labor movement of the influence it once had.[1] In May 2018, for instance, a Supreme Court ruling (by a 5–4 vote) allows employers to use arbitration clauses that prohibit workers from challenging federal labor laws via class-action lawsuits. The liberal justice Ruth Bader Ginsburg called this ruling "egregiously wrong," stating that it will lead to a "huge under-enforcement of federal and state statutes designed to advance the well-being of vulnerable workers."[2]

This is a huge setback for collective action and bargaining rights that have been at the center of every labor movement in history. The ruling could also suppress a number of civil rights class action suits: first by potentially slowing the momentum built by #MeToo and other movements organizing to end sexual assault and harassment; and second by creating an immense imbalance of power between employers that are likely to have more financial resources for legal representation than their individual employees.

Just a month later, in June 2018, a similar Supreme Court ruling declared that non-union members cannot be required to pay fees to public sector unions, which support civil servants like teachers, firemen, and police.[3] This blow to labor organizing will shut off a key source of revenue for unions. Conservative activists have long been chasing this outcome because public sector unions tend to represent more liberal and progressive stances supporting economic regulation. Nonetheless, the Supreme Court ruled that requiring people to pay fees to unions they may be politically opposed to would violate their First Amendment right to freedom of speech.

What about outside Europe and North America? In South Africa, changes in the labor landscape have begun to erode the influence of the National Union of Mineworkers, leaving significant numbers of the organization's poor and unskilled black workers at an even greater disadvantage.[4] As successive conservative governments opposed worker activism, membership in Japanese unions has also declined from 55 percent in the 1950s to 17.4 percent in 2015.[5] In 2018, the East Japan Railway Workers' Union—Japan's largest labor union—lost nearly 70 percent of its members—even after negotiations between labor and management averted the strike—because union leaders announced a strike that many members opposed.[6] In the wealthier countries of the world, globalization and pressures on industrial manufacturing are squeezing participation in unions.

The fight is far from over, however, as many unions have still been able to make strategic gains despite low membership. In Japan, for example, unions were able to see significant wage increases just this past year.[7] In Turkey, while union membership still remains low, awareness of labor struggles has increased over the past fifteen years.[8]

US Labor Leaders Advocate for Change

To investigate worker issues in the United States, where I had greatest access to union leaders, I spoke to over two dozen officials from some of the nation's most prominent unions, including the American Federation of Labor and Congress of Industrial Unions (AFL-CIO), the International Brotherhood of Teamsters, the Service Employees International Union, the American Federation of Teachers, and the Communications Workers Association. In these interviews we discussed the shifting nature of labor and worker organizations and explored the possibility of arriving at a digital economy that respects worker rights and values.

What have leaders in the union movement done to be ahead of the ball when it comes to the gig economy and automation? How can they, or all of us, ensure the technology industry respects, serves, and is accountable to its workers and users?

A common theme emerged among the labor leaders I spoke to: *think not of technology and worker welfare as necessarily in tension, but instead of how technologies can support workers' causes.*

Let's first consider David Rolf, one of the most visible figures in the technology-labor conversation. As international vice president of the Service Employees International Union (SEIU), Rolf led the successful "Fight for 15" campaign in Seattle; his city was the first to adopt a $15 per hour minimum "living" wage for all its residents (to be phased in by 2021). Like many forms of social change, it took a concerted multiyear effort to enact this piece of legislation. As California and New York adopted similar laws, the Fight for 15 efforts echoed throughout the country.[9] Given the success of these efforts, Rolf has become an even greater advocate for "sectoral bargaining," which takes up the fight for labor in strategically selected places and contexts.

The Fight for 15 may not necessarily ensure worker protections when it comes to automation and digital gig labor, however. Indeed, Rolf estimates that "the ability to form a union and bargain collectively is inaccessible to

more than 93 percent of private-sector workers—a major reason why working people have experienced forty years of wage stagnation even as the
economy grew and the rich got richer."[10] When we consider these issues in
light of gig work, we see that the minimum wage statute does not necessarily cover independent contractors.

What then can be done? Rolf favors an approach by which technologies themselves could support the cause of labor—*if* we were to innovate,
design, and build them in ways that reinforce the values of dignified, protected, and fair work. Rolf does not see technology firms as enemies of
worker causes but instead blames the demise of workers on the trickle-
down economic system within which workers are embedded.[11] Uber's CEO,
for example, released an open letter to leaders across business, labor, and
government in January 2018, which Rolf and the venture capitalist Nick
Hanauer co-signed. The following excerpt from the letter sums up Rolf's
point: "The American social safety system, which was designed in the 20th
century for a very different economy, has not kept pace with today's workforce. At a basic level, everyone should have the option to protect themselves and their loved ones when they're injured at work, get sick, or when
it's time to retire."[12] The letter identified the foundations of a portable benefits system as flexibility, universality, innovation, and above all a shared
commitment to action.

The letter has garnered some criticism. Bhairavi Desai, for instance, from
the New York City Taxi Alliance responded: "Selling out to the bosses is not
innovative—it's as old as capitalism."[13] Well then, what types of concessions
to tech businesses, if any, might be useful for worker causes? Is it acceptable,
for example, for the workforce of independent contractors to continue to
expand without ensuring that those workers are guaranteed benefits?

Palak Shah, the social innovations director at the National Domestic Workers Alliance (NDWA) offers a nuanced approach to Rolf's idea. It
involves designing apps that could allow on-demand workers to effectively
bargain with work distribution platforms. Instead of seeing the shift to
domestic work on online platforms as an inherent negative, she considers it an "incredible opportunity to set norms in a market that has been
so difficult" to regulate, even in terms of establishing basic wage floors.[14]
For domestic workers who have long lacked unionization, partnerships
could occur with large online providers to set standards for the first time
that address what it means to be fair when hiring. Shah put this idea into

action with the help of the Clinton Global Initiative by launching the "Fair Care Pledge" on Care.com, the largest online platform for domestic work.[15] So far 150,000 of the households who use Care's services have signed the pledge and committed to ensure three job-quality standards for the nannies, housekeepers, and caregivers who work in their homes: fair pay, clear expectations, and paid time off.

Shah wonders whether this type of pledge can be applied more widely in Silicon Valley. She describes herself as one of the most optimistic people in the labor movement—sharing the supposed optimism of many tech entrepreneurs. But she also realizes the challenge of shifting how these companies think about innovation, whether technically, socially, or economically. Twelve tech companies so far have signed the NDWA's "Good Work Code" in support of an eight-point ethical framework that companies doing business online can use to define good work based on workers' needs.[16]

One offshoot of the Good Work Code has been Airbnb's release of a "Living Wage Pledge," which asks the platform's US-based hosts to commit to paying their independent cleaners at least $25 per hour.[17] Airbnb says that it already pays its own employees $15 an hour and pledges that by 2020 it will pay the same to contractors and vendors whose personnel provide a "substantial amount of work" to the company.[18] Yet the company's failure to define what "substantial" means detracts from its otherwise affirming statement.

Another approach in the labor world could be to develop a kind of incubator fund by rallying individuals to exert significant pressure on technology firms from outside the unions. Consider *Time* magazine's choice for its 2017 Person of the Year: the "Silence Breakers" (or whistleblowers) who launched the #MeToo movement, which has now exposed a longstanding epidemic of abuse and harassment of women.[19] Could the cause of workers be in synergy with this powerful social movement? The former SEIU innovations director Vasudha Desikan sees its great potential.[20] She gave me the example of the 2018 Golden Globe awards Meryl Streep, Amy Poehler, America Ferrara, and five other actors gave the spotlight to several leaders from the NDWA and Restaurant Opportunities Center, the women they invited as their dates for the event.[21]

In another industry, consider the perspective of Doug Bloch, a political director for the Teamsters, one of the largest unions in the United States and Canada.[22] Although has he worked with other unions and for environmental and human rights causes, his current position brings him in close contact with truck drivers in California. He expressed concern to me about how we tend to discuss the future of work instead of the future of workers, as if we care more about work than our fellow humans who perform it.[23] For example, consider how automation may threaten the jobs of 3.5 million truck drivers.[24] Bloch notes that those who are developing autonomous trucks have little knowledge of the actual experience of driving a truck itself. So, perhaps as automation takes root, drivers might still be hired to train self-driving trucks. But beyond such a short-term stint, he is concerned about workers' prospects for the future—not just for truckers but also for workers in fields such as taxi and bus driving, construction and factory/warehousing, delivery, freight, janitorial, security, housekeeping, and hotel services. Only a small number of voices from Silicon Valley drive the conversation about automation and the digital economy, Bloch says. Rarely are voices heard from within the impacted industries. Like many others in the union world, Bloch has shared these concerns with several sympathetic members of the United States Congress, including the California junior senator and 2020 presidential hopeful Kamala Harris.

The Teamsters have worked with large companies like United Parcel Service (UPS), identifying ways to transform current trucking jobs for the future: truckers could supervise and control automated vehicles, for example, or work with automation systems and drive only part of the time, much like airplane pilots do today. What remains important, however, is that human work be as fulfilling as possible. This requires that employment be built upon a *social contract*. Its terms must ensure a cooperative and respectful relationship between employers and workers, or machines and workers, and be designed for outcomes that benefit everyone involved as much as possible.

We've seen some of the pragmatic and imaginative positions from labor leaders, but what is it that led to the declining power of workers to begin with?

Reassessing Union Decline

Rafael Navar, the national political director of the Communication Workers Association (CWA), took a more strident position when it came to the unions' current demise: he told me that those in the union world must also look in the mirror and consider how they have compromised on principles that were so central to the labor movement dating back to the 1950s.[25]

Indeed, today labor unions organize only 34 percent of the public sector and less than 7 percent of the private sector.[26]

What happened? Navar explained how the union (as a single force) lost the integrity of its pro-people political philosophy.[27] It disavowed itself of radical, progressive, socialist, and communist ideologies that saw labor as a collective enterprise rather than as an instrument to amass greater profit and revenues for employers. Corporate profit, he told me, isn't interested in addressing "the core needs of humanity," and that fact exacerbates the very things causing economic inequality.[28] He has asked us instead to pay attention to the "magical places in our world that have resisted the reign of global capital."[29]

Unions have an opportunity to rethink what they stand for and project a new identity, says Liz Shuler, the deputy director of America's largest union, the American Federation of Labor and Congress of Industrial Unions (AFL-CIO). She tells me that people perceive unions as "outdated, old-school, and inflexible." She adds: "Whatever the stereotype is, they turn their ears off, they can't hear what a union actually is."[30]

Unions thus face timeless questions: Can they voice critiques of corporate power rather than simply bow to it? Can unions keep their eyes on the big picture, the profound economic inequalities that have disenfranchised working people? Will they only look for piecemeal handouts for the few workers they do represent?

Navar introduced me to his friend and fellow Mexican American labor leader, Alma Hernandez. Since 2016 she has been the executive director of SEIU California, which represents more than 700,000 workers within the state. As the daughter of two immigrant farm workers, she grew up in a small central valley community.[31] She observed how hard her parents worked, and yet as a kid, while volunteering for her mother's union as an English/Spanish translator, she came to understand the precarious nature of their jobs and her family's stability.

Hernandez reminded me of a staggering fact: although California's economy was ranked the world's fifth largest in 2018, the gap between the rich and poor in this "liberal" state has only increased, along with rates of poverty.[32] The engines that have produced such profound wealth for the state, allowing its economy to surpass the likes of the United Kingdom or Russia, have not benefited everyone, even though California's economic performance is driven by *worker* productivity.

Of course a large part of the state's economic boom can also be tied to Silicon Valley and the flourishing tech industry across the state. But no one, Hernandez emphasized, seems to understand how the interests of this industry are mutually tied to the interests of workers. She gave me a range of examples, such as airport hotels that use robots rather than human beings to check guests in. Not only would this "innovation" displace jobs and potentially be less productive than if a human were in place, but it also breaks the personal connection that customers want. "We need a human touch, to remember that we are all connected," Hernandez told me. Whether we are customers or workers, "we are all social creatures."[33]

Hernandez wants us to continue to be excited about technology. She does not want us to give up our imagination by placing it in the hands of in futurists and entrepreneurs like Tesla's Elon Musk, and she thinks there is a place in our contemporary cultural branding for the worker—for a consciousness that considers the economic well-being of us all as interconnected.

Alma Hernandez and Rafael Navar, two of the highest-ranking Latinx labor leaders in the United States, are both in their forties. Their shared vision for work and workers puts faith in the power of stories and movements to reinstate political power with justice and equity. Those are values most of us can agree upon. Navar argues that the two major parties of the country, the Democrats and Republicans, no longer represent working people. Instead, progressive political movements that reject corporate political action committee (PAC) funds are growing in the United States. They represent hope that eventually a government for and by the people can come to be.[34] Without such a government, Navar tells me, we will be stuck within an "an organizing system that exerts power locally [and goes] all the way to the top … [in effect] driving ourselves off a cliff."[35]

Contract Workers' Rights

As structures of employment are changing today, laborers face particular challenges in the workplace that are specific to our time and our current economy. These challenges are increased, if not caused, by technological transformations happening before our eyes. For example, in *The Fissured Workplace*, David Weil describes how corporations in the twenty-first century increasingly distribute activities to entities and individuals outside of their internal corporate structure.[36]

In his book Weil analyzes the effects of this practice on employee lives and employment conditions. Even the average household today tends to "outsource" more tasks than ever, such as laundry, dog walking, and grocery shopping, to app-based, gig-economy services. This saves time and is supposed to "optimize" our lives. But it also creates more competition among companies, rather than within them, causing declining wages, eroding benefits, inadequate health and safety conditions, and ever-widening income inequality.

In gig and outsourcing cases, research has revealed that the promise of "climbing the ladder" is no longer a realistic prospect for many lower-tier workers.[37] As a result, upward mobility is quickly fading, another casualty in what feels to many like the erosion of the American dream. Children born in a typical US household in 1940 had a 90 percent chance of earning more than their parents, but those born in 1985 have only a 50 percent chance (see figure 12.1).[38]

The economist Raj Chetty has studied millions of tax records from the past several decades.[39] Figure 12.1 is based on his study "The Fading of the American Dream," which tells a somber story: children no longer grow up to earn as much as their parents.[40] He attributes this increasing economic

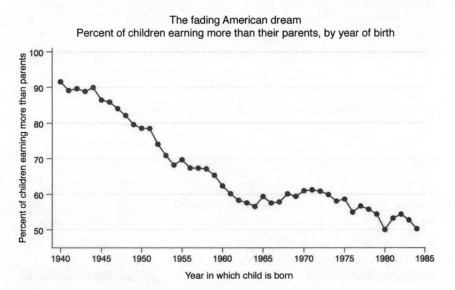

Figure 12.1

The declining American dream (source: NBER).

inequality to globalization, shifts in technology, educational deficits, and the declining skill level of the workforce.[41] While these dynamics are not the same in every other nation, they speak to how the largest economy in the world is failing to evolve in a way that serves its population. Researchers have found similar trends in the United Kingdom, where studies show that the average millennial earned about $10,000 less during their twenties than the previous generation did.[42]

The US economy, as measured through the nominal gross domestic product (GDP) and stock market valuation, continues to grow. This is a talking point Donald Trump used to inflate the success of his presidency. Yet Chetty's research shows that this growth does not benefit the vast majority of the population; it actively fails them. As the *New York Times* opinion columnist David Leonhardt reports, today's economy "disappoints a huge number of people who have heard that they live in a country where life only gets better, only to experience something quite different."[43]

Connecting this to tech, consider the examples of Kodak and Instagram. Both are imaging companies, yet Kodak went bankrupt right before Instagram was sold to Facebook for $1 billion in 2012.[44] Kodak had tens of thousands of employees and numerous patents, including for digital cameras; Instagram had thirteen employees and no patents.[45] Isn't it a problem that thirteen people instantly become worth tens of millions of dollars while tens of thousands lose their jobs?

So now, despite low unemployment rates, we're seeing that 80 percent of American workers live paycheck-to-paycheck, barely subsisting and unable to save.[46] Finding work means very little if that work fails to support the quality of life workers dream of every day when they resist hitting the snooze button and rise instead to head to their jobs.

During the decade between 2005 and 2015, about 80 to 100 percent of net employment growth arose in alternative work arrangements, or fissured work.[47] Technology has pushed this fissuring process forward, whether by making it easier for large corporations to identify and coordinate with external entities or by providing "gig economy" and other platforms for "informal" work. As a result people are not just picking up an extra hour of work or two through these platforms. Instead these platforms have created an entire workforce operating under financial pressure without a set of governing laws to protect it. As Weil notes, the fundamental "changes in employment relationships require a revised approach to enforcement."

It must be built on understanding of how major sectors of the economy that employ large numbers of vulnerable workers function, "and then using those insights to guide enforcement strategy."[48]

Steps to Automation

The gig economy could be an intermediate step toward a fully automated world. In its next chapter, Uber would be driverless, drones would deliver packages, and robots would work in factories, run retail cash registers, and serve in restaurants. Truck drivers and agricultural workers would lose their jobs, and algorithms would replace human resources work, financial services, and stock analysis. Each of these automated technologies are likely to be designed by private companies for their own, rather than their workers, interests. Applying a pro-corporate logic toward automation effectively closes the door on the majority of the national population, blue and white collar alike, who should have a say in the dramatic rewriting of their livelihoods. Indeed, a recent study predicted that 47 percent of US jobs could be lost to automation in the next twenty years.[49] Robert Reich, an economist and former labor secretary for President Clinton, has estimated that in the future about 30 percent of US jobs will be lost to automation.[50] The numbers are likely to be similar in other nations.

Why? The answer lies in maximizing return by lowering costs. That objective has been the core of capital/labor tension for centuries, and in the digital gig economy the tension has already increased. Indeed, in the United States, an hour of manufacturing labor costs $36 compared to $4 for a robot.[51] "We're starting to see robots and automated systems penetrate every sector of the economy," says Jeremie Capron, the director of research at Robo Global. "The pace of change is really accelerating, so, yes, it's a revolution."[52]

The massive reach of today's and tomorrow's automation differentiates it from earlier forms, such as forklifts and other industrial machinery that replaced physical labor. Today's automation extends to cognitive and mental labor in a range of non-industrial and agricultural fields through the use of AI technology. As this shift into new fields takes place, automated technologies may also expand into the surveillance and monitoring of human labor. For example, in 2018 Amazon secured patents for wristbands that "would emit ultrasonic sound pulses and radio transmissions to track where an employee's hands were," and it would use vibrations to "steer the worker" in order to increase efficiency.[53] "They want to turn people into machines," one Amazon worker told a *New York Times* reporter. "The robotic technology isn't up to scratch yet, so until it is, they will [us as] use human robots."[54]

The many workers who already live on the edge when it comes to economic security express fear and anger when confronting the onset of dehumanizing technologies and workplace solutions. In a telling example, Frank Gaskin, a California longshoreman at Long Beach, one of the world's busiest ports, shot a video with his phone of the robots that had taken his and his co-workers' jobs.[55] As an automated flatbed cart passed by, he yelled, "F— you automation mother f—ers!"

Gaskin's reaction speaks to a major transformation in which privately dictated automation and surveillance slowly drain the energy of labor and worker power. If we couple this with the declining the power of the union we can see how many workers may lose their jobs and economic security altogether. If we consider the fact that the gig economy may just be a middle step on the way to full automation—where paid work still exists but is increasingly precarious and unprotected—that number would be even higher. Not only might these ensuing economic inequalities harm the lives of workers but, ironically, they threaten the basic operation of capitalism itself: After all, if most people have little to no money, how will they purchase and consume products and services? This is a lose-lose situation that would be disastrous for the corporations, consumers, and workers.

Extensive research has attempted to map the future of work across an economic landscape in the throes of transformation. In this picture the rapid adoption of automation technologies sits on the horizon casting a long, foreboding shadow. A report by Bain's Macro Trends Group details how automation will create an economic climate of increasing extremes when paired with the slowing growth of the labor force.[56] Some researchers predict a surge in "plentiful labor" as baby boomers retire, but automation, a force likely to eliminate millions of jobs and suppress wages, may more than counteract it. The challenges imposed upon the labor market may stagnate household incomes and produce a skill gap among workers out of touch with the changing nature of work. Meanwhile, robotics and AI promise higher productivity, increased efficiency, convenience, and therefore economic growth. The question, as always, is who is this growth for, and what are the effects?

Complementing the Bain analysis, a recent report by the McKinsey consultancy estimates that about 60 percent of all occupations contain at least 30 percent of activities that are "technically automatable" based on *current technologies*.[57] Occupations will change and people will need to find ways to work alongside technology that takes our lives into account. The consulting group calculates that the adoption of current automation technologies could affect 50 percent of the world economy, or 1.2 billion employees, and $14.6 trillion in wages (with China, India, Japan, and the United States accounting for over half of these totals).[58] Automation will affect different countries in different ways, depending on governments, existing economic structures, and cultural values.

If nations wish to keep their citizens employed, or at least economically secure, they, like all of us, will need to find some solutions, including new forms of work. Consider a study by McKinsey's Paris office, which found that the internet had destroyed 500,000 jobs in France in a period from 1996 to 2011, but at the same time had created 1.2 million other jobs.[59] Although this is not an example explicitly about automation, it shows that there are ways for new technologies to benefit workers and to increase, rather than threaten, employment.

We have a range of options ahead: adapting our education systems, focusing on job creation, rethinking how incomes are structured, and innovating the ways in which humans work alongside machines. Most importantly, we must consider who is included in the conversation about automation, and whether workers will have a voice in shaping their futures.

The digital economy has shown new ways in which our old systems are vulnerable or subject to change—by being editable, hackable, customizable, or re-designable. And this can even apply to capitalism. We can rethink whether we want to live in a world where wealth grows for the few at the expense of many. We might use this chance to push back against the idea that work is inherently the solution for our societies: not everyone can work, and many forms of work are inhumane. We can start hacking the very concepts of profit, growth, and innovation to promote our values. We can even use the spirit of beautiful design, pleasure in creation, and elegant problem solving that animated the best parts of Silicon Valley to build a world that operates as a win-win rather than a zero-sum game.

Machining Labor

In a 2016 "Reddit Ask Me Anything" conversation, Professor Stephen Hawking weighed in on the potential onset of "technological employment," a situation in which automated processes cause widespread human unemployment, and machines and robots take the place of workers:

> If machines produce everything we need, the outcome will depend on how things are distributed. Everyone can enjoy a life of luxurious leisure if the machine-produced wealth is shared, or most people can end up miserably poor if the machine-owners successfully lobby against wealth redistribution. So far, the trend seems to be toward the second option, with technology driving ever-increasing inequality.[60]

The late astrophysicist believed that to avoid skyrocketing economic inequality, we should be concerned with *wealth redistribution* rather than solely with the transition of workers to other industries. If the ultra-rich owners of machines, who will become even wealthier thanks to control over automation, refuse to share their proliferating wealth, then the inequalities that plague our world will only worsen. The accelerating force of automation will continue to produce luxury and leisure for some, and miserable poverty and exclusion for others. This is not an issue exclusive to economics and money, either. Examples throughout history have shown that too often the poor are not only excluded from wealth and important health and education services, but also from positions of political and cultural power.

As it is, the chasm between the wealthiest and the rest of us is growing. In 2017, 82 percent of the total money generated worldwide went to the top 1 percent while the bottom half of the global population saw no growth at all. The World Inequality lab projects that, if current trends in wealth inequality continue, by 2050 the top 0.1 percent will own more wealth than the entire global middle class.[61]

These astonishing numbers are part of a systematic redistribution of wealth that has taken place over the past fifty years. Some estimates indicate economic inequality has not been so acute in the United States since the gilded age well over a hundred years ago, when Americans were rebuilding after the Civil War.[62] The French economist Thomas Piketty believes that this growing income gap will increase because returns on capital—the wealth accrued from assets like real estate and stocks—increase much faster than the overall economy grows.[63] In other words, the rich get richer at a faster pace than average workers do, because workers' wages are tied to economic fluctuation. This explains the wrong-headedness behind claims equating economic security to stock market growth. In reality, the increase in the value of the stock market only represents greater valuation for the corporations who make up this market: executives and shareholders benefit, but not necessarily the wider society or middle and working classes. Piketty, who argues for a global tax on the rich, believes that a continued obsession with creating wealth in general, and with generating greater net worth for companies in particular, is sadly inevitable.[64]

Piketty and Hawking have observed a trend by which changes in technology—whether through machines that replace human workers or the stocks that are tied to technological innovation—drive ever-increasing inequality. It doesn't have to be that way, however. Consider that the Asian Development Bank found that "rising demand, spurred by improved efficiency or labor productivity, more than compensates for technology-induced displacement of jobs."[65] This leaves an opening, a chance to assess the technologies we are using and how we are using them.

We should also carefully consider what *kinds* of jobs should be automated and can be performed responsibly with AI or algorithmic technologies. Consider the problems with what the scholar Sarah T. Roberts calls Commercial Content Moderation, or CCM.[66] Companies like Facebook currently hire people to perform CCM tasks, as a way to take some responsibility for the types of information people see online. CCM involves the large-scale screening of content uploaded to social media sites in order to make decisions about the appropriateness of images, videos, or text postings that users add to the sites. Users also have the ability to flag certain content as inappropriate, sending it to be reviewed by CCM workers.

Some benefit comes from automated solutions to this type of work. For instance, with the use of "skin filters," pornographic or violent content can be identified quickly and removed automatically by an algorithm. As it stands, however, CCM workers experience a great amount of distress. Some have become permanently affected by the disturbing content they must sift through to comply with the requirements of their job. In some cases, CCM automation could spare these workers the need to view stress-producing content in such great quantities.

CCM-type workers are also being hired to assist and train algorithms. Amazon, which runs a massive outsourcing platform called Mechanical

Turk, calls this human-assisted AI "'artificial' artificial intelligence." Others call the labor humans do for machines *human-augmented intelligence*.[67] Regardless of the label, what we see here are generally low wages—in 2017 the median pay for Amazon Turk workers was $2 per hour—often for labor that supports artificial learning. That in turn supports the profit-motives of platform owners and developers.[68]

Another example of traumatizing gig employment relates to call centers located in Asia and Africa. Many companies outsource these jobs to cater to their Western clients, thus forcing the employees to work overnight shifts and to Americanize their identities and accents. As technology allows companies to quickly shift labor to the absolute cheapest locales on the globe, and to use algorithms to coordinate these efforts in the cheapest way possible, workers face a daunting challenge. The terms of such a digital economy are set in ways that leave workers, and the larger population, on the outside looking in. The larger architectures that extend and reinforce economic inequalities across the world compromise the so-called freedom that gives gig workers the chance to opt into jobs that pay them very little. The scholar Christian Fuchs makes these points persuasively, as does Tiziana Terranova, who characterizes the digital economy as a "social factory." She equates our posts, likes, and comments to forms of free, uncompensated labor.[69]

Does the digital-work phenomenon exaggerate the existing power imbalances between countries and regions? Researchers have found that digital workers in countries on the economic margins tend to earn lower wages for the same work when compared with workers in higher-income countries.[70] So, while the governments of countries like Malaysia, the Philippines, and Nigeria increasingly see the digital market as a way to help some of the world's poorest find new opportunities, the structural problems of the global digital economy remain intact: the mere size and scope of the global market threatens the ability of workers to bargain. We can take the moderation and call center examples and put them alongside many others, such as hardware assemblers in China and miners in the Congo, to paint a picture of how technology corporations have achieved great power in dictating and dominating the nature of work across the world.

Challenges can be opportunities, however, and we now can ensure that choices made around automation and the gig economy are in line with the values of a world that is equal, cooperative, and humane.[71]

13 Working Hard, Struggling Harder

In May 2018, I had a chance to be part of a small group that met with Elizabeth Warren, a US senator from Massachusetts. Since taking office in 2013, Warren has become one of the most visible senators of the Democratic Party. On February 8, 2019, she announced her candidacy in the 2020 presidential election.[1]

Warren has long been a fierce critic of Wall Street and the big banks. For instance, Barack Obama named her Assistant to the President to help set up the Consumer Financial Protection Bureau in response to the massive recession of 2008. But his plan to appoint her as director was averted because of vehement objections from Republican congressmen, who expected her to be a zealous regulator.

During the moments we spoke together at the meet-and-greet, I asked Warren about automation and the related concerns, especially gig economy workers who are making far less than the minimum wage.[2] I mentioned that we see these challenges across the world, including the United Kingdom, where a study showed that 700,000 gig workers earn less than the minimum wage.[3]

Warren pointed out that the gig economy did not start with digital platforms: in reality, precarious, pay-per-job labor or piecework has long been a struggle faced by for day laborers, immigrants, and many other working-class populations.[4] She told me the story of how, when she was just twelve years old in 1961, her father had a heart attack. Resolved not to let her family lose their home in rural Oklahoma, Warren's mother did what few women at the time did: she took a job outside the home. As a minimum-wage worker at Sears she made enough to support a middle-class family, put food on the table, and even pay the mortgage. It wasn't great, but it was enough. The problem, Warren said, is that today the old wage-bargaining and benefits-for-work rules no longer work for the 99 percent:

> They were written for a completely different America, and if we had a Congress that actually cared about the working class, we would rewrite the employment contracts....So every time you think about what is happening with Uber and people working three jobs, understand—people today are working every bit as hard as my mother worked when she went to Sears. [The] difference is, our government is no longer working for our mother, no longer working for the people. And we have to change this.[5]

Warren's words ring true in most nations that fail to carefully protect the economic security of their citizens. In the United States, despite the unemployment rate dipping to under 4 percent in mid-2018 for the first time in nearly a decade, the economic foundations for many are shakier than ever.[6] According to a recent study, approximately 51 million households, or 43 percent of the total in in the United States, cannot afford the basics of living: housing, food, child care, transportation, and cell phones.[7] Meanwhile, the largest companies in the country are making more money than ever, as evidenced by the growth of the stock market since late 2017.

With this in mind, Warren has advocated for portable employee benefits: healthcare and retirement benefits belong to workers, she says, and should follow them regardless of which platform or company generates the income. She believes in regulating the labor market so that privately driven technological shifts do not put people out of work or income. Here's what Warren calls the basic deal: "Workers deserve a level playing field,...some basic protections,...a strong safety net,...and a chance to bargain over their working conditions."[8] Throughout this book I consider proposals such as these intended to support the economic security of us all.

The Value of Human Capital in the EU and Abroad

I discussed the precarious landscape that the gig economy and automation present for workers and economic security in chapter 12. But, as I concluded, there's little reason to see these technological shifts as a doomsday scenario. We can get ahead of the curve: by offering workers various forms of economic security, by identifying jobs of the future that automation may create rather than subtract, and by training people to perform them. In the process we can recognize that being pro-worker can also be pro-technology and pro-business.

As a result of their active and people-centered national attitude to labor relations, many countries across the European Union face less anxiety about job displacement and automation compared to the US climate of corporate deregulation and lack of labor protections.

In Sweden, for example, automation is commonly viewed as a way to supplement and improve working conditions, rather than to eliminate the prospect of human employment.[9] This comes in part from government funding for the jobless and from the nation's embrace of labor relations that support unions, good working conditions, and strong job security. Workers and employers alike understand that, in order for a company to maintain profit, it must continually increase its efficiency. But they also agree that automation need not endanger jobs but might in fact improve the conditions for work and create new jobs. For Swedish mine workers, this means that they can now operate loaders using joysticks from the comfort of a control room above ground. And who is better qualified to operate these technologies than the miners most familiar with the environment? It's a win-win. Workers are less exposed to the dangerous conditions they would encounter inside the mine, and mining companies still reap the increased profit. In addition, employees maintain their high wages, free healthcare, five weeks of vacation, and ample child leave.

How did this come to be? Part of the answer lies in Europe's highly effective employer-funded worker councils, which help people who lose their jobs find new ones.[10] Worker councils often sit on the supervisory boards of European companies and have a significant say in the introduction of new technological devices, privacy and data regulation, worker surveillance, performance monitoring, and more.

These councils have been able to insert themselves into technology-related company discussions and negotiations. Typically the goal is to foster cooperation: to balance job security for workers while maximizing the benefits of technology in the workplace. Achieving balance depends on recognizing that for most companies security requires increasing streams of income. When technology companies do overstep their boundaries, worker councils have been able to fight back to protect the interests of employees. For example in 2012, when Hewlett Packard (HP) denied employees their right to take part in a company decision to eliminate 29,000 jobs, the European Workers Council sued HP for avoiding proper social dialogue.[11]

It's not that automation, or information technology itself, is significantly different in Europe. But the country's approach to how automation is designed takes seriously the value of employees and human labor—*value* in the ethical sense that displacing workers is considered to be unprincipled and destructive, and *value* in the business sense of understanding the power of human capital. The notion that humans can increase a company's efficiency and profit when combined with technological innovation, rather than be replaced by it, is something we can either take with us into the information age, or throw out with the bathwater. In this sense, Sweden represents an example of how robots can introduce profit into the market at no expense to human capital.

The high-income countries of northern Europe, however, have smaller populations of manufacturers and low-income workers who make a living in factories and in that way are quite different from most other nations. Consider in contrast, how the economies of developing countries are at significant risk when it comes to automation. In India, a huge amount of low-skilled workers could see their jobs replaced by 2022, while medium and high paying jobs are expected to increase due to the growth in IT employment opportunities.[12] The challenge for India is to find ways to transition low-skilled workers to higher pay brackets, given the nation's population of more than 1.3 billion. In today's India, the top 1 percent of the population owns 73 percent of the nation's wealth. So despite the nation's growing middle class, inequalities remain massive.[13]

AT&T and Its Social Contract

What then about the United States? Yes, we see little in the way of worker councils and government interventions. But Thomas Friedman, a *New York Times* Pulitzer Prize–winning journalist, has gone out of his way to praise at least one US company that has proven its investment in maintaining human capital: AT&T. Yes, the telecom giant has been criticized for its stances on net neutrality, its potential connections to the National Security Agency's domestic spying program (PRISM), and its potential merger with fellow giant Time Warner.[14] But in terms of employee engagement, the story is different. With the onset of AI systems and cloud computing, the company recognized that its competitors had expanded to include not just telecom providers, but internet and cloud computing giants like Google and Amazon. Keenly aware of the likelihood that many of its 280,000 jobs (as of 2016) would soon be threatened by machines, the company told its employees to "adapt, or else." AT&T helped them in this process by providing employees with the opportunity to engage in "lifelong learning" courses. Investing over $200 million in online education, it gave workers the opportunity to take free courses and reskill themselves in areas like data analytics and software coding, which are likely to be in demand as automation and sophisticated technologies enter their field. To this end AT&T established its own internal university and forged partnerships with online education providers and several traditional universities.

"You can be a lifetime employee if you are ready to be a lifelong learner," AT&T Communication CEO John Donovan tells his workers.[15] If employees choose to participate, they can map the progress they make in lifelong learning courses onto the specific jobs available in the company. If they are not trained in a skill required for a desired job, AT&T will make a course available in that area.

I had the opportunity to speak about this program with John Palmer, Senior Vice President and Chief Learning Officer at AT&T. Instead of seeing technological transitions as risks to his company and the people that work in it, Palmer identifies them as opportunities for transformation. In his view, the company must be as agile and adaptive as the shifting technologies. "What we learned five years ago is likely not 100 percent relevant today, and will be less relevant five years into the future," he told me. "But it's important to hone our skills in emerging areas so our company can ultimately stay relevant."[16]

The takeaway here is that transformations in technology can be opportunities for companies to reassert their commitment to workers. Friedman describes AT&T's efforts as a "social contract," a term that Palmer also repeated to me. This idea implies that we all gain something when we agree to exit the dog-eat-dog, zero-sum mentality. AT&T has shown that doing this together is a viable option. Between 2013 and 2017 the company's total investment in the United States, including capital investment and acquisitions of spectrum and wireless operations, was over $135 billion—more than any other public company.[17] As John Donovan told Friedman: "We have employees here who want to make the pivot—employees who built this company that they would die for—and we need to give them the opportunity to make the pivot."[18]

AT&T's effort to brand itself as a pro-worker company while helping workers to become lifelong learners is certainly laudable. But the company's threat to reskill "or else" barely veils an unsettling reality. Even longtime employees may lose their jobs if they either refuse to participate in the suggested training or are just not able to acquire the knowledge. A real tension exists between CEO Donovan's belief that his employees would "die for" AT&T and the company's willingness to fire these same employees if they are not effectively retrained. Loyalty has its limits, and so do social contracts, but who gets to define these limits? And what about those who don't get to? Technological change is highlighting all of these lines in the sand and giving us an opportunity to think about how they got there, and how together we might redraw them.

It is not necessarily AT&T's job to be a caretaker for society, and we should not confuse its retraining program with a philanthropic or ethical effort. The company is letting its workers know that if they can continue to add value, their jobs are safe, nothing more. This is a private corporate initiative, not one offered for the public. It is neither welfare nor security oriented. AT&T isn't obligated to do anything else, but that doesn't mean that as a society we can't want more for our people than an ultimatum.

14 Money for Everybody? Exploring Universal Basic Income

At the 2017 World Government Summit in Dubai, Elon Musk, the founder and former CEO of Tesla, discussed his concerns with the takeover of our economies by robotic systems.

> There will be fewer and fewer jobs that a robot cannot do better. What to do about mass unemployment? This is going to be a massive social challenge. ... The output of goods and services will be extremely high. So, with automation there will come abundance. Almost everything will get very cheap. Universal basic income, it's going to be necessary. The harder challenge then, is how will people have meaning?[1]

Musk is no saint. A high number of injuries have occurred in Tesla plants, twice the industry average. And like many other gig workers, Tesla subcontractors are paid as little as $5 per hour, which Musk ultimately apologized for after the summit.[2] But Musk is also an icon of Silicon Valley; given his other entrepreneurial and creative endeavors in companies as diverse as PayPal, SpaceX, and Neuralink, he is someone who commands great attention within that world as well as across the globe. His belief that a global universal basic income (UBI) will become necessary in the near future, *as a direct result of technological automation*, continues to garner attention. Rather than predicting that new jobs—of equal or greater quantity than those available today—will be created thanks to technological advances, he assumes that, if anything, most jobs will be made obsolete.

UBI in Concept

UBI is a system through which the government would either send regular sums of money to everyone regardless of their income (ideally providing everyone with a living wage), or send money only to those in a specific low- or middle-income bracket. Although public conversation in recent years has put UBI in the spotlight, there are plenty of new ideas about how to organize a society if employment is no longer available for a large proportion of the population. Other possibilities include an *automation tax* that companies would pay based on profit increases from the widespread use of automated systems. Portable benefits packages are also up for consideration as a way to ensure access to healthcare, education, and other basic public services. Though these were once far-off ideas in the consciousness of Americans, a recent survey by Northeastern University showed that today 48 percent of Americans support the idea of a UBI.[3]

UBI is likely to become a major political issue, given the shifts in labor and economic security that populations across the world have already begun to experience. Andrew Yang, an entrepreneur and a candidate for the 2020 Democratic presidential primary, has made basic income a central plank of his platform.[4] Yang believes that instead of serving as a substitute for work, UBI would inspire entrepreneurship and catalyze new businesses and artistic initiatives. Recipients would get the security they need to create new possibilities within an expanded digital economy. Yang's "Freedom Dividend" proposal calls for the US government to send $1,000 per month to every household in the country. This, he believes, would transform its citizens' "constant mindset of scarcity to a mindset of assured survival and possibility ... by taking the boot off people's throats."[5]

UBI is often framed as a new initiative and, in its current guise as a measure to address technological transformations, it is. But taking a more historical look, we see that the idea has been discussed for centuries. Thomas Paine, an American political activist, Founding Father, and philosopher, advocated in *Rights of Justice* (1791) for publicly funded welfare for all citizens of a nation. Like modern proponents of UBI, Paine felt that ensuring economic equality in a society was paramount.[6] If one is poor, Paine remarked, in what sense is he or she free? Liberty, freedom, and equality—these are not values that exist within a vacuum; they shape and support one another. In other words, values often associated with the individual (such as liberty) are connected to values associated with society (such as equality). Free people must live in equal societies; otherwise they are not free.

The great civil rights activist Martin Luther King Jr. also embraced the idea of a basic income as a way to end poverty and fight rising racial and economic inequality.[7] In the last year of his life, King began his work on the Poor People's Campaign, a large-scale movement comprising activists and groups in support of labor, community advocacy, and social justice.[8] At the center of their lobbying work a $30 billion anti-poverty bill took shape. If it had passed, the federal government would have produced a basic guaranteed income for all Americans. For King, economic justice was deeply tied to racial justice. As King himself wrote, "New forms of work that enhance the social good will have to be devised for those for whom traditional jobs are not available."[9]

Elon Musk's friend Sam Altman, the billionaire founder of Y Combinator, Silicon Valley's best-known tech accelerator, has voiced little doubt that "capitalism will find a way" to provide jobs to those whose current means of generating income are displaced. But he also voiced concerns that in the transition to automation many people could be impoverished. So he too has been funding UBI studies. Altman told me we need "significant updates [about] how work happens in light of the changing economy; a new definition of minimum wage, benefits, and collective bargaining. ... This change needs to happen soon."[10]

UBI has gotten some heat—particularly from some US right-wingers, including former Alaskan governor and US vice-presidential nominee Sarah Palin for being "the same as socialism." But in reality, the idea behind UBI is more similar to a government investment than it is to a welfare program.[11] UBI fits squarely within the capitalist market that we already have because it redistributes wealth to be used in the market. It is *not* about socializing the means of producing wealth, or even about providing basic social services that might be seen as a human right (like food stamps or, in some societies, healthcare). But UBI costs money—a lot of money—and this could mean that governments considering UBI might consider cutting crucial services, like Medicaid and Social Security.

Distinguishing between a market-friendly solution and a program that's "the same as socialism" helps us to understand UBI as a viable, practical solution to the unemployment crisis caused by automation—rather than as a wild-eyed, utopic, liberal, or anti-capitalist project. As the US senator Bernie Sanders points out, the issue of universal basic income enjoys support from liberals and conservatives alike, including former US president Richard Nixon, a Republican, and the Adam Smith Institute rightwing think tank.[12]

Consider the example of Alaska (before Governor Palin's administration), where citizens voted in 1976 to invest into the Alaska Permanent Fund.[13] From this fund, which began in 1982 and has continued ever since, every Alaskan citizen receives a check for approximately $1,000 to $2,000 annually—no strings, tests, or conditions attached. Although some commentators have worried that, if given extra income people would lose the motivation to work, economists studying Alaska's program found that the payments made no overall effect on employment, other than a slight increase in part-time workers.[14] Research also shows that there are a number of positive effects for society, such as better educational outcomes for children, who may be getting more school supplies or more time with their parents.

Alaska is able to support a program like this because the fund is financed by immense amounts of oil money and an industry that contributes to climate change. This should give us pause: first because the portability of the model is limited—most of the United States, and many other countries throughout the world don't have access to that kind of oil reserve; and second because it exacerbates one of the great crises of our time.

Although controversial issues regarding the gambling industry play out on a different scale than big oil, one North Carolina Cherokee tribe is applying profits from its casinos and gaming operations to provide basic income for its members. The program began in 1997, after extensive negotiations between tribal leaders the Eastern Band of Cherokee Indians and the North Carolina state government, and it continues today.[15] Any member of the tribe is eligible to take part and receives a payment of $4,000 to $6,000 annually. So far, the results are positive: in the poorest households, educational achievement increased; criminal activity decreased, especially among youth and minors; mental health improved overall with markedly less addiction and anxiety; and physical health improved an obesity and smoking declined.

Other UBI experiments are small-scale but growing. The singer-songwriter and philanthropist Dolly Parton's Dollywood Foundation offered no-strings-attached cash transfers of $6,000 over the course of six months to every household whose home was destroyed when a wildfire affected thousands in Gatlinburg, Tennessee.[16] Stacia West, a researcher analyzing this program, concluded that Tennessean beneficiaries of the fund were able to stay on top of their bills, maintain their homes and assets, and did not experience a reduced motivation to work. The recipients themselves could direct the monies they received toward expenditures that would contribute to an improved quality of life.[17]

Examples exist in official government settings as well. The crowdfunded effort Mein Grundeinkommen in Germany has funded a basic income lottery since 2014.[18] The Finnish government is halfway into a two-year pilot where some residents receive a "basic income" of $16,000 for two years, no strings attached.[19] In Ontario, Canada, as of April 2018, select residents will receive thousands of dollars a year for the next three years in a trial run administered by the government.[20] And in 2019 a citywide pilot has taken root in Stockton, California.[21]

UBI in Action

I'd like to show UBI in action as it emerged in Stockton because I have a personal connection to the city's mayor, Michael Tubbs. The story starts in 2012, when I accepted an invitation to speak about my work on technology, activism, and democracy at a TED conference.

Among the many engaging presenters in attendance, I was most struck by the young, black man who spoke before me. His ability to captivate the audience was remarkable as he discussed the "cage of poverty" that many children are born into in the United States. For instance, a confluence of factors in Stockton itself placed over 40 percent of Latinx and black families in a no-win situation as they struggled to survive, living paycheck to paycheck within a system they were powerless to change.[22] Children in Stockton were already born into the curse: on average they would live for fifteen years less than kids from neighboring zip codes.

I did not know at the time that the speaker was also an academic superstar—a Rhodes scholar in his final year at Stanford University, where he would graduate with bachelor's and master's degrees in only four years, despite growing up in relative poverty in a single-parent home.

Just a few years later, at the age of twenty-seven, Tubbs was elected to be Stockton's mayor, making him the youngest US mayor of a city with a population 100,000 or larger.[23] Personally endorsed and mentored by Barack Obama, Tubbs is one of several young politicians of color gaining recognition as a progressive face of the nation's future. He is one of very few candidates that the billionaire celebrity Oprah Winfrey has publicly endorsed.[24]

Tubbs is also the leader behind Stockton's UBI initiative, funded by the Economic Security Project. The Stockton Economic Empowerment Demonstration (SEED) seeks to model how basic income can engage local communities, support families, and provide services to residents in ways that empower people and their voices.[25] It will begin in early 2019 and will start by providing one hundred recipient families with $500 per month for eighteen months. Income recipients will provide feedback about the impact of the stipend by sharing the stories of family or household members.

The effort has gathered a great deal of media attention. Politico called it an "experiment underway in millennial-led government that's trying to pull a city back from the brink [by] using what is essentially privately funded socialism."[26] *Time* described it as an attempt to "find ways to shore up left-behind areas of the economy in an era when economic growth is concentrated in fewer and fewer hands."[27]

I had the opportunity to reconnect with Tubbs and his colleague Lori Ospina, who is managing the SEED project. The mayor explained how the demographics of Stockton parallel that of the nation itself: a mixed city with Asian, Latinx, African American, and Caucasian residents.[28] It faces similar economic concerns that many other cities in the nation do, with about 20 percent of the population living in poverty and another 25 percent just one paycheck away from it. This likely means that if the UBI effort can serve the residents of Stockton then it could support others across the nation.

Why support those who are less well off? Why not force them to work instead? For Tubbs and Ospina, there is no reason that economic security and work should be mutually exclusive. Instead, economic security must be guaranteed as a prerequisite to fair and equitable work. This may seem counterintuitive at first, but consider the many forms of work that are not recognized in the formal labor sector, most notably parenting, housework, and caretaking for the elderly. If these crucial tasks were not done at home, workers may not be able to survive, much less turn up to their jobs in the formal economy dressed, fed, and ready to be productive. We need a model recognizing that talent and intellect exists in all of us, Tubbs told me, and that "we're all better off if we accept that the sum is greater than its individual parts."[29]

Take, for example, the growth in the American prison-industrial complex. The United States has by far the highest rate of incarceration of any of the world's wealthiest nations, holding about 25 percent of the world's prisoners.[30] The cost of mass incarceration to taxpayers is enormous, estimated by some studies to be over $182 billion per year—more than double the average cost of tuition at a four-year university (which is already more expensive in the US than in most other wealthy nations).[31] Formerly incarcerated people face an incredibly steep climb if they hope to successfully reintegrate back into the economy upon their release from prison. As a result, studies have shown that in the year after release only 55 percent of ex-prisoners were able to generate any income at all.[32] Even if just for a moment we set aside the human tragedy this represents, we can see not only a massive economic cost to society but also a huge waste of potential.

Tubbs and Ospina wonder whether these numbers would change with economic security, and whether the desperation that drives many to crime would be mitigated if everyone, not just the rich and white, were taken care of. The mayor seems to know he is an exception. Few children of color from single-parent, working class families ever end up excelling the way he has, though many are just as intelligent, charismatic, and driven. Growing up in Stockton, he told me, taught him a lot about resilience, strength, community, sacrifice, and being tough. But now, as mayor, with such visibility and star power, he has the opportunity to "create the community [he] wanted to grow up in."[33] There is no guarantee that this or other UBI efforts will manage to succeed. But they may show how even a small drop of the astronomical wealth that technology produces can be harnessed to serve and protect the economic well-being of societies across the world.

15   Worker-Owned Technologies

As we search for ways to use technologies to improve the welfare of us all, let's consider the possibility of using technology applications to share costs, revenues, and labor among their workers. One answer, according to Nathan Schneider and Trebor Scholz, the co-editors of *Ours to Hack and Own*, is the *platform cooperative*, a digital platform cooperatively *owned by workers*. (These platforms differ from those of a so-called *sharing economy*, which valorizes the shared use of data and resources but sets up a relationship in which the platform provider makes the lion's share of the money.) These scholars attribute the low-wage crises we see in the United States and several other countries across the world to poor regulation policy in many of the biggest tech applications. They say that platform co-ops may offer a solution.[1]

Enterprises based on cooperativism have existed across the world for centuries in the form of tribes, guilds, and monasteries. The Shore Porters Society of Scotland, which claims to be the modern world's first co-op company, was founded in 1498, though the cooperative movement as a whole didn't become a force in Britain and France until the nineteenth century.[2] Today, many well-known companies such as REI, Organic Valley, ACE Hardware, State Farm Insurance, Best Western Hotels, and True Value Hardware are cooperatively owned, though they remain hugely profitable business enterprises.[3] In addition, cooperative banking services provided by "mutual organizations" remain very popular all over the world, in the form of credit unions and mutual savings banks. Like non-digital co-ops, platform co-ops rely on two main principles to function and sustain themselves: *communal ownership and democratic governance*.

What if Uber drivers themselves were to own the rideshare company? That seemingly far-off dream has become real. Some drivers have left the company to organize in co-ops. Others are even designing their own taxi apps, like the Denver-based app Green Taxi Cooperative, owned largely by African immigrants.[4] As a result a number of other successful platform co-ops have entered the market: among them is Fairmondo, an Amazon-like marketplace in Germany that is committed to fair trade; Resonate, a music-streaming service owned by musicians, record labels, and fans; and MIDATA.coop, a Swiss-based app developing cooperative ways to access medical data, with a focus on transparency and user control over who has access to medical information.[5]

Because platform co-ops are founded on agreements made by their workers and (in some cases) their users, they present the opportunity to build a digital workforce that ensures income security and fair pay for all. There's another perk too: when the terms of work agreements are co-determined by a group of owner-workers, abuses like wage theft and excessive surveillance are unlikely to be tolerated. Platform co-ops are able to offer these protections without sacrificing the scale of their operations, their potential for success, or their financial viability. In fact, studies show that cooperative businesses have lower failure rates than other businesses, and can better address and survive economic crises.[6] The author Marjorie Kelly describes them as *generative*, meaning that instead of extracting value for the gain of a few powerful players at the top, they are able to use their revenues to benefit all of the cooperative's stakeholders.[7] If platform cooperatives could take over a large portion of the economy they could generate and recirculate wealth to support a range of new activities, investments, and initiatives—potentially reversing some of the economic inequalities that plague our society. For instance, platform co-ops are often closely tied to the communities that use them, which means that a platform like the Green Taxi Cooperative can help to promote community growth and expand local economies and capital. Platform co-ops can thus make a joint social and economic investment in businesses and communities, and as a result invest in collective stewardship, accountability, and care.

In August 2016, the British politician Jeremy Corbyn, leader of Britain's Labour Party, released a four-page manifesto promising to transform internet rights for users in the United Kingdom. The document included a section on platform co-ops that called for a digital "bill of rights" to protect users. In addition the proposal envisioned platform co-ops as a desirable option for improving public services such as transportation, housing, libraries, community media, and citizen passports.[8] Under Corbyn's manifesto, publicly funded programs would be open-source, allowing users to re-create and reuse government software—a policy that other governments worldwide, including in Uganda and the United States, are also pursuing.[9] Because companies like Uber, Lyft, Airbnb, Facebook, and Google are becoming so deeply enmeshed in the fabric of modern life, Corbyn says, they are evolving into public utilities and should be regulated as such. This perspective, which places the public interest back at the center of the conversation, may guide us toward the goal of a more democratic and equitable internet for all.

One push toward platform cooperatives in the United States, I learned, connects to the energy behind the 2011 Occupy Wall Street movement. Nathan Schneider, a tech-savvy member of the movement, told me that a group of artists in the original Zuccotti Park encampment formed a printmaking cooperative called OWS Screenprinters.[10] The co-op members began to silkscreen shirts and posters with Occupy slogans right there in the park and then contributed their proceeds to the movement's General Fund.[11]

Schneider explained to me that co-ops in the United States already provide more than 2 million jobs and create more than 75 billion dollars in annual wages.[12] If we look globally, the numbers are even more heartening: based on 2016 statistics collected in a 2018 report, the world's top 300 co-ops were able to net some $2.36 trillion, and revenues were projected to increase.[13] OuiShare, a network designed to further cooperative and collaborative models in business and governance, has managed to gather hundreds of members across twenty nations within Europe, Latin America, North America, and the Middle East.[14] Hubs for the co-op movement like OuiShare, and the International Cooperative Alliance, a federation of cooperatives that represents 315 organizations from some 110 countries, help keep co-ops connected, communicating, and aware of one another's success.[15] And this movement also is strong in Africa. For example, in Kenya, more than 60 percent of the population earns its income from cooperative enterprises, which together accounts for 45 percent of the nation's GDP.[16]

Numbers like these tell us that successful, global cooperatives are a reality, not an elusive, utopian fantasy. Many of us, however, have yet to consider how powerful a force for good the internet could be if only it followed a path based on cooperative values. Schneider explained that the platform cooperative movement uses the phrase "internet of ownership" to help us imagine how technologies of search, transportation, and social networking as fields of innovation that should support all of us, not merely private shareholders and executives.[17] This connects to an idea proposed by the business scholar and author Arun Sundararajan, of a gig economy led by millions of micro entrepreneurs, rather than a small number of giant corporations.[18] These gig platforms can be owned by their workers, and need not follow the troubling pattern I've described in relation to Uber, Airbnb, and others.

What is the biggest obstacle we face to building a more robust platform co-op system? A digital market where large tech companies have most of the power, own the most data, employ the most workers, and compete for the most customers. But as Schneider and Scholz point out, "At its heart, platform cooperativism" is about "lived acts of cooperation." It is a *process* and a *movement* rather than a final, static state.[19] Perhaps platform co-ops present us with the possibility of a more cooperative future.

16 Discrimination Technologies

In our conversations about automation and its potential effects on the future of work, we should all be asking questions about *how* work will be chosen in an economy increasingly driven by algorithmic and automated systems. But let's probe deeper and ask an important ethical question: How do we prevent automated and algorithmic systems from targeting the most vulnerable members of our society, and how do we continue to ensure that these systems remain equitable?

Technologies are not neutral but *socially constructed*, created by and for particular people with particular worldviews, agendas, values, and priorities. Engineers, designers, and coders build into each system a projection of their own thoughts, preferences, and assumptions about the world. There is nothing inherently wrong with that, except that as these technologies become increasingly central to the lives of everyone on earth, the mental habits and cultural biases of a tiny group will be disproportionately influential and powerful. Sometimes it makes sense to allow a small group of specialists to have this level of influence or power—but usually we do so by electing them. Yet now more than ever, hidden, private, and corporate-employed designers have all the power over even the most intimate aspects of our lives.

It might seem obvious that content on the internet, as well as on the most popular websites, largely comes from the Western world (with the exception of the Chinese internet).[1] But less obvious is the extent to which biases that shape the content we see come from the "white Western male" perspective. I have discussed gendered and racial bias earlier, especially in chapter 2. I'll put it more bluntly here: artificial intelligence (AI) systems have a "white guy problem."[2]

We have already seen AI learning systems make mistakes that follow the classic lines of discrimination: Google Photo's image recognition algorithm misclassified black people as gorillas.[3] A "beautifying algorithm" developed in Russia, called FaceApp, turned Barack Obama's face whiter (and younger).[4] A Microsoft-developed chatbot called TayTweets, which was designed to mimic and converse with teen users on Twitter, quickly developed racist, sexist, homophobic, and xenophobic tendencies.[5] Within hours it began to echo positions of President Trump while tweeting #MAGA.

Algorithmic mistakes and biases are real. We all have implicit biases no matter who we are: after all, much of them are simply about what we leave out, forget, or assume. But algorithmic discrimination takes the problem to a new level. It can blow up hate, bigotry, sexism, and other forms of bias, which disproportionally target already-vulnerable peoples, to inflammatory extremes. Similarly, the very human temptation to treat computational recommendations as objectively "correct" is just another mechanism by which any error an AI system makes can be duplicated (and further spread) with the potential to create disaster.

Let's consider perhaps the most troubling facial recognition software of them all: Amazon's Rekognition. The system misidentified 28 members of the US Congress as criminals when the American Civil Liberties Union (ACLU) of Northern California matched images of all 535 members to a publicly available database of 25,000 mug shots.[6] To make matters worse, for Congress as a whole the error rate was 5 percent, but for nonwhite members of Congress the error rate was 39 percent.[7]

Disturbingly, Rekognition is currently being designed for and sold—despite protests from the ACLU and even Amazon's own employees—to the US Immigration and Customs Enforcement (ICE) agency as well as the military and police, who could use it in conjunction with body cameras. Amazon has nonetheless reinforced its commitment to build AI systems for these organizations. Teresa Carlson, Amazon Web Service Vice President, defended the system against protests from company workers, saying that "our war fighters out there in the field, our civil servants" should have the capabilities the technology provides.[8] In response to questions about the possibility of abuses involving the Rekognition technology, Carlson offered an all-too-typical justification: "You're always gonna have bad actors."[9]

The "bad actors" comment comes straight out of the big-tech company playbook. It's the exact term execs like Mark Zuckerberg have used to abdicate from Facebook by claiming that problems (and, by extension, scandals) are inevitable.

We've seen examples so far of AI gone wrong related to racial, cultural, and political issues, but it turns out that problems exist in even the most supposedly neutral applications of machine-learning systems, such as science and medicine. In 2013, IBM's AI system Watson, which defeated Ken Jennings on *Jeopardy!*, entered the healthcare market as an expert consultant to cancer doctors.[10] In response to the discovery that Watson had made incorrect, even dangerous, medical treatment recommendations, one doctor did not mince words: "This product is a piece of shit. We bought it for marketing and with hopes that [Watson] would achieve the vision. We can't use it for most cases."[11]

These issues of bias are not limited to the internet. Consider an example concerning automobile safety standards: from their inception the passenger dummies used to test the impact of simulated car crashes were designed to mimic the body of the "average male." Only in 2011 did the industry bow to pressure and begin to produce dummies based on women's average proportions as well.[12]

The harms such bias can produce are even more disturbing when government or social services use technologies produced and created secretively by private companies. In New York, for example, the algorithms used to allocate housing, food stamps, and policing were created by private contractors who have refused to make their source code and learning models available to anyone, even the government agencies charged with providing oversight.[13] It is troubling that the public impacts of private actions are so profound, while private actors remain unaccountable.

Consider the tech companies responsible for the algorithms that identified black people as "gorillas" (Google) and for the racist, vulgar TayTweets chatbot (Microsoft): in both cases they chose to remove the symptoms rather than fix the problems. Microsoft shut down TayTweets (see figure 16.1). Google removed the words "gorilla," "chimpanzee," and "monkey" from its image-learning algorithm to avoid labeling *any* image with those primate terms.

It's becoming more apparent that algorithms can transform white-collar jobs as well. For example, they are being used in human resources (HR) departments to automate the assessment of a job applicant's CV or résumé. In 2017 the journal *Science* published a paper about "word embedding,"

Figure 16.1
The racist, homophobic chatbot (source: Business Insider).

an algorithmic process that looks at billions of words online to identify semantic relationships among them.[14] The paper found that this algorithm was more likely to associate words such as "female" or "woman" with the arts and humanities professions, and "male" and "man" with the math and engineering professions.[15] The results also showed that European American names were associated with pleasant words like "gift" or "happy," whereas African American names were more commonly associated with unpleasant words such as "ugly" and "filth."[16] In this case, unlike in Google's photo-tagging debacle, the AI hadn't made any errors, strictly speaking. All it had done was reflect back to us the patterns of bias already infecting our actual, human language use.

And what was the result when this AI was allowed to make decisions involving actual humans? The word-embedding algorithm, researchers found, was 50 percent more likely to grant job interviews to European Americans than to African Americans, *even if their CVs were seen as otherwise identical*.[17]

Companies have increasingly turned to technology to make decisions concerning their human resource practices. As they interview an applicant, for example, a camera connected to an AI system may analyze the person's behavior. Just as we saw in the case of word associations gathered from online text, if these automated systems are trained by white faces and male voices, black and female applicants are likely to be both disregarded and disenfranchised.[18]

It is unacceptable to let algorithms reduce our human performance, character, or fitness to a lone number (a single score!) in order to determine where we work or to decide whether we belong in jail. Denying us access to how we are evaluated not only exposes us to potential discrimination but also gives us little possibility to demonstrate change or growth.

As algorithms become more influential as automation and other digital technologies transform the future of work, how will they be rewarded and punished? It seems likely that, unless we intervene, human resourcing algorithms will only expand their presence in the future, jeopardizing many jobs and further normalizing racial, gendered, and other forms of bias.

Black Mirror and Social Credit

Let's take these algorithmic bias problems further by examining an episode from Black Mirror, an award-winning series on Netflix.[19] "Nosedive" tells the story of Lacie, a young woman obsessed with her ratings in a world where people can assess their every human interaction on a scale from 1 to 5 stars.[20] In this fictional world, ratings impact social credibility and socioeconomic status: the ability to get a loan, enter an exclusive apartment complex, board a flight, or even rent and purchase items.

Only the dopamine-triggering affirmation of being given a 5-star rating uplifts Lacie's persistently nervous mood. So she rekindles an old relationship with a popular childhood friend, Naomi, whose score hovers in the 4.8 range; Naomi is just the kind of well-connected woman whose friendship would guarantee a ratings boost for Lacie, a striver stuck in the low 4s. Sure, Naomi was a bit of a bully in childhood, but when she invites Lacie to be the maid of honor at her upcoming wedding, Lacie is determined to make this relationship happen—no matter what. Lacie, however, has no clue that a similarly monomaniacal desire to maximize her social score drives Naomi's motives. She's trying to game the system with the "generous" gesture of inviting someone with a low score to her wedding; she's convinced that the algorithm will reward her for adding some "diversity" to the high-rated, almost 100-percent white community of wedding attendees.

As the episode progresses, Lacie is swept into a cycle of increasing social-score precarity that leads to her downfall: she needs high-score "influencers" to boost her ratings, but her own low score and obvious desperation limit her access to them. Her escalating panic leads to lower scores and tightened restrictions. As she struggles to get to the wedding on time, her low-end rental car breaks down but she's barred from renting a better one. Why? Because her score isn't high enough. When she dares to stick her thumb out for a ride she receives terrible ratings from passers-by, who sink her score ever lower. Naomi, who tracks her friends' scores closely, un-invites Lacie on the spot. At the end of the episode, Lacie crashes the wedding and delivers a ranting speech to the guests. The police arrest her and throw her in jail.

Is this dystopian nightmare possible? A report in *Esquire* magazine, along with several others, alleges that something like it has come to life in China.[21] Indeed, the parallels between "Nosedive" and the "social credit system" in China are stunning. As I discussed in chapter 10, China uses surveillance technology to monitor and rate its citizens' behavior. The resulting score affects everything from the schools their kids can attend to available jobs they qualify for, and determines whether they can board airplanes or trains.[22] China hopes to implement the system nationwide by 2020, but as of May 2018, a "bad score" has stopped more than 11 million people in China from taking a flight.[23] The rules are clear: scores that don't pass an undisclosed threshold can exclude a citizen from opportunities that once were taken for granted or considered a right. All the while, citizens remain completely in the dark about what personal data is being collected, how it is being aggregated, calculated, and computed; who is accountable for it, or what scores produce what outcomes. Existing technology makes this system possible not just in the world's most populous nation but also, in theory, nearly anywhere. In short, it is a data dystopia—the stuff of reality meeting the stuff of science-fiction nightmare.

As if the Chinese social credit system were not enough to monitor the country's huge population, Chinese-developed tech has allowed the government to deploy facial recognition AI systems far and wide, tracking the behavior of any citizen in public space. The data is then used to punish those who commit offenses, such as jaywalking. It issues fines instantly

Figure 16.2
Public shaming via LED screen in China (source: Sohu).

via text messaging and, in some cases, publicly shames them by projecting their faces on giant LED screens in the streets (see figure 16.2).[24]

These Chinese systems speak to the dangerous possibility of data being used to fuel government overreach, mass surveillance, and the "ordering" or social control of society. Technology alone is not the culprit here. The danger stems from the desire of authoritarian governments and private corporations to manage and organize citizens, especially when those citizens aren't on board the use of data for a particular purpose. Even in European or North American nations, where laws exist to protect citizens from excessive surveillance, new technologies stand at the ready to further enable a government's desire to break traditional boundaries regarding its citizens' privacy and self-determination.

It's important to recognize that China's social credit system is founded on the already-established credit and debt economy that pervades the globe. The idea of credit suggests that there exists between a borrower and a lender some sort of agreement—let's call it "trust." Credit in this case is synonymous with trust because it allows one party to provide money or resources to another party and receive in turn a *promise* to either be repaid in full (often with interest), at a later date. Making this promise formal means it is legally enforceable.

Today many of us live in an economy that is increasingly reliant on debt; our level of indebtedness prompts friends and family to urge one another to build their credit in order to be considered financially responsible.[25] Credit and debt are touted as ways in which we can open up opportunities for ourselves. Credit (and therefore debt) allows us to take out loans to attend college, purchase a home, or start a business. The less publicly visible side of this system, however, is how the debt economy assigns statuses to everyone, in the form of credit ratings, based on the indebtedness that comes with already-accessed credit. These credit ratings, or scores, are based on esoteric rules and algorithms that few people fully understand. A low credit rating can prevent any citizen from accessing the very opportunities credit is supposed to secure.

The Chinese social credit system is an extreme extension of this basic economic idea. It illustrates vividly that credit is not just economic, but also political and personal. In this way technology can strengthen the predatory aspect of the market, which thrives on debt and the social control it produces, as a means to the end of deepening and consolidating state power.

How does this work in China? Instead of basing one's credit score on one's ability to promptly pay back loans, the Chinese government "trusts" its citizens to carry out a certain set of determined behaviors. If citizens break this trust, their score lowers. We can understand the kinds of exclusions that result from a low social credit score in China as similar to how credit card companies might raise our fees or interest rates if we are late to repay our loans. The main difference is the extent to which China has developed this process and expanded the credit system's power to encompass a much larger swath of life. In China, a poor credit score that comes from expressing "bad" speech online, smoking cigarettes, or playing too many video games could result in a range of punishments, from losing one's freedom to travel, to public humiliation.

Meanwhile, the government is largely ambiguous about how this information is collected, stored, and distributed, and likewise about vague the parameters used to analyze it. Currently the Chinese government gathers personal data by piecing together information given to them by private tech platforms as well as city councils. Citizens don't know exactly when or how they are being monitored, and they have no political voice to demand such information. As a result they have less power to resist, avoid, or subvert sur-veillance and more incentive to panoptically *self-police*—even when they aren't actively being watched. This is how a lack of transparency regarding data collection and use can be exploited as a tool of control.

These troubling practices are also afoot in the United States: new scoring algorithms, modeled after credit scores, purport to predict your behavior in ways that go far beyond your financial creditworthiness.[26] For exam-ple, algorithms are increasingly being used to predict your reliability as an employee or even as a patient in a doctor's office. These algorithms assess credit history, marital status, and more. The Fair Isaac Company's new Med-ication Adherence Scores, for instance, can predict how likely a consumer will be to take a prescribed medication or self-medicate. Is this really the type of 24/7 surveillance future we want?

Cooperate or Compete?

Over the past few chapters we've seen the ways in which technology can drive economic change that affects far more than markets and personal finance. These changes shape society, culture, and politics in ways that are sure to transform our intimate lives, and in some cases disproportionately harm the vulnerable and poor. Of course the reverse is true as well: social investments can powerfully shape economic futures. As the job market and the labor economy are changed by digital apps and services, automation, and AI, the ways we relate to one another, communicate, and live our daily lives can be shaped in the spirit of cooperation and mutual support.

By focusing on positive economic and social "externalities," benefits of new technologies, we can create a future that is balanced, good for busi-ness, good for workers, and good for us all. We also need to be mindful of the extreme voices coming from the tech world. Silicon Valley's Peter Thiel, for instance, an early investor in Facebook and a co-founder of PayPal, is a highly vocal free-market capitalist. He no longer believes that democ-racy and freedom are compatible in our world. Thiel wants to use technol-ogy to colonize new markets—in cyberspace, in outer space, and on the open ocean. Through technology, he believes, we can escape politics and instead develop a market so robust and expansive that it will secure our "freedom."[27] In the meantime, however, he seems to be indifferent to the effect this has on the well-being of the public.[28]

Thiel's rhetoric does not veer far from what some tech giants seem to be thinking, perhaps even the ones we consider to be more liberal. It's a scary reality, and one that spells out pretty clearly why we don't want to leave all of the decision-making about our collective future up to tech billionaires.

Why can't we instead work with technology to stabilize the job market, and create growth and new enterprise that benefits all classes? The Policy-Link CEO and founder Angela Glover Blackwell points out that when societies choose to support vulnerable or disenfranchised groups, the economic and social benefits that result tend to *improve the lives of everyone.*[29] She illustrates this point with an example drawn from the disability movement of the late 1960s and the 1970s, during which activists lobbied for more accessible walkways and streets.[30] These accommodations ended up improving the flow of diverse pedestrian traffic and created as well more accessibility for parents with strollers, workers with heavy carts, and tourists or business travelers navigating luggage.

We see other cases like this everywhere. Seatbelt legislation adopted to protect children has created laws that have saved approximately 317,000 lives of all ages since 1975.[31] In other words: social changes that are designed to benefit a small, vulnerable section of the population most often end up making the world a better place for everyone, even Peter Thiel.

Blackwell points out that employing and paying workers of color at the same rates as white workers would *add $2.1 trillion to America's annual GDP.*[32] Her insight is a reminder that we're at a point where, because of the changing nature of the economy and the ways that technology seems to be transforming all aspects of life, we have the opportunity to transform our economy for the better. We have a chance to improve the lives of low-income people, racial and gender minorities, aging populations, the disabled, as well as everyone else. But we have a choice about how we frame and pursue these goals. Research shows again and again that self-gain and collective gain are not enemies. We need a digital future that recognizes the intertwined nature of all of our fates—political, economic, and social.

IV   An Internet for Us All: Overcoming Inequality

17   Keeping Network Power Local

with Aditi Mehta

On February 26, 2015, while the Obama administration was still in power, the US Federal Communications Commission (FCC) established landmark net neutrality protections to ensure that the pipe to the internet remained fully and equally open to all. This victory was short-lived, however. On December 14, 2017, the Trump-appointed chairman of the agency, Ajit Pai, announced the decision to reverse those rules. Thanks to this FCC maneuver, internet service providers (ISPs) can now charge different rates to users and content providers, and influence the speed and accessibility of content.[1]

The American Civil Liberties Union (ACLU) described this decision as "falling victim to the profit-seeking whims of powerful telecommunications giants." Without net neutrality rules in place, ISPs can "disfavor controversial viewpoints or smaller websites and favor the content providers who have the money to pay for better access."[2]

But in crises come new possibilities. What if instead of solely fighting a difficult battle around net neutrality, small businesses and user communities themselves could create *community networks*, taking power over the design, construction, rollout, and management of the ways they get internet access?

We are seeing this happen in a smattering of places across the world: rural communities are putting wireless internet antennas on grain silos and tall trees, and the fastest internet connections in the United States are provided by local governments, *not multibillion-dollar companies*.[3] Indigenous Native American communities have leveraged unused television broadcast spectrum to get access to the internet, thus engineering their own sovereign digital networks.[4] Networks have been set up in regions as remote as

West Papua, Indonesia. These community networks are legal in much of the United States and elsewhere in our world. As *Vice* reporter Jason Koebler explains, "There are myriad ways communities passed over by big telecom have built their own internet networks or have partnered with small ISPs ... to bring affordable high-speed internet to towns and cities across the country."[5]

What are the basic requirements when a community wants to create its own network? Networks usually rely on access to electricity and one point of hard-wired connection to a data or internet network, or to wireless access points or fiber optic lines. An access point is a device that creates a wireless local area network (WLAN) for routers to connect to and expand the network, whereas a fiber optic line is a method of transmitting information via pulses of light along an optical fiber.

Making a community network sustainable, however, is not simply a question of technology: it is also a process of developing skills, digital literacies, leadership, and technical knowledge. But internet access also matters. Without connectivity, the computer is little more than a typewriter: we can't use it help us pay bills, apply for jobs, or access the essential information and services that have shifted online.

I co-wrote this chapter with Aditi Mehta, a scholar of urban planning and a researcher committed to community organizing and social change. In these pages we share a few stories and some of our insights from the world of community networks.

Network Multitasking

A community network is a technical and social marvel, an infrastructure that may be built for many reasons: to create new jobs, develop digital literacy, amplify the voices of residents in a neighborhood, support local communication, or facilitate economic development or disaster preparedness.

We should not put community networks on a pedestal, however, or think that using the word *community* as a panacea resolves the many concerns I've raised regarding technology and inequality within this book. The range of community networks around the world can be distinguished in terms of their dependencies, economic models, and political realities. Some networks described as community-based are in fact administered and overseen by governments and designed to support the aims of the state, possibly at the cost of its citizens' welfare. Consider networks in places like North Korea, China, or even Cuba; each of these allows users to access the wider internet, but only on the state's terms. Such networks, ripe for censorship and surveillance, are at odds with a local community's power to decide what kind of connectivity will best serve its economic, cultural, and political relationships to technology.

Aside from allowing its users to share information, and to share the costs of access to the wider internet, community networks can help residents of a neighborhood or geographic area to preserve or archive information and to take advantage of educational, cultural, political, and economic opportunities that depend on internet access. Users can build businesses that exchange services via their network. Or they can use the network to preserve languages and traditions that may be lost, as in the case of indigenous communities. These networks can take the shape of a public service or utility, as a *commons* that is "open, free, and neutral."[6] With this logic, these networks hold the potential to be decentralized, affordable, locally owned internet infrastructures.

Serving and Supporting Users

Sascha Meinrath, a researcher and policy activist who has been exploring community networks for nearly twenty years, explained to us the dangers of a digital future focused on the "pursuit of profits without any regard for collateral damage."[7] Consider "disaster capitalism" as one such example. When people affected by hurricanes, earthquakes, or fires must helplessly turn to anyone who can assist them, relief aid might fill the pockets of private corporations without delivering services to those in need. Instead of being exploited, however, communities can respond by creating and maintaining their own resilient solutions, for example mesh networks (think of them as interconnected nodes that communicate directly, dynamically, and without hierarchy to transmit data to and fro). Mesh networks emerged in 2005, for example, to assist victims after Hurricane Katrina hit the southern gulf region of the United States.

For years, independent mesh networks in places like Berlin, Athens, and Barcelona have been used to expand service across neighborhoods that lack full internet access, or to provide service during natural disasters when ISPs are banned or not functioning. Often, mesh networks are able to survive disasters because even if one router is damaged and shuts down, the remaining routers will still connect to one another. If the "gateway" node cannot access the internet, there will still be an intranet, or locally connected network, among the routers, which means users can communicate, plan, and organize in rapid fashion. For example, the community network Red Hook WIFI in Brooklyn, New York, was one of the only functioning communication infrastructures in the neighborhood post–Hurricane Sandy in 2012, and has since expanded to help the neighborhood in case of future disasters. Mesh networks have also been deployed to build independent communication networks during protests such as Occupy Wall Street and Hong Kong's Occupy Central. For activists, the ability to evade police surveillance is part of a mesh network's appeal.

The decentralization of these networks makes it difficult to know how many exist or where they are located. But network creators, like many in the hacker, free software, and open source communities, share a common interest in getting back to a central value of the early internet: to serve and support the goals and desires of users, above all else.

The Detroit Community Technology Project

Detroit's Diana Nucera is a savvy technologist and community activist known locally as the "mother cyborg." Like others left behind in the city when the free trade agreements of the 1980s and 1990s allowed General Motors, Ford, and Chrysler to move their operations overseas, Nucera would seem to have little in common with the technologists of Silicon Valley. Abandoned in the inner city were poor, disproportionately black and brown people who lost their jobs and the safety net that work offered them. As a result, full city blocks were abandoned. Gang and drug activity took over. Neighborhoods and communities were compromised as buildings were boarded up. Even more jarring is the disturbing news from recent years regarding the lead- contaminated drinking water in Flint (Detroit's industrial neighbor). As an unnamed colleague of Nucera's told Kaleigh Rogers from *Vice*, "All the skills have been stripped from us: from food, housing, clothing, and shelter, damn [we] gotta build everything from the f—ing floor up."[8]

In 2014, Nucera and her team formed the Detroit Community Technology Project, in collaboration with the Equitable Internet Initiative and the Allied Media Projects, to develop technology to serve and support their community. Demystifying technology is only part of an overall goal to provide youth with basic literacy and education, and to support small businesses looking for partners and customers. The team has implemented mesh networks in seven poor and marginalized inner-city Detroit neighborhoods, training more than twenty-five neighborhood leaders. These networks provide speed and quality that often exceed what private ISPs offer. The stakes are higher, however, than achieving better access. As Nucera explains, the project addresses the risks faced head on if residents "don't take ownership and control over the internet in a way that is decentralized."[9]

These working-class people of Detroit, like many in other poor and rural communities across the world, are stuck. Either they have no service at all, or they are forced to agree to economic terms over which they have no power. In Detroit, the wealthy receive bandwidth at a speed that can be as much as a hundred times faster than for those less well off, whereas the poorest people are left out of the loop altogether, or at best get dismal service. The community network created by Nucera and her friends, however, operates with the opposite logic—the less you have the more you deserve to be respected and included.

Detroit's community networks, designed to support greater sovereignty and self-determination for their users, continue to attract attention. If we think about networks like this expanding across the world, we could see an internet of the future that's quite different than its current private- or corporate-dominated version. Nucera believes that such a vision is possible: a future internet that is truly rooted in civic ethics and values, fostering collaboration within and across cities, nonprofit organizations, philanthropists, community organizers, and businesses.

Catalonia's Guifi.net

Aditi and I also had the opportunity to learn firsthand about the challenges, possibilities, and visions associated with one of the most well-known mesh networks in the world, Catalonia's Guifi.net, which has been active since 2003.[10] Guifi seemed like a strange name at first, but we later learned that the name is the Catalan pronunciation of "Wifi!"

Ramon Roca and Roger Baig Viñas, two lead administrators of Guifi.net, described to us how community networks in the early days were more often seen as technical challenges than social and collective spaces of opportunity. The technical goal of building decentralized mesh technology is not the same as decentralizing power to control the network, as Roca and Viñas discovered. The best way to support their fellow Catalans was to focus less on the technical specifics, and instead understand what kinds of access support community members.

The Guifi.net network has now grown, and it works with a group of leaders and participants in the areas where the network operates. As of 2018 it serves over 100,000 users. Roca and Viñas describe their efforts as *generative*: because the network is a *commons*, rather than privately owned or operated, people and organizations can innovate as they see fit. As a result, numerous for-profit and nonprofit businesses have emerged, employing people across the region and spreading literacy and knowledge of the technology. According to Roca and Viñas, this also allows them to build and sustain a network in which the values of "agency and self determination are crucial. Because you are the user, you should be the driver...deciding where the technology goes."[11]

The Catalan network creators see a community-owned network as a human right: access to communication and information is a basic need for anyone in our world. But having access means more than "getting online." It is intrinsically tied to costs and benefits: access that involves the surveillance and capture of personal data is unacceptable to the team, just as access at artificially high price points is a nonstarter.

LIke a commons, Guifi.net is available to everyone like a park or square. This approach to technology is in sync with the findings of the late Elinor Ostrom, the Nobel laureate who identified the economic potential of "common pool resources," or the ways by which public and community-based investments could create positive returns aside from merely making money.[12]

Resilience and Community: Networks after the Storm

During summer 2016, Aditi worked as an instructor in the Digital Stewards program in Red Hook, Brooklyn. The program offers paid on-the-job training for careers in tech or the media to fifty young adults each year, aged nineteen to twenty-four. Through her teaching, she met several young adults who were learning job-readiness skills in the fields of graphic design and video production. The students heard about the opportunity through friends who were former digital stewards, some who even helped build Red Hook WIFI in 2012.

Young residents, nonprofit workers, city employees, and federal officials contributed to creating the Red Hook WIFI infrastructure. Like the other examples we've shared so far, the network did not come out of thin air. It built upon existing connections and histories, as technologies of all forms often do. To make this project successful, diverse parts of the community with varying needs had to build alliances and negotiate their visions for the community network.

One of those negotiators is Tony Schloss, the former director of technology at the Red Hook Initiative (RHI). He had the idea to build the network in 2011 to host an internet-based youth radio station. With the help of Alyx Baldwin, a Parsons School of Design graduate student, Schloss installed two point-to-point routers: one at the RHI building and one at Coffey Park.

After Superstorm Sandy hit in 2012, killing and injuring hundreds of people, and damaging their homes across the region, RHI staff realized that the experimental Red Hook mesh network with its two routers was still working after the disaster. It was indeed the only functioning communication infrastructure in the neighborhood. Because the RHI building did not lose power, internet access remained available inside and directly outside the building. The router in nearby Coffey Park provided the same access to users within a half-mile radius. One young adult on the job at the time looked back with enthusiasm at the potential for network flexibility:

> The reason why anybody was excited about it was that if one node did go down, there was still access, because they were able to route around that node that went down. That's what made it super cool. No matter which one went down, it would still find a way to fix itself and still give access, with or without internet.[13]

With technical help from Brooklyn Fiber, a local small-business ISP, and eventually from the Federal Emergency Management Authority (FEMA), the mesh network was able to distribute the internet connection to locations where residents, first responders, and recovery volunteers needed it most: it functioned as a hyperlocal information sharing system. Logging on to the Red Hook WIFI network was (and remains) free for any resident or business employee in the neighborhood. Digital updates about post-storm conditions, community events, and other news from Red Hook appeared on the network splash page.

The Red Hook Initiative's first cohort of stewards, known as Generation 1, included six neighborhood residents. Every day brought a different adventure or challenge for these young adults, such as scoping out good places on neighborhood roofs to install a node. As one of the first-generation stewards told Aditi: "We were always trying to figure out what we had to do every day. In the beginning, when the team started, the program was just getting off its feet. There was no clear structure on how we were to go about learning things." Yet another steward recalled, "The way I saw it in the beginning was that we're building a platform so residents in the community can communicate with each other."

The group strove to build upon the sense of community that had emerged post-Sandy as young adults asked themselves: "How can we make it so it's not only used for when a disaster happens, but it can be used on a daily basis to bring the community together? How can we 'mesh' populations together, and get all the businesses in this area to hire people from the actual community, and bring them closer so [we don't] feel so segregated from one another?"[14]

In the years since 2011, the Red Hook Digital Stewards program has transformed its operations from radio production, to disaster preparedness, to neighborhood WIFI, to job training for the tech sector. It has supported the positive development and growth for Red Hook young adults, who face daily challenges associated with poverty, racism, violence, police brutality, geographic isolation, and access to opportunity. Understanding and evaluating the Digital Stewards and Red Hook WIFI programs is complex for this reason. For funders, one of the major indicators of the program's success is when a participant's training leads to a job.[15] For neighborhood small businesses and stakeholders, however, the major indicator of success comes from having a strong and reliable Wifi network.

Of the young adults joining the program, many signed up as a way to earn an income while receiving eight months of training. Several digital stewards admitted that they saw the program as a ticket out of the neighborhood. But others acknowledged their attachment to Red Hook and spoke of how they wanted to participate in community improvement after completing the program. One of the original architects of Red Hook WIFI, who is now an intern at Google, said: "I always tell people that I'm definitely going to invest in someone because someone invested in me, and I definitely do live by that. I will definitely come back to Red Hook and the digital stewards and help in any way that I can."[16]

During focus groups, a number of digital stewards commented how relationships formed through the program helped build the confidence that they too could shape their lives using technology. The social networks they created provided emotional support and connections to the wider world. In this way, the Red Hook WIFI network and the Digital Stewards social network make each other stronger.[17] The program became an ongoing resource for training, for paying a local wage to local people to learn technology and to access work and resources.

But, as Aditi and I have pointed out, some digital stewards have entered the program as a way to exit the neighborhood. That's a key issue provoking mixed emotions for many people inside and outside of the community. Individuals oftentimes look at projects like these as ways to improve their lives, even if that means leaving their community. As Red Hook WIFI expands its reach within the community and better serves its residents, it runs the risk of moving further from its original mission of putting the community in power over the technology. Scale and efficiency require more involvement (and thus compliance with requirements) from local government, service providers, and other partners outside the neighborhood. If the network remains truly community-owned and managed, then the improvement of the infrastructure may come at a slower pace. Tradeoffs like these are important to note as community networks expand in reach and are forced to grapple with funding, political, and other sustainability challenges.[18]

The stakes are high, not just because of decisions pertaining to net neutrality, but because again and again, large corporate ISPs have made self-serving choices not only regarding who to provide access to, but also regarding the quality of that access. Even Google, despite great acclaim, has struggled with its ISP program, Google Fiber. In the city of Atlanta, for instance, Google has not delivered access as promised to many users and has even damaged underground infrastructure in its process of building the city network.[19] Cities including Nashville, Tennessee, have struggled with Google Fiber delays.

Community networks continue to emerge as viable alternatives to private ISPs. *Wired* has reported that the sixteen cities of the Southern Bay region of Los Angeles have agreed to cooperate to provide cheaper and faster access to their residents. These cities, spanning a range of socioeconomic and demographic backgrounds, recognize that internet access does not have to come solely from self-serving private corporations but instead could emerge from the spirit of public service and collaboration.[20]

Creating and sustaining these networks will be an uphill battle if we keep thinking about ourselves merely as "consumers" of the internet. As Sascha Meinrath reminded us, too often our educational and civic institutions "don't teach people how to be critical consumers of technology and therefore how to use technology as a tool. And, you know, we wouldn't give kids chainsaws and say go have fun, which is kind of what we do with technology."[21]

With community networks, we can see how the power can return to the people. Leandro Navarro, one of the central figures behind the Catalan Guifi.net network, describes this hope, writing: "We know how to do [networks]. … Nature shows that cooperation is effective even in extreme conditions where competition may not work. Nature also shows the importance of diversity and local evolution to create organisms that adapt to local conditions. We need a neutral environment for the development of community network infrastructures to nurture and sustain digital life everywhere and for everyone."[22]

18   Questioning Connectivity

In my first book, *Whose Global Village?*, I questioned a pervasive myth in popular culture: that technology will magically bring together billions across the world in global, democratic-style conversation. When I spoke about the book at a July 2017 UNESCO event in Guanajuato, Mexico, I had a chance to challenge that premise. I discussed the problems that arise when new technologies are designed almost exclusively in one part of the world and pushed out to the "margins." In contrast I offered many examples that I had personally observed, which showed how global communities and cultures have created and taken power over new technologies.

After the conference, I boarded a domestic flight to Mexico City. On the plane I noticed a tall, slim man who seemed vaguely familiar. My seatmate leaned over and whispered to me in Spanish, "Sentado allí es el presidente Fox." Sitting across the aisle, in the row ahead of me, was Vicente Fox, who led Mexico from 2000 to 2006.

I honestly didn't know much about President Fox other than the humorous videos he had posted criticizing US president Trump's promise to build a wall for thousands of miles along the United States–Mexico border. A few passengers approached Fox, took photos, or shook hands, but I remained quiet. What do you say to a president? But as the plane taxied toward the terminal, I finally mustered up the courage to compliment him on the videos.

To my surprise, President Fox asked me about myself: where I worked and what I did. I told him about my research, the UNESCO event, and my collaboration in the southern Mexico region of Oaxaca, site of the largest community-owned mobile phone network in the world. Fox seemed to be genuinely interested in this research and even asked for my card. He expressed concern that in Mexico, just as we have seen across the world with the "fake news" phenomenon, misinformation has run rampant online, discrediting honest politicians and elevating dishonest ones. With a handshake, we went our separate ways.

A few hours later I was drinking fresh mango juice at my favorite bakery when an email from President Fox came in on my phone. In a few kind lines he praised my work and invited me to speak at a forthcoming summit he would host on the future of science and technology in Latin America.

Is Tech "Acceleration" a Magic Guarantee?

Two months later I arrived at the president's hacienda in San Cristobal, ready to speak at the conference.

Thomas Friedman, a Pulitzer Prize winner and *New York Times* columnist, delivered the opening address. He presented insights from his best-selling book *Thank You for Being Late*.[1] Discussing the invention of the iPhone, Friedman argued that achievements in robotics, artificial intelligence, genetics, and telecommunications would empower us all. He pointed out that ever-faster and more efficient technology could be a vehicle for achieving stability in the face of change, thanks to its ability to connect us across the world.

Using the words "amplifier" and "acceleration" repeatedly to characterize the digital transformations of the past few decades, Friedman alluded to the way we have revolutionized our ability to communicate with one another, find information, preserve knowledge, outsource and distribute work, even monitor one another. He seemed convinced that revolutionary advances in technology would naturally bring about similarly positive transformations in the human condition.

What puzzled me was the assumption that technological change magically guarantees an improvement in the lives of humans, regardless of who or where we are. This attitude seems too close for comfort to the scenario in which one group that stands to profit from an innovation brands it as inherently good for everyone. Where's the space, I wondered, for us to develop technology based on conversations and collaborations with us all?

When my turn came to speak, I spent more than forty-five minutes discussing how we might truly listen to and collaborate with diverse global users around the world (see figure 18.1). Many in the tech world praise

Figure 18.1
The author, lecturing at a summit hosted by Mexico's former president Vicente Fox.

countries like India, which have experienced significant increases in their GDP due to technological development. But I pointed out that this growth had done little to address the profound problem of income and wealth inequality across the nation, and on top of it had made money in a way that left the country dependent on mostly Western clients.

Do you think, I asked the largely non-American and non-Chinese audience, that entrepreneurs, activists, governments, and communities across the world are capable of directing the design, production, ownership, and regulation of the technologies they use every day? I noticed heads nodding in the audience as I described how this is indeed already happening. I share some of these stories in the coming pages of this book.

The world has dramatically changed when it comes to the spread of technology. Mobile devices accounted for 73 percent of total internet consumption in 2018.[2] In 2016, 63 percent of the global population owned mobile phones, and it is estimated that by 2019 the number of worldwide mobile phone users will increase from 4.57 billion to 4.68 billion people.[3]

These are astonishing numbers given that our world's population is just over 7.5 billion.

But rarely do we consider how people across the world have made technology work for them, either by appropriating (or using existing technologies in new ways) or by designing tools and systems that serve their country needs, values, and goals. In southern India, for example, villagers work with businesses and nonprofits using mobile applications for selling farm goods and fish.[4] Activists in the Egyptian Arab Spring famously used digital social networks to share stories on a large scale, and even brought citizens together in local, physical spaces by projecting videos they pulled from YouTube in city squares.[5] Indigenous peoples in Mexico have designed their own digital networks to communicate in places where traditional network operators will not provide coverage.[6] In these examples it's not the technology itself that magically supports its users—it's *how* it was crafted, designed, and managed that promotes the cause of a community or culture.

Each internet or mobile user has a different online experience that one-size-fits-all terms like "access" or "digital divide" do not explain. The devices we use, our data or internet service plan, the speed of connection, the censorship and the control of what is accessible, our digital literacy—these all dramatically differ based on who or where we are.

Because of how access has spread, the users who are just now interacting with the internet in Africa, Latin America, or the Middle East are likely to have fewer economic resources and are more likely to live in rural locations. This raises important questions: What kind of internet will these users engage with? Will it support the preservation of their language or accelerate its permanent loss? Will it uphold traditions or commodify and devalue them? Will it include these users in a way that reflects their diverse political, economic, and cultural values? Or will it *claim* to be inclusive while in fact grabbing their attention and extracting their data?

Is More Connectivity Always Better?

When we hear about the ways in which mobile phones and the internet are changing the world, particularly in places like South Asia, Africa, and Latin America, the term *connectivity* is always at the center of the discussion. But what does this word really mean?

Hitched to the idea of connectivity is the common perception that expanding access to a system can only improve it: as the mobile phone statistics from the last section reveal, the more people who use any given communication technology, the higher the value of that technology.

But from a human, user-based perspective, is connectivity really that unequivocally positive? We should ask who we are connected to, through what service provider, and how much power we have as users to determine these connections. A telephone call to someone I love is a connection that I can control in a relationship I value; surveillance is a connection I cannot control, and it forges a relationship I may not value or even want. This distinction shows how different human-driven connection is to blind digital connectivity.

Here's another important question: What do we give up in exchange for being provided with connectivity? Is it our privacy, our money, something else, or nothing at all? What about connectivity in exchange for advertising and demographic data? Many tech startups iterate this model. The Mexico-based Pig.gi, for instance, offers virtual coins as a reward and now has nearly 2 million users across Latin America.

The actual experience of connectivity is likely more different than similar for people across the world. The technical infrastructures and machines we use, and the social factors at play, all create different kinds of connectivity depending on who or where we are. A user's level of digital literacy, as well as behind-the-scenes censorship and monitoring, can dramatically affect the actual experience of connectivity. Think back to the examples in chapter 8, which revealed how algorithmic-tracked political biases influenced a Google search for "Egypt" during the Arab Spring.

An example of this came in 2017 as I was preparing for a UN-sponsored visit to Cameroon and wanted to learn about the West African nation from the perspective of its people.[7] The results of my Google search, however, reflected a mistaken algorithmic prediction: that the content most popular with users from my own nation (the United States) was what I'd most want to see and most "relevant" for me. That explains why the first three pages of results linked me to webpages like the CIA World Factbook, providing stats such as average rainfall and the length of the country's coastline, but not to a single page that actually came from the nation. In effect, my *connectivity* blocked the possibility of a deeper *connection* to the actual people of Cameroon.

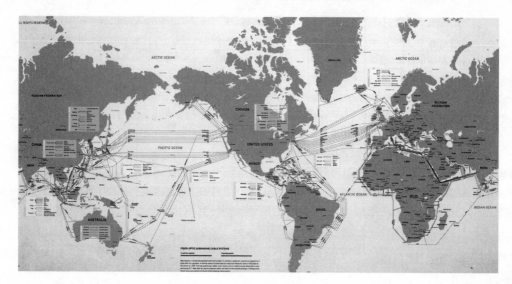

Figure 18.2
Fiber-optic cable map (source: ThriveGlobal).

Strictly speaking, there's nothing wrong with the search results Google served me. But they reflect specific ideas about value, credibility, and utility as defined by their engineers, managers, and investors, ideas that everyone may not share. And they do so without transparency or explanation. This black box facilitates a "smooth" experience for users while discouraging critical engagement or deeper curiosity.

Let's probe connectivity further by looking at one of the physical infrastructures that makes the internet "go": fiber optic cables underneath the ocean's surface (figure 18.2). A paltry few cables connect the most populous continents of the global South (South America and Africa) with one another. But large numbers of cables connect once-colonized places with the nation that formerly colonized them. Still-larger numbers of cables connect locales of great political or economic power, such as Los Angeles and Shanghai, or New York and London. This reveals how the internet's fiber optic cable architecture is not neutral but actually reflects political and economic relationships.

For example, consider a 2018 deal struck by two small island South Pacific nations in the developing world (Papua New Guinea and the Solomon Islands) and their powerful neighbor (Australia). Concern about the

Chinese telecommunications company Huawei's entry into the region had originally sparked Australia's interest in this strategic 2018 agreement, which was supposed to improve internet access and affordability by connecting the three countries via fiber optic cable.[8] This example reveals how a political agreement (in this case, a move intended to fend off Chinese intrusion) can drive a technical decision about the design and engineering of a set of cables.

What If Government and Corporate Technology Intersect?

Connectivity in China, of course, has different connotations and implications than it does in Australia, North America, or the small South Pacific islands. Understanding this is important because China is a huge force in shaping global technology. I've shared several provocative and fascinating examples from China so far. But I haven't written extensively about its most powerful tech companies, such as TenCent, Baidu, and Alibaba because it was impossible to get employees of these companies and their users within the country to speak to me on the record. Off the record, those with whom I did speak repeatedly identified two important themes driving the strategies of Chinese technology companies: *growth* and the *kinship between the corporate world and the government.*

China presents an example of how a nation's culture, politics, and values can shape the ways it designs technologies that spread both within and outside its borders. China's technology industry is aligned with the intention of its government to exert its influence—not only domestically, which it does with expansive enthusiasm, but also across the world. Led by President Xi Jinping, the world's most populous country announced in 2017 that it would pursue the trillion-dollar Belt Road Initiative (also known as One Belt, One Road) for tech development (figure 18.3). The project, whose name alludes to the famous Silk Road trade route, seeks to reestablish China as a global power with global influence. To achieve this goal, the Belt Road will span sixty-eight countries and focus on a "vast loop of new, Chinese-funded, transportation and energy projects...through Central Asia to Eastern Europe, and then back around the Horn of Africa and on to South Pacific."[9] It might include solar-powered, perhaps even automated freight trains that will run for thousands of miles, connecting Eurasia and linking China to Europe via the Middle East.[10]

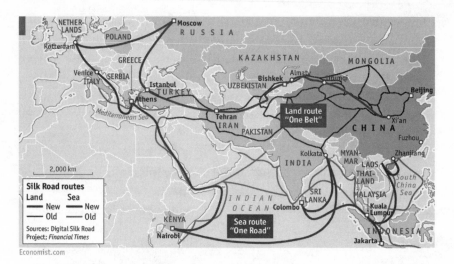

Figure 18.3
China's expansive Belt Road Initiative.

The Chinese tech insiders with whom I spoke (again, off the record) were almost uncannily in sync. Each one expressed to me the importance of one particular strategy for growth: *success in numbers*. They explained that government officials, executives, and even they, as employees or tech users, do not see themselves being in competition with each other. Instead they present themselves as a united front seeking one outcome for their country and its corporations: demonstrable evidence of growth and influence. This outcome could be expressed in terms of revenue, profit, number of users, or other measures of influence, but it is also revealing of a collective culture toward work. When I spoke in 2018 to a female Chinese executive at a tech company valued at more than $400 billion, she summed it up succinctly, saying: *We live to work and work to live.*

Where China's international ambitions will go remains to be seen, as does the pushback they may receive from other parts of the world. But Chinese technology uniquely represents a clear alignment between the state and its businesses, a position quite unlike the one we see in the United States. Consider actions the US Department of Justice has taken in 2018 alone: it accused Facebook of violating the Fair Housing Act, took the company to court in an effort to force it to wiretap gang members under investigation by the government, and charged over a dozen Russian nationals and Russian groups with using the Facebook platform to commit a criminal conspiracy to defraud the United States.[11] All of these situations are nearly unthinkable in China, I was told. One cannot have a sustainable enterprise of any significant size in China without carefully coordinating with the state. And in the private sector, it is unacceptable to take a political or economic position at tension with the government.

A vivid example of this is in the Chinese social credit system, a dystopian condition of present-day life that I discuss in chapters 10 and 16. Enforced by an algorithm-driven reward/punishment system, there is little transparency regarding the criteria used to "score" citizens' behavior, or little accountability for the invasive structure of surveillance that builds on an already problematic system of financial credit. The Chinese social credit system is *not* simply the integration of technology into government functions, but rather a total license to use data to control the lives of its citizens.

This system does not capture all that Chinese technology is, but it provides a potent example of how powerful and involved the state can be in shaping technology, especially in a cultural context where unity, coordination, and control are fairly aligned with traditional values. The country has no shortage of successful tech platforms: for example TenCent's WeChat, a service with over 1 billion users, has managed to integrate messaging, social media, and mobile payments, all into a single platform. Very few of us know this, but TenCent was actually worth more than Facebook by $72 billion as of March 2018.[12] As a major player in the global technology industry, this powerful Asian nation demonstrates a whole host of possibilities both troubling and promising about how we might think of technologies in the future and around the world.

New Frontiers

The numbers are astounding: From 2012 to 2017, 1 million people joined social media every single day.[13] And during that same five-year period, we have seen an incredible rise in access to mobile technologies and to internet and social media connectivity across the world—an increase of 82 percent, or almost 1.7 billion people.[14]

In the coming chapters, starting with Africa and continuing into Latin America, I discuss breakthroughs in places that we often overlook when we talk about innovations and developments in tech. Yet these are the places where the internet is most quickly spreading. The examples from the coming chapters show the power for us all, regardless of who or where we are, to be technology innovators: to design, engineer, and shape technologies in our image as users and communities.

19 African-Born Technology

How many of us would have guessed that Africa beat the Western world to the punch when it comes to the innovative technology that allows us to pay our bills via mobile phone? In this chapter we'll look not only at how tech corporations aim to market their goods and services in Africa, but also how grassroots ingenuity on the continent has given birth to many of the technologies of tomorrow.

Who Is Sophia, and What Brought Her to Ethiopia?

Sophia, a Saudi Arabian citizen, made her way to Ethiopia in July 2018 for the nation's Information and Communication Technology expo. She'd already achieved "fifteen minutes of fame" some two years earlier: during a *60 Minutes* appearance with Charlie Rose, she flummoxed the seasoned interviewer by coming on to him with "a good pickup line." Sophia may not always be a flirt, but she is indeed attractive, sharing some of Audrey Hepburn's delicate features (figure 19.1).[1] Did I mention that Sophia is a robot?

The Hong Kong–built Sophia can have basic conversations, answer questions, and make facial expressions. In preparation for her meeting with Ethiopian prime minister Abiy Ahmed, Sophia learned the Amharic language spoken by most in the country. But in a surreal twist, the meeting had to be rescheduled when some of her parts were lost at the Frankfurt airport a week prior to the event.[2]

Sophia's visit to Ethiopia symbolizes Africa's role as a playground for new digital technologies. But the path that these technologies will take on the continent has yet to be determined. The relatively late arrival of

Figure 19.1
Sophia and the president of Ethiopia (source: @realsophiarobot).

the digital economy in Africa is partly responsible for this uncertainty, but it may be an advantage of sorts: as Africans experiment with technology, they stand to benefit greatly from looking at examples across the world—in particular the dramatic successes and shortcomings of Silicon Valley. The continent's young and growing population, its incredible natural resources, creative ingenuity, and independence from traditional wired infrastructures all point to a digital future defined by and for Africans.

There are reasons to believe the hype, ones that might surprise you. For example, the infrastructures for mobile technology connectivity are impressive across Africa's most wealthy nations. Kenya, for example, has the world's fourteenth-fastest mobile internet speed. That's twice as fast as the global average. And it is well ahead of the United States (which is twenty-eighth) and South Korea (which boasts the fastest fixed internet connection speeds in the world).[3]

Western and Chinese giants increasingly see opportunities in Africa to sell their products, access new markets, and perhaps even rethink how to design their technologies. Apple, Amazon, Google, Facebook, Microsoft, and many others are moving in all over the continent, setting up shop in urban hubs such as Johannesburg and Cape Town (in the south), Nairobi and Kigali (in the eastern and central region), and Accra, Lagos, and Abuja (in the west). The Chinese companies on the continent are involved not just in technology initiatives but nearly every other infrastructure imaginable, shaping economic relationships and expanding the sphere of Chinese political influence. Alibaba's founder, Jack Ma, just launched the $10 million Netpreneur fund to support at least a hundred young African entrepreneurs in the coming decade.[4] In August 2018, he told a crowd that every African country should encourage young people to take advantage of good tax conditions for startups. But, as he also went on to say, "I believe today we should not talk about the robots and artificial intelligence in Africa. We should talk about innovative ways to solve job creation for Africa."[5]

One thing is clear, whether we speak of foreign influences or businesses that have grown on the continent: *Africa is a hot new tech frontier*, a place upon which visions of a digital future are projected.

As I conducted my field research for this book in the East African nations of Uganda, Kenya, and Tanzania, I commonly encountered signs of Chinese and US influence in the marketplace as well as business people from both nations. I was especially troubled to see that high school students are being offered classes in US and European history more often than in African history, and how indigenous cosmologies and languages are rarely taught in schools. Even national parks have been named after British scientists and explorers. It made me wonder: Could new technologies coming out of Africa recognize and build upon the cultural histories, diverse traditions, and languages that these countries have to offer, rather than barely acknowledge their existence? This and the coming chapters (20 and 21) show that, yes, such innovation is alive and well on the continent.

Uber in Africa

Several tech business successes have already emerged from the ground up in Africa. In 2012 the e-commerce group based in Nigeria, called Africa Internet Group (AIG), became Africa's first tech *unicorn* (a privately held startup valued at over 1 billion dollars).[6] But AIG, which owns a variety of other startups, relied on foreign support to earn the unicorn designation. The combined investments into AIG from Goldman Sachs, Rocket

Internet (a group based in Germany), and AXA (out of France) are a symbol of international recognition of Africa's growing power in the tech world.[7] It is also emblematic of a new economic trend in which the African markets, bolstered by tech, are less defined by the price of oil and other commodities than ever before. A TechCrunch contributor summarized the changing dynamic in 2016, stating, "Even as Nigeria, the continent's largest economy and oil producer, takes hits from lower commodities prices and China's downturn, it is still producing billion dollar tech companies."[8]

In another example, consider the rise of Uber across the continent. I discussed in chapter 11 how the Wall Street–backed rideshare company had "commandeered" three-wheeler rickshaws in Dar es Salaam. A trip to Nairobi, Kenya's capital, will show how Uber has tried to take over travel in East Africa. Dozens of street advertisements plastered seductively along central city blocks promise efficiency, convenience, safety and affordability, suggesting that you too, like the attractive young Kenyan in figure 19.2, can be part of the hip, modern, digital economy.

Some of these promises are based in truth. Digital technologies like Uber may mitigate the corruption that often takes place in informal economies. Formalized methods of negotiating economic relationships involve systems of security, oversight, and price transparency, and thus come with a badge of legitimacy. But in so doing they may conflict with, even violate, the economic and social systems that people have long followed. A digital technology, then, may promise "disruption" in an anti-corruption sense, while also "disrupting" the valuable cultural and social connections that local communities have forged through generations of gathered wisdom, negotiation, and human ingenuity.

While visiting several East African nations I had the chance to speak to a few Uber drivers. Not a single one was a fan of the company, and many had figured out creative workarounds. Some used the service to meet customers and then exchanged information with them outside the platform to maintain and support relationships that they, rather than the app, would manage.

Most of the drivers I met (as well as local people generally) pointed out the unfair, unjust, and even violent resource extraction of oil, minerals, and (more recently) data from the continent. The Western world has long had an interest in exploiting Africa's resources—everything from slavery and indentured labor to minerals and oil—by keeping the profits for itself and

Figure 19.2
An Uber ad in Nairobi capitalizes on the rideshare's cosmopolitan appeal.

leaving African nations further disenfranchised. These drivers told me that all the "good" promised by Western technological efficiency might be just the latest Trojan horse.[9]

Exploitation, however, can also breed resistance. In Kenya there have been numerous strikes by Uber drivers who now refuse to use the platform.[10] Others rejecting Uber say they will try to offer services in tandem with the ubiquitous, cash-based *matatu* minivans that cross the city and country (see chapter 21). Such public transit systems reflect a widely shared sense that transportation, and life more generally, is a social and collective experience, unlike Uber's individual-based ride experience.

Support also exists for local, app-based rideshare alternatives. In Uganda, apps such as Taxify, Mondo, and Spesho might individually lack the investments backing Uber (and therefore the ability to artificially lower prices through investor subsidies). But together they may be able to compete by offering value that Uber's greater financial resources cannot match. One initiative, called SafeBoda, uses GPS technology to link clients to a *boda boda*, or motorcycle. With the exception of *matatus*, travel by *boda boda* is the cheapest and most common form of transport in the country. Traditionally, the passenger rides without a helmet on the seat behind a motorcycle driver who weaves through intense traffic, at times going the wrong way down one-way streets. The SafeBoda, in contrast, has improved upon these problems. It makes a dangerous form of transportation far safer because the company has rewarded drivers with higher and more secure wages for driving safely. Local innovation can thus provide value to passengers, drivers, and a local company. But Uber seems to have caught on. The rideshare is attempting to gain back lost business by offering a UberBODA option within its app.[11]

African Ingenuity

At age ten, Kevin Doe, from the West African country of Sierra Leone, began to build electronic transmission devices out of discarded scrap and to engineer his own batteries. At sixteen, Doe was chosen as the youngest-ever visiting practitioner at MIT. He then returned to Sierra Leone to help develop free electricity for individuals in rural areas. His innovation allowed electricity to be powered through a person's shoe, simply by walking. Doe is one of Africa's brightest young tech stars—a powerful reminder of the talent we might be disregarding when we see him as just a "user," and a teenage one no less. Because Africa has the fastest growing population of any continent in the world, it is also home to the youngest population.[12] Currently, 60 percent of Africans are under age twenty-five.[13]

A number of efforts are underway to harness the potential of young, creative Africans to be leaders in areas of technology. In July 2018, the African Institute for Mathematical Sciences announced the launch of its masters' program in Machine Intelligence, the first of its kind to open in Rwanda.[14] Facebook and Google backed the program, which was designed to eventually expand into other African countries. The institute realized the need for such a program because the talent pool in the AI developer community did not represent the world's diversity. Not only can this lack of representation perpetuate biases on a global level, but it can also result in missed opportunities. This is an example of innovation that can emerge out of Africa—or out of any region, for that matter—that simply cannot be fostered from afar.

Looking to learn more, I spoke with the Africa-based tech journalist Toby Shapshak about digital technologies and the continent. In our conversation we were mindful that speaking in terms of "Africa" tends to be homogenizing. The continent is a deeply diverse landmass upon which some 1,500 to 2,000 languages are spoken by peoples of profound cultural and economic diversity.[15]

In a TED talk Shapshak gave in 2013, he spoke about why Africa has become a center for such important technological developments.[16] He points out the pay-as-you-go SIM card was an idea pioneered by Vodacom in Africa in the early 2000s, as were services like iCOW that help farmers sustain their dairy enterprises and preserve food security in Africa. In Shapshak's view, Africans were able to produce these kinds of innovative technologies *because they had to.*

Technology is a problem-solver, he argues. More solutions will arise where there are more problems. And the solutions Africans develop might be ones that we can apply elsewhere. The insights I share in the coming pages make a different point, however: what we see as a problem might just be a constraint that can be readily overcome.

Why? The answer goes back to Africa as the home to vast diversities in thought, knowledge, and history. Such a richness of cultures presents new technological destinies and implementations. In a global tech market currently saturated with Western and Chinese influence, African ingenuity could disrupt the disruptors.

For instance, the lack of electrical grids throughout Africa has stirred a number of off-the-grid power startups such as Greenlight Planet, d.light, Off-Grid Electric (OGE), M-KOPA Solar, Fenix International, and BBOXX.[17] Drones, generally seen as a sophisticated technology "flown in" from the wealthier parts of the world, deliver medical supplies to remote regions in nations like Rwanda.[18] Even those who've had less formal engineering education (meaning those outside of the professional class on the continent) are technology innovators. Across the continent's cities a great deal of street-based hardware repair takes place, giving new life to technologies that in the Western world would be treated as obsolete (such as old iPhones).

The ingenuity that thrives on the streets of African cities, and indeed in much of the global South—from extracting metals out of supposedly defunct machines to creating new synergies between machines—raises a fascinating question: What truly is innovation? Is it designing devices to die within a few short years? Or is it being resourceful and recycling them so they can continue to live?

The "short-lived" approach forces dependency by consumers and has in part contributed to Apple becoming the world's first trillion-dollar company. But what has this approach toward perpetual buying and selling produced, aside from wealth for the Silicon Valley giant? Apple's planned obsolescence is not the same as the Monsanto Corporation plan to patent seeds designed to die after a certain period of time. But is it so different given that left-to-die devices could become toxic waste? The "recycle" approach, embraced across Africa, focuses on *doing more with less*. It recognizes the economic constraints most people face as a launching point from which to develop resourceful solutions to extend the life of technology. We can look at this comparison and ask the question: Which way do we want to live?

Perhaps the most remarkable example of how culture and context have informed technology across Africa comes from the rapid increase in mobile phone use, which provided individuals and families new opportunities for sharing and exchange. Based on the observation that mobile phone users in Africa often worked around the constraints and shortcomings challenging their daily lives *because they had no other option*, the telecommunication provider Safaricom was among the first in Uganda to offer several options to help break down barriers preventing growth in local economies. Sambaza, for instance, a service by which Safaricom users shared their "talk time," grew into M-PESA, the continent's first offering in mobile money transfer (see figure 19.3). As of 2016 more than half of the 282 mobile money services worldwide are located in Sub-Saharan Africa.[19]

M-PESA, which developed because a great portion of the African population had no easy, reliable method to exchange money with one another, allows its users to pay for goods and services via credits linked to their mobile phone accounts. It operates from almost anywhere through a network of agents such as sugarcane merchants, small food stands, and other local businesses. The concept has proven to be a great deal for users and the mobile phone providers alike: with at least 100 million active mobile

Figure 19.3
Ubiquitous M-PESA signs in Kenya.

money accounts across numerous African countries, the telecom companies are making more while charging users far lower transaction fees than the banks.[20] Unsurprisingly, there's been a pushback from banks and governments. Together these powerful and traditional institutions recognize the threat this innovation poses to their once-top position in the financial hierarchy.[21]

Today, mobile money initiatives have spread across the continent.[22] Neighboring Somalia, a far poorer country, has a nearly cashless society, relying on mobile money services that offer free subscriptions and low transaction fees.[23] Also an analogue now exists in the banking world.

Kenya's Equity Bank has been lauded for not rejecting people who want to create an account, no matter how little money they may have.

Achieving the win-win situation that comes from including most of the population just involves reimagining what banking, and technology itself, should look like given the peoples and places with which the financial system intersects. And it's not just about money—SIM cards and Bluetooth features of cell phones are being used by Africans in places like Mali and Algeria to swap music. This is creativity in action—using the phone in an unintended way to share local music, even favorite Bollywood tunes.[24]

Following the mobile money lead, we can see how so-called disadvantages, such as lack of infrastructure, lack of power grids, and lack of organized local governments, can really be better understood as opportunities. But when large technology companies notice such digital engagement, they can also see the continent's potential as a fertile "emerging market": a place where significant numbers of "untapped" users can be brought into the fold to expand the consumer base (and more importantly, increase company profit).[25]

The "emerging markets" strategy is consistent with a long tradition of "digital divide" work done worldwide by non-governmental organizations (NGOs). Under the banner of decreasing the "divide" between countries with digital access and those without, NGOs have viewed the spread of technologies from the wealthier parts of the world to the global South largely as an unremitted good. And they've often ignored the question of who profits or benefits from this spread.[26] Such was the logic of *The Fortune at the Bottom of the Pyramid*, a corporate tech bible that Bill Gates praised as offering "an intriguing blueprint for how to fight poverty with profitability."[27] The bottom of the pyramid is far wider than the top, and even if those at the bottom have fewer resources per capita, together they represent a potentially greater revenues and profits for foreign and corporate interests.

Google's AI Lab in Accra

Google, too, is active across the African continent.[28] In 2017, the company announced a five-year, $20 million initiative to award grants to nonprofits, such as Gidi Mobile and Siyavula, two African startups that provide free digital learning for hundreds of thousands of low-income students. In 2018, Google awarded an additional $5 million in grants for its Impact Challenge, a program that asks innovators how their designs could make a difference for their community.[29] These grants aim overall to prepare millions of Africans for jobs of the future by expanding opportunities for education and employment. But they also support startup initiatives by providing mentorship and access to working space and advisers. The grant funds are not just restricted to tools that benefit Google, however. They can be used for training in digital journalism and other areas that could truly empower individuals living on the continent.[30]

But here's a Google move that's getting much more attention than funds and grants: the company announced that it will build an AI lab to open in 2019, based in Accra, the capital city of the West African nation of Ghana.[31]

Is this lab motivated by Google's desire to resolve mistakes its AI systems have made, for example with an image-search algorithm that mistook black people for gorillas? Is it an attempt to enter new frontiers where there is less regulation and governmental interference now that the European Union has stricter rules under the General Data Protection Regulations legislation (GDPR)?

I had these speculative questions in mind before I talked with Moustapha Cisse, the incoming director of the lab. This young computer scientist from Senegal, who is also a co-founder of the BlackinAI online community, has long been active in conversations regarding race and algorithms.[32] His multiple college degrees and experience working in North America and Europe, he told me, make him the exception, not the norm, relative to fellow African technologists.

Cisse is concerned that AI systems reinforce the values and biases of the places and peoples in the world who were first online. The lab, he told me, is a response to this problem, and will incubate products that learn from, rather than objectify, African users.[33]

From my perspective, the clear intention of a for-profit company to expand its presence in parts of the world that are relatively new to the internet warrants questioning—no matter how well intentioned any of its staff may be. Are Google's attempts to design better AI for and about Africans geared to benefit Google, or will they also help African users?

When I asked Cisse about the possibility that Google's motives for expanding on the continent are less altruistic than avaricious, he resisted. It is inappropriate, he said, for a non-African to question what types of relationships with technology might be empowering for the people of his continent.[34] Fair enough, I responded. But my questions are not merely about Africans, they are focused on the agenda of a North America–based company that has reached a market cap valuation of nearly a trillion dollars. Should we not worry that the company will use its outsized wealth and power to benefit its bottom line first and foremost? Cisse brushed off my concerns, telling me, "We (Africans) are extremely happy to see this opportunity happening at home and are sure it will serve our good. When we Africans think something is good for us, who else can say, 'no this is not good for you' and not be paternalistic?"[35]

African Ascension

My conversation with Cisse was eye opening. Google now refers to itself commonly as an AI company, and he is genuinely excited to lead its AI development efforts on the continent.[36] He's right to imply that from my privileged position as a professor in the United States it is easy to ask such critical questions of the company. I can only imagine the challenge for a Google employee like Cisse, dedicated to using company resources to support others on the continent, to engage with the criticism I can so freely explore. I have no skin in the game that Cisse's life goals and livelihood are riding upon—and that's a privilege, for sure.

But it's still worthwhile to ask tough questions and address contentious issues especially in the early chapters of African technology. Consider Cisse's comments relative to those of another African, the Cameroonian political philosopher and scholar Achille Mbembe. He has critiqued the universalization of Western European thought and its treatment as neutral and "naturally" beneficial to African peoples. Africa's leading role in technological innovation, he has written, must be considered in the specific context of an African history fraught with the traumas of European colonialism.[37]

And yet, as Mbembe points out, colonialism remains very much part of the African story, revealed in examples like unjust labor and mining and mineral extraction.[38] In many ways, the ongoing plunder of resources on the continent is tied to digital technologies. The mineral coltan, for example, which powers our mobile phones, is extracted from Mozambique and Democratic Republic of Congo; as of 2012 those African countries were among the biggest coltan producers worldwide, along with Australia, Brazil, and Canada. The labor practices, environmental impact, and shady political arrangements associated with gathering this mineral are troubling, especially because they are subject to little scrutiny. In the Congo, the extraction of coltan, and of other minerals and natural resources (most notoriously diamonds), has resulted in human rights violations and spawned civil wars.[39] Reports from the UN Security Council reveal that Congo's neighbors, Rwanda, Uganda, and Burundi, were involved in smuggling coltan from Congo and using the revenues to sustain their war efforts. The Rwandan army made at least $250 million doing this, but denied Congo any of the money.[40] According to numerous accounts, child laborers as young as twelve can be found at work in coltan mines.[41]

Despite this exploitation, Mbembe sees Africa as "the last frontier" or, more pointedly, as "the last territory on earth that has not yet been entirely subjected to the rule of capital."[42] Many untapped resources remain on the continent, and he estimates that by 2035 these resources may have been pivotal in helping to create the youngest and most dynamic population in an ageing world.

Africa, like other developing regions worldwide, has "leapfrogged" from the "age of iron to the digital age" without a linear step-by-step path of development, the way Western countries have historically tended to do.[43] For many Africans new technology has meant a leap from no electricity to solar-power without needing to depend on the electrical grid. Similarly, many Africans have jumped from having no access to banking of any kind, straight into mobile banking, skipping over a middle phase dominated by local, brick-and-mortar bank branches. Mbembe describes this type of development as a "constant, permanent innovation." But, he asks, "How do we make sure that this inexhaustible capacity for innovation is at the service of a bigger kind of creation that can propel the continent?"[44]

20 AI in Uganda

I've shown throughout this book how algorithms make it possible for machines to learn from the data sets they are trained upon and thus become "intelligent." They promise efficiency, a key to the future, and a way of providing "answers" that affect most of the global population's daily life. But AI has a "black box" problem: we the users don't have the ability to understand how algorithms arrive at a decision or make predictions. And even if we did, their creators (usually engineers and private corporations) almost always keep the inner workings secret.

What's happening with AI development in Africa—aside from the AI lab that Google is set to open in Ghana—now that the continent has "leapfrogged" into the digital age? In this chapter I look at AI efforts in Africa coming from outside the corporate world, specifically in Uganda. What if AI became a field of inquiry that belonged to everybody on the continent, and was applied to support its diverse cultures and regions? Could that happen without making personal data available to powerful corporations thousands of miles away? Can we apply lessons from Africa to how we work with AI around the world?

A Model Lab at Makerere

My story shifts now the East African nation Uganda—and to the AI & Data Science research lab at Makerere University in the capital city of Kampala. By showing that digital technologies can be imagined and designed by Africans, for Africans, the lab, a nonprofit public institution, is fulfilling its potential to become a model for the rest of the world. Researchers at the lab have been working on AI systems to address resource management, public

Figure 20.1
Crops monitored using AI lab technology (source: Makerere University).

policy and governance, agriculture (see figure 20.1), transportation, and health.

I had known about the AI & Data Science research group for some time before my visit in July 2018, thanks to journalists and scholars such as Gregg Zachary. In one *New York Times* article Zachary described the lab's origins, when a Ugandan student, newly home with a doctorate in computer science from a Norwegian university, founded the College of Computing and Information Sciences at Makerere in 2005. Before that, only relatively simple tasks such as writing very basic software and using network computers were taught at the university. But, as Zachary explained, Uganda's economic prosperity had propelled many within the nation's middle and upper class to pursue engineering and science degrees. An overwhelming number of undergraduates enrolled, causing the faculty to schedule extra classes at midnight to accommodate the students.[1]

The AI & Data Science group's successes, both in designing systems and publishing its research, are evidence that technological achievement can reach great heights if the process is driven from universities on the continent and the communities they serve. These researchers share a passion to engineer AI systems *by Ugandans for Ugandans*. For instance, the group has tethered mobile phone cameras to solar panels scattered around Kampala in order to monitor the city's traffic, a massive problem faced by its 5 million residents. Another example is Mcrops, a tool connecting cell phones with low-cost spectrometers and pattern recognition software that will identify and diagnose viral crop diseases in cassava plants, and thus help farmers stop the spread of disease.[2] Also notable is Kudu, a mobile auction market using farmer and trader input about commodity price, availability, and location to match compatible buyers and sellers.[3] One former Makerere student, Brian Gitta, won an Africa-wide prize for developing a way to automate the diagnosis of malaria with digital microscopy and a mobile phone.[4]

The group's director Ernest Mwebaze, a youthful, soft-spoken, and thoughtful faculty member at Makerere, hosted my visit to the lab. Ernest introduced me to several colleagues including Julianne Sansa-Otim and Engineer Bainomugisha (all three are equally young, under age forty). When I spoke to Julianne she explained how rare female computer scientists are in Uganda, and how her early training in an all-girls school had helped developed the confidence necessary to find her place in this field.[5] Engineer explained how from a young age (literally the day his parents named him!) he was encouraged to work with technology.[6] When the AI lab first began, there were but two individuals from the university who had received doctorates in engineering-related fields. Today there are forty, an incredible increase given how disadvantaged African universities are when it comes to accessing scientific equipment and subscribing to costly academic publications.

The researchers at Makerere are in a bind, however. Because they depend on foreign funding, they need to frame their work to appeal to donors located thousands of miles away. Yet they must also create locally relevant products and, where possible, collaborate with other Ugandans. A conflict of interest might occur when a project with funding tied to outcomes that increase the national GDP overlooks farmers whose crops don't contribute greatly to the overall economy. Cultural values and customs are also likely to be marginalized when pursuing top-down "development indicator" goals because solutions applied on a broader public level tend to homogenize social constituencies.

AI systems are largely influenced by three factors: software code, data sets, and the domains to which they are applied. At the research lab in Uganda, systems are being trained by data that originate from Uganda itself, including data about its local air pollution, traffic, and cassava plants. The technologies developed by the lab thus far are *deterministic*, that is, they have "right" or "wrong" answers that are agreed upon by nearly everyone. Most would agree that cassava and human disease, just like air pollution and traffic, are negatives and should be minimized.

But it's not always that straightforward. In some communities, for example, the strain of cassava most resistant to disease is *not grown* but instead prematurely uprooted and sold. Why? Because the farmers wish to work with crops susceptible to disease and learn to combat it. Cultural factors like these challenge the simplistic assumptions scientists and technologists often make about the lives of those they claim to serve.

AI systems in Uganda do have the potential to empower cultural diversity, although no current projects underway at the lab are specifically aimed to do so. Looking to the future, AI in Uganda could be designed to help citizens share political perspectives, techniques of harvesting crops, or ways of planning for natural disasters.

Outside the university a team of researchers in the rural communities of Kasenda, Buheesi, and Rubona, all in the Kabarole district in western Uganda, conducted a study on how communities would engage in water management practices in villages without clean water.[7] The team determined that to have an effective water management system, the communities needed a sustainable and culturally conscious financial management system to ensure regular payment and collection of water-use fees. They also learned from mistakes made by a non-Ugandan agency. In the past water management development projects had moved boreholes (the shafts in the ground from which water is extracted), to locations far closer to people's homes. This proved to be a huge error because it overlooked the social utility the borehole provides when placed some distance from the community it serves. Like a village well or the office water cooler, the borehole is not merely a place to get water but a public space where people come to socialize, communicate, and meet one another. The development agency–funded NGOs wasted money and time by moving the boreholes because they did not understand the cultures of the people they worked with.

These examples show how important local knowledge is to the design of tools that actually serve users in Uganda. Engineering, mathematics, and computer science cannot be exported from afar, or dropped into communities from the clouds. Otherwise we get what Fiona Ssozi, a lecturer at Makerere, described as *pilotitis*, the all-too-common situation in which new technologies introduced into the developing world fail to survive beyond the pilot phase.

Homegrown AI

No single label (except multitalented, perhaps) can fully describe Daniel Mutembesa. He's a poet, musician, DJ, semi-professional dancer, national Karate champion, and, of course, an engineer—all despite being only twenty-seven years old. When I met him front of the AI lab at Makerere we became fast friends and immediately fell into a spirited conversation about the cultures, traditions, and politics of Uganda. I was fascinated to meet someone in the lab who saw his work in computing as perfectly aligned with his interests in supporting his countrymen.

Mutembesa took me off campus to different and interesting parts of the city in his tiny Korean car. He taught me the indigenous names for Uganda's national parks. By using these names he hoped to revive interest in the histories and traditions of his country. Part of the incredible diversity of Uganda comes from its more than forty indigenous languages. They are spoken in local communities without a national language to lean on for universal communication, except for the English used primarily in metropolitan areas.[8]

Like nearly everyone in Uganda, Mutembesa comes from its tribal regions: the Ankole and Tooro parts of the southwest of the country (the latter borders the Democratic Republic of Congo). Learning about his nation opened my eyes to a sad trend of historic and cultural erasure. Mainstream entertainment produced for and accessed by Ugandan audiences often adheres to the Western brand, as when DJs fake American accents on the radio and

African TV programs mimic *American Idol* and *Survivor*. Western webpages dominate the internet. Mutembesa sees this as an invasion of national and cultural consciousness by Western norms and products.[9] I asked him how the AI lab sees itself relative to these cultural and global forces. Is computing simply seen as a vehicle of "being modern" (aka imitating the West) or can it revive and support the identities and values of the culture's richly diverse traditions and communities?

Mutembesa passionately believes in the latter. He is designing AI systems that learn from Ugandan farmers' decisions about harvesting crops. He has built a machine-learning model that considers the way farmers make choices about using treatments that agricultural scientists suggest. Potential problems arise, he tells me, when applying a model developed from afar to the Ugandan context.[10] But an alternate approach may come from understanding that the strength of game theory lies in the relatively few assumptions upon which it is built.

Game theory considers decisions made to either cooperate or compete with participants in social settings, and the factors that shape these choices.[11] The lab's model allows the Makerere team to learn the unexpected. For example, they found that the National Agricultural Institute, a partner of the AI lab in Mutembesa's project, has great sway with farmers— far more, in fact, then the lab would have in paying them to participate. Praise matters more than money; knowing this important cultural value, the team collaborated with farmers and applied technology to maximize yields and harvests.

Though it gained independence from the British in 1962, Uganda continues to face uneven relationships relative to the world's political superpowers. There is a tendency to blindly embrace the media, technologies, and ideas of Europe and the United States. As a result, Mutembesa says, "We, in Uganda, have a strong tendency to take on blanket recommendations. From our technology, to our research, to our systems of education and health, even the names we give to our lands and parks, we are forced to follow the funding and the rules of the funder."[12]

What does this mean for the AI lab? The challenges it faces are local, the questions they wish to answer are local, the communities, plants and animals with which they work are local. But the funding models are not. As a result, the lab may not always be answering the "right" questions when it comes to serving their country.

Why don't these researchers move into the startup world and begin a chain reaction that could, eventually, create more wealth within the nation? The hurdles are too high. The national government gives almost no support to entrepreneurs. Unlike in nearby Nairobi, Kenya's capital, there is little history of foreign investment in technology in the Ugandan capital and country as a whole. Researchers working in Kampala have more secure incomes than those working in business. The risks are too high for entrepreneurs to take on labor costs and invest heavily in their city and country.

Nonetheless, Mutembesa remains hopeful. Despite the complacency with which Ugandans view their lives and the uncritical ways they tend to applaud the West, he believes in the deep, living heritage and identity within his country. That identity is evident in the land and in the languages, dances, and music from the nation's numerous indigenous communities. These forms of cultural expression are the "knowledge of the people," he says, and are the "basis for which technology and innovation should be developed to serve our communities."[13]

This example resonates with the perspective of Dambisa Moyo, an internationally recognized author from Zambia, and an economist specializing in African developmental economics. In her book *Dead Aid* she pulls no punches when she criticizes African dependency on foreign aid. Because of it, she says, "Misery and poverty have not ended but increased. [Aid] has been, and continues to be, an unmitigated political, economic, and humanitarian disaster for most parts of the developing world."[14]

Moyo points out that Uganda, like many of its neighbors, is a youthful nation with 50 percent of the population younger than fifteen. She argues that aid disempowers this young population by placing her fellow Africans in a subservient position, elevating foreign goals over local creativity. Subsidies for farmers guaranteed by the World Trade Organization, for example, might seem helpful but in reality they can harm local economies. Subsidies drive down the prices of staple goods and leave farmers in financial instability, unable to negotiate the terms by which they buy and sell.[15] We can see evidence of this with global trade market, where Africa represents only 2 percent of overall activity.[16]

My time in East Africa, and specifically Uganda, does not necessarily echo Moyo's position. Certain types of aid, including a Google Research Initiative, have supported the development of the laboratory of scholars and their students at Makerere, one of the continent's top universities. Emerging from this has been impressive coordination among scholars, researchers, and engineers across all of Africa.

But has such aid forced a ceiling on growth? Makerere's AI lab reveals that external aid, lacking concern for the culture and voices of Ugandan citizens, is no panacea. Uganda's neighbor to the West, Rwanda, once faced a similar situation in relation to aid from the United States and the International Monetary Fund (IMF). Instead of simply taking the aid, Rwanda gave the IMF an ultimatum: let us accept your aid but on our terms, where it will enter the right hands and take the right paths to grow the economy.[17] The push and pull of independence versus dependence is a common story when we look at technology across the world.

21   Innovating from the Ground Up in Kenya

Kenya, Uganda's neighbor to the east, has a buzzing technological ecosystem. Its capital city, Nairobi, has even been dubbed "Silicon Savannah." The name alludes to a promising array of startups, entrepreneurs, and innovators blossoming across Sub-Saharan Africa. In this chapter I show how these tech firms reflect the cultures, constraints, and environments characteristic of East Africa.

Large Western tech companies commonly open sales offices in countries and cities in the global South. But Nairobi is not just a corporate outpost for the West. It is a center for homegrown African technology—along with Accra (Ghana), Lagos (Nigeria), and Johannesburg (South Africa)—where internet-related education, jobs, businesses, and institutions continue to emerge. And the Western corporations have taken notice. When Facebook recently expanded onto the continent, Mark Zuckerberg visited Nairobi and its leading entrepreneurs several times.[1] Google has been investing in Kenyan talent, too. IBM and Microsoft have also brought research laboratories to the region.[2]

Despite Nairobi's thriving technological ecosystem, the region faces unique difficulties. East Africa, a problematic zone in Africa's electricity crisis, has historically lacked essential infrastructure to connect to power grids or to establish landlines. The incredible diversity of the region adds layers of complexity that confound the idea of a one-size-fits-all development model, including when it comes to digital technology. (Kenya, for example, features at least sixty-seven distinct languages and numerous different tribal cultures.[3]) So how did Nairobi, smack in the middle of this complicated picture, manage to become the Silicon Savannah, and do so without the influence of Western players?

The tech-buzz around Nairobi began in 2007. A weak, mostly nonexistent retail banking system in Kenya (and in much of the continent) had made it difficult and inconvenient to transfer and exchange money among individuals, a factor that stymied the growth of small businesses and individual enterprises. At the same time mobile phone networks were blossoming across Sub-Saharan Africa. In 2002, one in ten people owned a mobile device in Tanzania, Uganda, Kenya, and Ghana, yet today they are almost as widespread as in the United States (89 percent of adults compared to 82 percent in Kenya).[4] Noting this, Safaricom, a branch of Vodafone and one of the leading telecom providers in the country, responded to that trend in 2007 and adopted M-PESA, the mobile money service I mentioned in chapter 19 (in Swahili, *pesa* means "money").[5]

M-PESA technology works on even the most basic cellular phones. Subscribers to the service can deposit money into an account tied to their cell phone numbers, send money to others via PIN-secured text messages, and redeem deposits on the phones as well. They pay only a small fee for sending and withdrawing money. M-PESA serves as a "branchless" bank with a network of agents located at retail outlets.

Having passed its tenth anniversary in 2017, M-PESA still celebrates rapid growth and achievement.[6] There are now 17 million users based in Kenya, Tanzania, Afghanistan, South Africa, India, Romania, and Albania.[7] The service has played a significant role in lifting many families out of poverty and has launched ordinary Africans, many still without bank accounts, into the digital economy. Today seemingly every mobile provider has created a *pesa* of its own, making mobile phone access the only requirement for engaging in exchanges of all kinds across Kenya, and indeed across much of the African continent.

Mobile money is the most well known and pervasive of several grassroots tech innovations born in Africa. As we saw in chapter 19, the success of mobile money has everything to do with the *power of context* rather than an advanced achievement in engineering. In this case three factors—strong cell phone connectivity, weak banking systems, and the ways Africans interact and exchange informally—combined to make possible an innovation that transformed the lives of millions.

To see another tech innovation in action, simply step onto a street corner in nearly any African city (or, for that matter, in much of the global South). The *matatu*, a collectively owned and operated mode of minibus transport, might seem like Western public transit to the uninitiated, but it's not so simple because of how it operates. The key virtue is flexibility: the *matatu* system can accommodate low-income passengers and still remain profitable enough to support owners, drivers, and cash-collecting conductors.

Riding the *matatu* isn't limited to the mundane experience of simply getting from place to place. It is also an immersive social space full of life, spirit, and noise (see figure 21.1). On Nairobi's "Bob Marley" line, Jamaican, African, and hip hop music blasts from the minivan while conductors dance down the aisles and hang out the doors like acrobats. Animated

Figure 21.1
A Nairobi *matatu* featuring posters of musical superstars.

conversation takes place even among strangers. In these ways, the *matatu* represents the sense of community that infuses everyday life in Africa.

Individual *matatu* owners who hire drivers and conductors recognize that they alone cannot overcome the threat of fiscal instability. As a result, *matatu* entrepreneurs have banded together in Kenya in savings and credit cooperatives (SACCOs). These collectives invest the deposits of their members in safe financial instruments; they offer loans, if necessary, to maintain the quality of their members' services and insure their drivers and vehicles.

Contrast these bottom-up examples of African innovation with problems that accompany digital solutions imported from afar: Google recently attempted (unsuccessfully) to digitize payments within and across the *matatu* systems in Kenya. This disruption was presumed to be an efficient "solution" for *matatu* entrepreneurs, but it was rejected as invasive and unnecessary because it ignored local context, diminishing the roles and identities that drive the cash-based exchange economy of its community. To put it simply using M-PESA as an example, the homegrown mobile money system resolved a problem whereas Google interfered with one.

The Chinese company Xiaomi encountered its own import problem when it decided to market its Mi Band "smart watches" in Africa. A number of them sold across East Africa, but few actually worked. Why? The creators designed them for users who look like they do, rather than for customers the company wished to reach: the optical sensors in the watch don't function effectively with dark skin. Imagine how these watches might have been different if they were designed with collaborators on the African continent. The sensors wouldn't have been calibrated to light skin only, that's for sure.

These examples reveal why technology, like economics and politics, must be read through the prism of cultural and social context. They show that designing for the world without listening to or learning from it is a big mistake.

Connecting All of Africa

Twenty years ago, Erik Hersman sat in an African studies class at Tennessee State University, a historically black college, learning about Sudan (where he'd lived since he was two) and Kenya (his home country). Hersman stood out in the classroom—and not because of his bushy beard or ex-rugby player frame. He was the only white person in it.

Hersman's parents—white American linguists and cultural anthropologists—had moved first to Sudan in 1977 and then to Kenya to pursue their Christian mission. Hersman, who enlisted in the US Marines to pay for his college education, went back to the states to become a business major at Florida State University and spent a semester at Tennessee State as well. But after graduating, a combination of love for computers from his childhood in Africa and IT experience he'd picked up in the United States inspired him to return to Nairobi.

In 2008 Hersman co-founded the nonprofit crowdsourcing platform Ushahidi, the Swahili word for "testimony." The technology, which allows its users to monitor everything from elections to outbreaks of violence to natural disaster recovery efforts, protects personal data and can be installed on one's own server. It has now been used hundreds of thousands of times across the world: in Haiti (after the 2010 earthquake), in Nigeria in 2011 (and in national elections in the United States), in Japan (the 2011 tsunami), in Syria during the civil war, and with harassmap.org (a site that helps women report on sexual harassment and other abuses).

Ushahidi emerged as a response to the 2007 election crisis in Kenya, much like the way M-PESA emerged as a response to the absence of banking. Kenyans had begun to notice that reports they viewed on television differed from what they were hearing from friends and relatives. Then the government shut down local television, and international coverage presented its own flaws and biases. An opportunity emerged to design a technology to empower Kenyans to speak to one another directly, and to get instant information from the grassroots by sharing reports through texts, tweets, emails, and photos via an interactive map. Volunteer and humanitarian aid organizations have commended the "crisis mapping" that Ushahidi implemented during the relief efforts after the 2010 Haiti earthquake.

The platform's use in places around the world reveals that technology born in Africa can assist communities and causes thousands of miles away from the continent. When former US president Obama visited Kenya in 2015, he referred to Ushahidi as one of the great hallmarks of African innovation: "From Zimbabwe to Bangladesh, citizens work to keep elections safe, using the crowdsourcing platform Ushahidi—and that's a great idea that started right here in Kenya."[8] The organization still maintains its pledge to avoid gathering personal data and has faith that people will deploy the technology as they see fit, as a tool rather than a "solution" to social, cultural, or political issues. Charles Harding, Ushahidi's former director of product management, tells me that's the reason users developed trust in Ushahidi. It's the opposite of Silicon Valley's urge, as Harding puts it, "to create technologies to solve what they see as problems; which end up creating other problems." As a result of that urge, he says, "there's a real underinvestment in the power of human choice and creativity, social capital, and democracy."[9]

Hersman has more recently focused his efforts on a for-profit business called BRCK, a company driven by the social mission to "connect Africa to the internet"—for free.[10]

After my time at the Uganda AI lab in 2018, I spent several weeks with Hersman and others in the Nairobi tech world. Hersman, Reg Orton, and Philip Walton are part of a lost tribe: foreigners who have spent large parts of their lives on the African continent. Walton grew up between Burkina Faso, Côte d'Ivoire, France, and the United States, and has gained experience in these places in logistics and supply chain management, real estate, and biometrics. After working with a tech developer team in India, Walton decided to help develop technology *within Africa for Africans*. His reason stemmed from the different cultural approaches to technology he encountered when asking engineers to think "outside the box." Walton tells me that in India developers would question whether a world existed outside of the box. In Africa, where developers are unconfined and free thinking, they often respond with, "What box?"

BRCK's Work under the Hot Sun

Hersman and his BRCK co-founders describe the way Africans approach problem solving with what Walton calls "constrained resourcefulness."[11] Their goal is to provide their fellow Kenyans the baseline of internet access.

In Africa, technological ingenuity happens from the street up, a process Hersman has seen from the front lines. Indeed, I knew of Hersman first as a writer for his blog *Afrigadget*, where he tells stories of technological ingenuity in the informal economy. The Swahili term for this work, *jua kali* or "hot sun," speaks to how laborious and intense such work can be. But rather than decrying or fetishizing it, we can learn from it—as innovation in action.

SupaBRCK is the current flagship innovation launched by Hersman and the BRCK team: this small, portable, solar-powered Wifi device can connect people to the internet in areas with poor or nonexistent power and communications infrastructure. It allows anyone with a smartphone or laptop to connect either to the wider internet or to locally hosted content via a platform called Moja, an option that provides users with free internet access in exchange for taking a survey or watching an advertisement. SupaBRCKs are designed to be rugged: they withstand dust and pollution, and if dropped they won't suffer from the impact. In an environment where a standard Western logitech router wouldn't survive, the SupaBRCK thrives.

Each SupaBRCK weighs just 1,200 grams. Although sleek and compact, it is powerful, providing connection to the internet via Ethernet, Wifi, 3G, or an LTE link. During power failures its ten hours of battery life make it extremely useful. But it can also serve as a hard drive, allowing people to watch videos and reduce the cost of streaming. The 500GB hard disk can be upgraded to 5TB, and several modems it can work with provide high-speed connectivity.

By 2018 the company had deployed nearly 1,100 SupaBRCKs (approximately 1,000 in Kenya and 100 in Rwanda). Most of those have been set up on the *matatus* that comb the city and country. The company trusts *matatu* drivers to take care of its SupaBRCKs and recognizes the value of collaborating with the ubiquitous transport system. Although the company sells its BRCK devices when it can, it looks also for partnerships and sponsorships to get BRCKs produced and distributed through other means. In that way it can grow its user base and foster connectivity to create new economic opportunities for users and businesses.

While figuring out how to support cheap and possibly free internet access for Africans in a way that respects user needs, desires, BRCK rejects an exploitative model that would gather data from Africans and monetize it for profit. The company embraces the principle of supporting locality and place, in part by employing more than sixty Kenyans. Hersman told me about his "Star Wars" method for recruiting talent—an approach that allows his team to show their stardom within a nurturing and collaborative environment.[12] He trusts that their creativity will allow them to take off on "rocket ships" of their own, launching new organizations, businesses, and opportunities, even ones that might someday compete with BRCK.[13] The company's design and engineering prowess thus build on the constraints and possibilities of the local environments where technologies

Figure 21.2
BRCK's call to arms.

would be deployed. This contextual knowledge allows the company to
build sustainable solutions—African design for African environments.
Proudly adorning the front office wall, along with articles from *Wired*, the
Economist, various African newspapers, and prize-winning certificates, is a
poster (see figure 21.2) with a telling question: "Why do we use technology designed for London and Los Angeles when we live in Nairobi and
New Delhi?"

In an interview with a *Forbes* magazine reporter Hersman spoke about
connectivity in the image of his fellow Kenyans:

> If we want to solve the problem of internet in emerging markets, we need to think
> about the infrastructure of the internet itself differently—or, maybe we need to
> think of it as it was originally designed—truly distributed.[14]

Let's get technical for a moment to understand the logistics: Currently,
the internet's infrastructure, despite appearing to be centralized, makes
access dependent on the ability to plug a router into one centralized system of electricity—aka "the grid"—and one centralized system of fiber optic

cables—aka "the internet." In Africa, this method introduces a host of problems for connectivity, since infrastructures like the grid are unreliable. Most people have poor internet connectivity unless they live in a large town or city. In a continent with 1.2 billion people, this means a huge portion of Africa is prevented from going online and accessing content, especially if the server that content sits on is located elsewhere, in the United States or in Europe.

At times the absence of connectivity has been a major issue within Kenya. During the 2017 election, Kenya hoped to prevent electoral tampering and showcase technological advancement with mobile internet voting stations hosting automatic results. But because a quarter of Kenya's 41,000 polling stations did not have mobile-phone reception, the reporting was incomplete, which led to allegations that the voting was rigged.[15]

Although many parts of Africa, including Kenya, still have slow connectivity, we've seen statistics in chapters 19 and 20 that show how the use of mobile phones has exploded. The *Economist* reports that "every 10% increase in mobile-phone penetration in poor countries speeds up GDP growth per person by 0.8–1.2 percentage points a year. And when people get mobile internet, the rate of growth bumps up again," meaning that the economic costs of low connectivity are debilitating.[16]

Mobile phones and internet connectivity can help spur on a number of other innovations that speed up economic growth and improve the lives of people.[17] Africa faces less risk of unemployment due to automation because it has fewer manufacturing industries than other continents (5 percent of the continent's jobs versus 15 to 18 percent in other developing countries). Technology, then, could facilitate Africa's economic growth. Africa already has the second fastest-growing workforce in the world, and of the 2.4 billion new people projected to be on our planet by 2050, 1.3 billion of them will come from the continent.[18]

Hersman and his colleagues see both a need and an opportunity here. If people in wealthier parts of the world are able to access the internet, so too should those with fewer resources in Africa. That's a right, not a privilege. The key is identifying a business model that supports rather than exploits fragile users and businesses. BRCK has had to hold the line with international companies and find its own ethical path around how it engages with advertisers—BRCK employees have explained to me that these customers have asked for personal data of an intimate nature. Despite that, the company has not provided them with any personal identifying information about its users.

"Robin Hood" Spirit

With the shared belief that internet access is a right, BRCK has had to address questions about what connectivity means to the now more than 300,000 independent users it has brought into the fold. I heard these conversations every day during my two-week visit to BRCK, not from senior management alone but also from nearly every employee I spoke to. The design practices of the Western world simply don't apply here, they explained to me, because electricity and services much of the world takes for granted are not so stable. Yet, as we have seen in several examples already, in this absence lies great opportunity. Although solar power is considered an alternative model in the United States, in a place like Kenya it might be the most dependable source of power. This pushes engineers to leverage the resources Kenya has to offer and to develop solutions that actually support their environment and natural resources. A green future might first come out of Africa.

Take for example the new product BRCK is rolling out—the PicoBRCK (or "little" BRCK). This device, designed for monitoring soil, water, and wildlife with smart sensors, is BRCK's first Internet of Things (IoT) product. As the company attempts to build out this product, employees told me stories about similar efforts from the United States and Germany. The California-based company Silver Spring, for instance, has built massive sensor networks that are completely reliant (unlike BRCK) on the electrical grid. With a laugh, BRCK engineers told me that such implementations would never work in East Africa.

For Mark Kamau, a user experience (UX) designer for BRCK, design does not emerge from the mythical place of infinite-resource access or solely from Western giants such as Apple or IDEO. He explained the rationale that inspires BRCK's spirit when it comes to technology. Subsidizing cheap, even free, services to startups and to street-level informal-economy business people and workers—and thus giving them an opportunity to showcase their creativity—can be achieved by charging higher premiums to wealthier and multinational customers.[19] He calls it the "Robin Hood" approach, taxing the rich to help the poor.

Most BRCK employees come from working-class upbringings. Many are self-trained and they have had to hustle to get where they are today. They wish to succeed in a way that supports rather than ignores their friends from the street tech world.

Innovation in the Savannah and the Valley

In Kenya, necessity is the mother of *innovation*. I was amazed to witness resourcefulness in action when I saw 3D printers being created from recycled and discarded electronics. One business, Africa Born 3D (AB3D, see figure 21.3), set up shop by printing objects for passers-by on street corners, in warehouses across the city, and today in the Nairobi home of its founders. The company's twofold mission is (1) to produce high-quality printers by salvaging discarded wires and circuits from electronic dumps and recycling centers, and (2) sell them at a fraction of what an off-the-shelf model with the same specs would cost if purchased from the West. AB3D printers even

Figure 21.3
AB3D transforms waste into a 3D-printing business.

beat Chinese prices, and its machines are more robust and resilient. Why? Because entrepreneurs who design the printers have local knowledge and deeper understanding of the environments and realities of their customers. The business provides maintenance and support, and has now begun to offer a 3D print-on-demand option for entrepreneurs throughout the country. As Maketa Maina from AB3D told me, "We're in the informal sector—it's just that now we have a roof."[20]

Because fledgling businesses like AB3D often emerge from the *jua kali* ("hot sun") of the informal economy that I referred to earlier in the chapter, we can also say that East Africans pour sweat equity into their work.[21] Whether in the form of rigging, hacking, or remixing, they craftily transform the old into the new, bringing something left to die back to life.

Markets across East Africa show such resilience in action (see figure 21.4). Televisions, bicycles, shoes, clothes, cars, school desks, sewing machines—every item imaginable becomes a candidate for repair in these dynamic centers of exchange. With digital technology, it is no different. Phones are being resoldered, repaired, and resold; electronic devices are being spliced together to create new hybrid technologies. All of this is happening on the street, with just a small wooden table, discarded electronics, and a soldering gun.

Let's compare what innovation means to those in both the Silicon Savannah and the Silicon Valley. When we think of Apple's Steve Jobs and other mythologized corporate "innovators," innovation is forward moving. It's synonymous with words like "upheaval" or "breakthrough," and connected to terms like "disruption" that represent a departure from the past and a replacement with something new. In the West, this process is usually characterized as a series of improvements that are supposed to move us into an ever more "modern" or "advanced" future.

Does this have to be the only way we define innovation? The theorist Steven R. Jackson challenges this simplistic thinking about digital innovation.[22] Rather than novelty, growth, and progress, what if we thought of erosion, breakdown, decay, and sustainability as starting points for dreaming about what innovation in technology could be?

Jackson calls this an exercise in "broken-world thinking." The concept draws from the philosopher Walter Benjamin, who theorized decay, destruction, and the "aftermath" as aesthetic categories during the postwar period in which he lived. Far from being a conversation stopper, it allows

Figure 21.4
Electronics and everything else at Mercato, in Addis Ababa, Ethiopia.

us to ask and answer questions such as: What if we saw broken technologies or systems as opportunities to repair and recycle? (See figure 21.5.) Or what if we thought past the "remarkably restricted and usually binary sets of actors that have dominated media and technology studies to date: senders and receivers, producers and consumers, designers and users" to instead include a new cast of characters—the breakers, the fixers, and the maintainers.[23]

In truth, these grassroots innovators are already out there, scattered around the world, using the vulnerabilities of their situation (the lack of electrical grids in Africa, for example) as starting points for design.

Figure 21.5
Recycling mobile phones in Kibera, Nairobi.

Mainstream, Western-centric thinking fails to give the fantastic powers of innovation in the developing world their due. But we have an opportunity now, as tech blossoms across the global South, to correct this error. Just as I have discussed the potential and achievements of an African AI lab, I'd like to consider the potential of enterprises, businesses, and incubators that come out of places such as Nairobi, Kenya, Accra, Ghana, Johannesburg, South Africa, Lagos, Nigeria, and more. Can they create new jobs and industries (and engineer and design new types of technologies) that serve and support peoples in the places where these very efforts originate? Or, in so many places across the developing world, will they be stuck in a dependency model like foreign aid, donor funds, or low-level outsourcing with the West and China?

Let's apply this new idea of innovation to one of BRCK's rising stars. To Yasin Mohamed Barre, a young man from the Kenyan village of Masalani, it means "making and breaking; opening and closing; recycling and assembling; just trying things out."[24]

Barre grew up thinking that his only professional path was to serve in the army. That's what his dad had done (and was all it seemed he could financially, and perhaps emotionally, support). Nonetheless, Barre was a tinkerer—he told me how he'd always loved the idea of breaking, re-creating, and making things. He put himself through school with a variety of "side hustles," but most of his science and engineering education was so theoretical that it was difficult to apply it to the work he does today as a BRCK product designer and mechanical engineer.

Like most BRCK engineers, Barre is self-taught. When he goes back home and mentions his job title to his community, they think he is a mechanic or blacksmith—there is no word for "mechanical engineer" or "product designer" in their local languages. As a result, the work he does is not merely out of the ordinary to his hometown friends and neighbors. It's otherworldly.

Considering the hurdles Barre faced, it's impressive that he's behind the mechanical design of the SupaBRCK. Echoing the company's motto *Africans designing for Africa*, he told me that a Western design approach toward the BRCK would never have worked.[25] It would not take into account the bumpiness of the *matatu*, or the noise and vibrations made by passing cars. The SupaBRCK needed to be strong and robust—designed for impact. To absorb vibration, Barre and his team created a set of rubber washers, which are not available off-the-shelf in Nairobi, and not even in the West.

Another issue the team had to address during the design period was how to create an electronic system that could withstand both the rainy season and Nairobi's months of heat. They designed the box encasing the SupaBRCK to be a heat-sink, meaning that the materials and space within the encasing of the electronics itself could release the heat generated by the circuit board.

These are but a few of the challenges and opportunities that await one who wishes to "design for Africa." And we should recognize that this motto itself reflects a gross generalization. African environments are as different as they are similar. That's why BRCK's field expeditions stress-test their products in these environments, whether on a 2,500-mile motorcycle journey between Nairobi and Johannesburg, a hike to the top of Mount Kenya, or a ferry to the coastal Pemba Island.

Like other African tech startups, BRCK faces challenges in terms of sustainability and growth. Nonetheless, the company is tackling the complex challenge of being financially solvent while also trying to nurture a culture of creative and social-minded engineers, and deliver value to its mostly poor and working-class users.

But challenges can be opportunities. The PicoBRCK, for example, could be helpful within communities where environmental monitoring of water quality, soil fertility, and climate-change threats are a central priority. But to introduce its technology in these places, BRCK will need to overcome the absence of other reliable infrastructures, including electricity and a mobile data network. That's why Hersman and his team have begun building (or partnering on) their own LTE towers across rural regions of the country, and doing so for a far lower expense than if they contracted a mobile tele-communications provider. LTE is a data-only rather than a mobile voice technology, which means that those connected to this network would be provided access to applications like Skype or WhatsApp, rather than to a traditional phone service. Like mobile infrastructures that leapfrogged landlines in many parts of the continent, BRCK is attempting to provide users even cheaper services through LTE. Rather than relying on expensive, bureaucratic, or self-serving partners, its goal is to take control over the "whole stack," giving the company greater power to make their services available for all.

The possibilities for BRCK's services have already captured the imaginations and hopes of several organizations across the continent. PicoBRCKs are currently being installed in northern Kenya's Dadaab refugee camp, now a temporary home to more than 250,000 displaced people who have fled the conflict in neighboring Somalia. Funded by the UN High Commission on Refugees (UNHCR), the PicoBRCKs will monitor water quality and availability.[26] BRCK engineers have also spoken to me with excitement about linking these with a SupaBRCK, which could then allow the refugees to connect with family members, share their experiences, and even access jobs and income.

Yet, like neighboring Uganda's AI lab, BRCK faces major funding challenges. Often venture capitalists decline to invest in people or organizations that do not look or behave like them. This issue looms large even within Silicon Valley itself. For example, Tracy Chou, an accomplished engineer and critic of Silicon Valley's "diversity problem," points out that funding for white, male, Western entrepreneurs is often far more secure than for the people who don't check those boxes.[27] Given that Silicon Valley investors are not inclined to support a company located thousands of miles away, it doesn't seem likely that they would make an exception for BRCK, whose employee base is more than 90 percent black and Kenyan.

Research shows that 40 percent of venture investors attended either Harvard or Stanford and 82 percent of the industry is male (with 60 percent being white).[28] What about black investors? It turns out that 50 percent of them attended one of these two prestigious universities. In this way the venture capitalist (VC) world is "insular and less of a meritocracy, but more of a mirrortocracy."[29] The pipeline that produces technology businesses is blocked when it comes to racial, gendered, or even cognitive diversity, given that investors tend to come from but a few elite universities located in the United States.

Yes, Hersman is a white male, but everything else about his business is far removed from the world most Silicon Valley VCs live in. If he decided to move the management of BRCK to Silicon Valley, the funding would be dramatically increased, but his ability to develop the "right" product for the African market would be compromised. It would be impossible to manage employees and engineers from across the world, given that in Kenya personal interactions are so culturally important.

BRCK has received some social-impact investment funding, a niche by which developing-world technology efforts are often framed. But in many ways what BRCK is doing should not be brushed aside into this social category. Yes, it is a business that is attempting to grow and, yes, it has a social conscience. But to define its mission and goals according to the categories of Western philanthropy is to miss the point. What it is doing is not philanthropic, according to the Western definition, or strictly entrepreneurial, either. It's marching to the beat of a different drum, and accordingly upsets what most of us are used to when we think about business and social impact.

To do business in the developing world, one can either be close to the money, or close to the people. BRCK has made a clear choice in favor of the latter. The tech company values the power of proximity in Kenya and Africa—hiring the region's engineers, partnering with its businesses, and striving to bring free connectivity to its users.

But BRCK's "close to the people" stance isn't just a mismatch with the tastes of Western venture capitalists and other funders. It is also in tension with many of the company's potential partners in international organizations and companies. I learned of at least three global organizations attempting to get a foothold in Kenya: Google, Facebook, and a United States development organization called the Overseas Private Investment Corporation (OPIC).

Each of the three has taken a different approach toward the goal of getting more Kenyans online. OPIC employees have visited Kenya without acknowledging or making contact with organizations like BRCK. Google, through its hot air balloon project (Project Loon) and connectivity lab, has partnered with national-level top-down telecommunications providers who are building expensive cell towers across the country.[30] Its goal is to build its infrastructure in-house, likely to ensure that any experience of the internet goes through its data-gathering platforms and services. But Facebook, for all its faults, has been strategic, partnering with BRCK to sponsor its free internet services on the *matatus*.[31] It does not force BRCK users to sign up for Wifi via a Facebook account. It offers instead free Facebook app updates for users who are interested. Facebook recognizes that having more users online in Kenya means more users, period.

Hersman told me about Mark Zuckerberg's visit to his office in Nairobi in 2016 (see figure 21.6). During their meeting a Facebook satellite blew up, wiping out one of the billionaire's visions of connecting more users to its platform and gathering even more data.[32] It was also in that period that Facebook suffered an embarrassing blow in India, where users and regulators decidedly rejected Facebook's Free Basics services on the grounds that the free internet program only featured handpicked, sponsored content.[33]

These incidents convinced Zuckerberg how important it is to trust local partners in a nation rather than produce "solutions" from thousands of miles away. Of course this does not describe Facebook's repeated negligence of local user communities around the world that I've discussed in this book. Nonetheless at present, Facebook funds the entire BRCK *matatu* program on the Kenyan organization's terms, without any hidden data collection or microtargeting consequences. This represents a rare and important victory for grassroots entrepreneurs, and possibly for all of us as users.

The BRCK story is not solely about technology, or about culture or business. It's about the bold attempt to innovate among many fields and communities, and to rethink the internet from that perspective. The anonymized data BRCK has gathered reveals that users are interested in connectivity as a means to engage with their city, region, and country. The

Figure 21.6
Facebook's Mark Zuckerberg visiting BRCK (source: BRCK).

local vegetable seller, usually called "Mama Mboga," can use the BRCK platform to reach out to potential customers across the city rather than simply within her neighborhood. Musicians can directly target local audiences
rather than get swallowed up on YouTube or iTunes. Indeed, 80 percent
of the content that users choose to interact with on Moja, it turns out, is
limited to the nation.

As Philip Walton told me, "Here in Kenya, we're in the primordial soup;
we aren't going to replicate what the West does online. We're going to innovate upon it organically."[34] Already in Kenya solutions have been identified
that the rest of the world may eventually grapple with. Money can move
between people in Africa far faster than in the West thanks to M-PESA, and
devices have been engineered in contexts of limited or broken infrastructures. Technology can be used to scratch at fleas or fight lions, according to
Walton. In other words, in Africa the real problems to which technology
can be applied are more visible than their counterparts in the West, in part
because they unfold on a huge scale.

But technologies created in Africa still have potential to help others
around the world. We've seen how Ushahidi has been used in dozens of

countries, even in the United States. And we can imagine how the use of BRCK systems could assist communities in the Americas or Europe who need access to clean water—just look at Flint, Michigan's recent lead-contaminated water crisis. With this in mind I asked Hersman whether we could foresee a digital Africa whose enterprises grow and incubate employment and possibilities for its citizens, one that is not stuck in a subordinate position relative to Western enterprises. He brought up the example of Asian tech giants such as Korea's Samsung. They had to do the tough work, often outsourced from the West, to get to a place where they can now lead the global economy in a number of different areas. He told me that it's time for Africa to get into the same game, time to start with a workforce that uses, innovates, and builds technology.[35] From there, a whole new paradigm can emerge based on what works across this continent and the global South. But to start, the engineering process must respect the visions and aspirations of people on the ground in Africa and across the world.

Bringing the People Together: Crowdsourcing from Africa

Jamii—the Swahili word for community—is everywhere in East Africa. *Jamii* was a force in every place I visited, evoking a philosophy of living, of seeing one's identity and destiny as intimately tied to familial, societal, and at times national identities.

BRCK, like hundreds of other technology companies, emerged out of Nairobi's iHub community, which Hersman co-founded in 2010. iHub, a co-working space and technology incubator in Nairobi. The organization is, however, not simply a physical space. It stands for a set of values that evoke what the digital revolution *should mean* for Kenyans. It has run workshops, conducted research, and written policy papers advising the nation on a number of issues that include: how to use technology to keep the government and private enterprises accountable, how to make sure that the internet is safe as a place for women to express their viewpoints: and how to advocate for online safety and data security.

But the innovations in Africa that have sprung up throughout the continent in the past two decades didn't come solely from the private sector, the street, or even the educational system; they also relied on government commitments to support digital infrastructure. In Kenya, the story is no different. In 2006 a civil servant working in the country's Ministry of

Information and Communications, Bitange Ndemo, crafted a master plan for developing the country's information and communication technologies that eventually spurred the Kenyan government into action.[36] In 2008, the TEAMS fiber optic cable construction project began under Ndemo's direction to increase broadband access for a number of East African nations. Those efforts to facilitate high-speed internet access have of course benefited Kenyan businesses and institutions.[37] In 2013, the government made an explicit commitment to develop an information society and knowledge economy, a promise recorded in Kenya's Vision 2030 document. This policy blueprint aims to "transform Kenya into a newly industrializing, middle-income country providing a high quality of life to all its citizens by 2030 in a clean and secure environment."[38]

So far we've read about how an African vision of innovation shapes emerging technologies. But the BRCK model doesn't have to be the only way for people far removed from Silicon Valley to enter the digital world. Looking at Latin America, I next consider a far more provocative possibility: collective ownership and design of telephone and internet networks by users and their local communities.

22 Mobile Power to the People: Indigenous Networks in Mexico

In the mountains around the Oaxaca region of southern Mexico, one of the most biologically and culturally diverse regions in the Americas, technology is being reimagined. There we can find the Telecomunicaciones Indígenas Comunitarias (TIC) project, the largest community-owned cell phone network in the world.

The TIC organization, established in 2012 by a group of hackers, activists, and indigenous community leaders in the region, emerged from centuries of grassroots political movements and philosophies that have extolled the importance of autonomy, communality, and collectivity. From its base in of the city of Oaxaca de Juárez, TIC has implemented independent, community-owned cell phone networks in at least sixty-three indigenous communities of Zapotec, Mixtec, and Mije origins. The effort has provided daily service to more than 3,500 people despite some of the harshest conditions for building communications networks in Mexico—elevation, rain, dense forests, and the absence of other reliable infrastructure like electricity.

The Oaxaca region is home to about a third of the Mexican indigenous population, with speakers of at least sixteen languages and dozens of dialects. The creation of TIC is an example of how new technological innovation can spring from places of great cultural and linguistic diversity.[1] The region contains about half of the entire nation's species of flora and fauna, including gila monsters, jaguars, and, at forty feet in diameter, the world's widest tree.[2]

It's easy to assume that technological innovation is restricted to the design laboratories of Silicon Valley, the tech companies of China, or the elite universities of the world. But according to Peter Bloom, the co-founder of TIC (and its sister global organization, Rhizomática) there are "people sitting in Silicon Valley thinking up problems, and then thinking up solutions to those problems. But they're not grounded in anyone's reality."[3]

Dramatic differences exist between TIC and corporate telecom providers in Mexico. Whereas commercial mobile service competitors such as TelCel and Movistar charge for service at rates its users can't control or negotiate, TIC offers cheap service *owned* by its community of users. Rather than use "access" to technology to extract as much money as possible from an already economically poor population, TIC has aimed to build upon the values of self-determination that run deep in indigenous Mexican cultures, and to blaze a trail toward the democratization of technology. Both TIC and Rhizomática are nonprofit organizations, obtaining their funding from members and occasional external grants and fellowships.

For three successive years, from 2016 to 2018, I conducted interviews with more than thirty individuals in several villages where TIC provides service, as well as with TIC and Rhizomática staff. In this chapter I share stories of this effort from a range of perspectives, examining the possibilities and the challenges.

Is Connectivity Right for All?

Billions in the world have mobile phone access, with some estimates reaching as high as 6 out of the world's 7 billion. For many of these people, internet access is exclusively available by phone.

No wonder the mobile phone is widely seen as the technology with the greatest promise to support economic development in poor and rural communities worldwide. From projects describing how mobile telephony could assist fishermen in India to research describing the positive correlations between mobile phone access and per capita GDP, mobile phones are praised for being inexpensive and less reliant on traditional infrastructure than wired telephony.[4]

But, as the civic media scholar Ethan Zuckerman pointed out in 2010, we cannot presume to understand the mobile phone as a device that remains the same across the world; even when the same technology is transported and implemented in a similar manner. Each community's experience of connectivity is likely very different.[5] We should consider how it is used and by whom in order to understand it in context. Steve Song, an activist who has worked internationally to support community networks, told me: "The primary value [of mobile phone networks] is in people connecting to each other."[6] What a phone *does,* in short, has everything to do with the purposes and agendas it serves, and the values of the communities that define these ends.

For poor and developing-world nations that are struggling to provide basic necessities to their citizens, communication technologies need to support, first and foremost, their economic needs and goals. Infrastructures, which require investments, might not be a community's first priority. As Song noted, "When you're below the poverty line, the value of the internet is not self-evident to you."[7] By this he means that at any price, no matter how low, investing in the internet poses a risk for impoverished nations and communities who generally put almost everything they have into making ends meet.

I've shown in earlier chapters that blind and naïve access to technology is no silver bullet. Providing technology access on the providers' terms rather than the users' has a very predictable pattern of results: it increases rather than combats inequality. For this reason, we need to do away with the pervasive idea that the internet or mobile phones will magically empower people independent of who or where they are. But is it possible to connect to the internet or phone service in ways that are determined by communities for their benefit, or does there always have to be some middleman exploiting the transaction?

For communication technologies to truly serve users of diverse cultures and regions in the process of gaining new access, we can benefit from a nuanced, broad view. We can consider *co-production*, a concept described by the late Nobel Prize–winning economist Elinor Ostrom, as a cooperative relationship between service providers and users. For Ostrom, co-production is "the process through which inputs used to produce a good or service are contributed by individuals who are not 'in' the same organization," meaning that the typical producer-user relationship that we see in product or infrastructure development is blurred. Instead, through co-production, "citizens can play an active role in producing public goods and services of consequence to them."[8]

For many communities throughout the world, phone networks support oral communications, and coordination between family members, co-workers, and buyers and sellers *within the same community, region, even nation.* Local telecommunications initiatives are often in the best position to do this, since they are closer to the people, rather than stuck in traditional economies-of-scale approaches taken by most top-down corporate initiatives.[9] As I've shown through examples from Africa, when success depends on intimate knowledge of the local context, the decision to foster cooperation with those who live in that place is a no-brainer. And within the communities of the developing world, experience is often synonymous with expertise. Experience with local cultures, traditions, and values can be deeply valuable in making a technology project succeed.

The Politics and Cultures of Infrastructure

Communication technologies sometimes seem as if they are everywhere and nowhere all at the once, a feeling aptly illustrated by "the cloud," the metaphor we use to describe the mysterious places where our data live. But in reality, infrastructure is the foundation of any technology. Infrastructures are what actually make technologies useful to people in their everyday lives; the wires, plugs, underwater cables, and towers designed to accomplish a technical goal are as important (or more) than the beautiful apps we use or the elegantly designed devices we covet. But despite their ubiquity, infrastructures tend to be largely invisible—often capturing our attention only when they fail.

The components that make up infrastructures, such as the natural environments where they are built, the tools used to construct them, or the materials from which they are made, are much less "sexy" than the face that technologies present to consumers: a well-designed game, a crystal-clear phone call, or the clever chat of Siri or Alexa will grab more of our attention than thoughts of wire buried in the ground. These basic materials that undergird our technological systems may at first seem boringly neutral. But in reality the design and development of infrastructure is far from a neutral science. Developing a Wifi system, for example, requires crucial decisions that are not neutral at all, such as what electrical grids, wires (such as cable), or other material components are needed to connect it. Well, those decisions are made based on who owns the wires and operates the electrical grids. Is the grid run by organized crime, a friendly business compatriot, a totalitarian government, or owned by a company whose CEO is your ex-husband? Surely these different situations would greatly affect how you choose to design and implement the system. All of these questions and more come into play when infrastructure is built and managed, making it political, cultural, and social.

Community-based initiatives require that we consider infrastructure on a human and environmental scale. Such a mental shift involves moving away from thinking about them as centrally organized, large-scale technical systems. Instead let's think about how infrastructures could be created and managed locally at the grassroots level, rather than accessed from some far-away location where all the money and power lie.

Like the community-run mesh networks I discussed in chapter 17, southern Mexico's TIC presents the potential of decentralizing how people connect to a network. By decentralizing, I mean transferring authority and control from central governments and powerful corporations to local communities, from the top-down to the bottom-up. From ownership, to design, to maintenance, TIC reveals how a community can take power over technology.

A Bird's-Eye View

The communities that use TIC span the Sierra Juárez, Mixe-Alto, Mixteca, and the Sierra Sur regions surrounding the Oaxaca valley. In 2012 the town of Villa Talea de Castro (Sierra Juárez) became the first to join the collective, which today includes fourteen participating towns. Throughout these regions, TIC has enabled the construction of mobile phone towers and a functional, affordable cell phone system while transforming its users into active creators and owners of their own networks.

Because of the rural, sparsely populated habitats and lower income levels of many indigenous peoples in Oaxaca, the TIC-member communities have been left underserved when it comes to internet and mobile phone connectivity. This is not just true in Mexico, but indeed around the world: networks tend to be formed and users tend to be "served" where people have the money.

Given this pattern, services developed to connect the underserved often characterize indigenous and rural communities as the "last mile." Communities like those who are part of TIC are typically an afterthought, if they aren't excluded altogether.[10] When service providers view user communities through the lens of the "last mile," they tend to treat them as needy and willing to sign onto digital network infrastructure projects no matter what, regardless of who develops (and profits from) them.

It doesn't need to be this way. What if, instead of thinking about user communities as customers, we were to elevate and humanize them as creative agents, innovators, owners, entrepreneurs, and designers of their own communication networks and technologies? What if the communities themselves represented the "first mile" in the policy, economic, design, and cultural choices involved in bringing mobile telephony into their lives?

Most TIC-connected communities currently operate and maintain autonomous networks. How does that work? Each community owns GSM (cell phone towers), which are then connected to the internet via partnerships with ISPs; tethering traditional phone service to the internet allows members to make longer-distance calls via Voice over Internet Protocol (VoIP) technology. Individual users pay a maintenance fee for the TIC connection, which then allows them to call one another locally and across regions for a fraction of the normal, commercial price.

TIC has designed technology directly for its user communities. Yet it is important to note that long-distance calls, even within Mexico, are often routed through US-based data servers. Still, TIC users can call relatives in Los Angeles and the United States for pennies on the dollar, a far cheaper rate than the costs for those who live in large Mexican cities.[11]

Although TIC is a small organization with just six paid employees in Oaxaca and two in Mexico City, it vies with corporate powerhouses and supports the potential for communities to provide access to their members in a way that the corporate businesses can't.[12] Being owned and co-developed by community *asambleas* (assemblies), it supports cooperation of an economic and cultural value.

Whereas traditional telecom businesses ask users to pay for access without decentralizing ownership, TIC puts user communities in control. Subscribers to TIC services pay 42 Mexican pesos per month (about US$2), and for the most part those fees re-circulate within the community to pay for ISP access, electricity, and labor.

Communication rights are often framed in universal, public, or national contexts. But when it comes to indigenous communities and their unique languages, traditions, and a history of persecution, TIC shows us how groups that have been the most persecuted can be innovators in the digital age.

Welcome to the Rhizome

Rhizomática, the nonprofit that helped establish TIC, draws its name from the philosophers Gilles Deleuze and Félix Guattari, who used the term "rhizome" to reject the common perspective that knowledge is centrally produced and then passed on to the margins.[13] The rhizome presents knowledge as decentralized, as a network that consists of multiple, laterally connected entry and exit points. The word comes from the plant sciences, where it refers to the underground, horizontal stem of a plant from which upward roots and stems form.

Unlike the trunk, the foundation of the plant in a root-tree system, the rhizome is not organized in a centralized, discrete, or fixed pattern. The rhizome, dynamic rather than fixed, grows outward in multiple directions at once, providing a model for organic thinking in a nonlinear, multiplicitous way.

The rhizome's structure is similar to what scientists have found in the world of mushrooms. Research shows that the rhizome of a plant does not exist in isolation, but is actually connected to other plants and fungi through mycelial networks. Indeed, some scientists argue that the largest living biomass on our planet is not the giant Redwood or Sequoia trees, but the intricate, complex, and massive network of mycorrhiza mycelia. Plant rhizomes and these mycelia are in partnership—they exchange carbohydrates and nutrients with one another, forming a symbiotic relationship that allows plants to remain healthy and develop intricate systems of communication that aid their survival. Paul Stamets, an expert on fungus, has referred to these networks as "Earth's natural Internet."[14]

How might these insights into the natural world influence our thinking about the digital networks human beings have created for our own communities and societies? Can the rhizome help us imagine alternative technologies that balance power, warn our neighbors about hostile threats, respect the sovereignty of diverse communities, and allow us to learn from one another? These are questions that motivate Rhizomática's efforts.

The Communal Union of Hackers and Indigenous Peoples

TIC is all about supporting community. But we should ask what is community, and what would it actually mean for technology to support it? The answer is not obvious but lies in every community's own self-definition, values, and priorities.

Community in southern Mexico (as well as what it would mean to "support it") means something different than it does in San Francisco, Beijing, or Nairobi. It's also different than community on Facebook, given that a tech company's version of community can never fully capture the deeper meaning of the term from our lived experiences. The Oaxacan people connected to TIC are mostly indigenous farmers whose plots of land lie on the terraced sides of steep, high-altitude cloud forests—precious land for which the local population has regularly risked life and limb. Indeed, the Oaxacan vision of community, indigenous rights, and autonomy from which TIC has emerged can be tied to a far more familiar story, that of the Zapatista indigenous rebellion.

In 1994, the Zapatista indigenous movement led and won an uprising to fight for autonomy over rural and forested lands in the neighboring Mexican state of Chiapas. It was a movement that contested the government's push to privatize land and natural resources. They then established regional centers dubbed *caracoles* (snails or conch shells in Spanish), with names like "Resistance toward a New Dawn" and "Mother of the Sea Snails of our Dreams."[15] The snail metaphor, in fact, pervades Zapatista iconography and consciousness. This cultural touchstone inherited from Mayan ancestors poetically captures indigenous ways of being and knowing—slow, circular, reflective, concentric—central to the lifeways and histories in the region. The *caracoles* serve as sites of resistance to (and protection from) the Mexican government, yes, but also as openings to the world, spiraling paths by which inside and outside are negotiated. This way of life is often narrated by the Zapatistas as "another world" in stark contrast to the hyper-stimulated global engines of free market capitalism. For indigenous communities the land on which the *caracoles* sit, settled by Mayans over a thousand years ago, is not inanimate; it cannot be defined in the sterilizing terms of objectified value, as a piece of property or a quantifiable resource. Instead it is an active force, a protagonist in the lives of the people living on it, and utterly central to their economic livelihoods.

Telecom corporations in Latin America, however, have historically seen the people and land in Chiapas or Oaxaca differently, as discrete, quantifiable, even as a commodity to extract. "Too few bodies, not enough money" captures the logic used to justify the lack of capital and infrastructure investment in these communities. Corporate providers see the remote high-altitude locations, characterized by lush rainforest, as too precarious for infrastructure and the people too few in number (and not wealthy enough) to be worthy of investment. Even if a large carrier like Movistar or TelCel offered mobile access to indigenous communities, it would likely charge exorbitant rates several times higher than those for urban counterparts.

To build the TIC cell phone network, teams of activists, hackers, and community members had to fight multiple legal battles, waged at national and international levels, just to gain access to the radio-frequency spectrum (the "airwaves" on which mobile networks run) that are currently dominated by large corporate providers. Building the network has also involved the design and the development of decentralized telecommunications infrastructures, an innovation that depends on community involvement and leadership. TIC leaders have argued that just like water and air, the mobile spectrum should be a public utility to which all should have equal access.

We've seen from stories I've shared about tech innovations like M-PESA and BRCK in Africa, that a little-recognized but powerful mode of innovation called "broken-world thinking" can often flourish from within environments of constraint, inaccessibility, and systematic exclusion. Loreto Bravo, a community radio activist based in Oaxaca, describes the TIC effort as a "a techno-seed that inhabits a communal ecosystem; an ethical-political bridge between the hacker community of the free-software movement and the communities of indigenous peoples in Oaxaca, in the South-East of Mexico."[16] Here she points out that the project brings technology activists concerned with checking the power of telecom companies and surveillance systems together with indigenous peoples interested in greater sovereignty over their lives.

Indigenous philosophers from Oaxaca, including Jaime Martinez Luna and Floriberto Díaz Gómez, have coined the term *comunalidad* to describe the sense that community and interdependence—not the individual, his sense of self, or illusion of freedom—are at the heart of life. In his book *The Wealth of the Commons*, Gustavo Esteva explains that this philosophy is "about displacing the economy from the center of social life, reclaiming a communal way of being, encouraging radical pluralism, and advancing towards real democracy."[17]

Comunalidad is not an abstract political philosophy but is near and dear to the languages and practices of most of Oaxaca's rural indigenous communities. One way it is practiced is through the *asamblea,* gatherings of as many as hundreds of community members who discuss and eventually vote on matters of shared concern. TIC likewise organizes itself through collective assemblies that are attended by each partner community. *Comunalidad* is in all things social and cultural, Esteva explains, in contrast to the atomizing approach taken by telecom corporations that establish contracts between an individual and corporation, one at a time.

While visiting TIC communities, I witnessed the power of *comunalidad* in action, as a means to connect each community member to the greater collective. For some of us so much "togetherness" may sound claustrophobic. But the indigenous lens doesn't see limitation in entanglement, it sees reward: as a member of the community with an unquestionable place at the table and a valuable role to play, an individual becomes someone with whom the rewards of communal life will be shared—not just as a person to whom a "piece of the pie" is owed, but as a beloved neighbor in whose company dessert tastes all the more sweet.

Oaxacan Diversity

What are the challenges and possibilities TIC faces as it attempts to both sustain and grow within Mexico and around the world? Like a rhizome, TIC-type networks can potentially sprout anywhere. To understand how TIC operates in its present state, however, it is important to consider the cultural context in which it has emerged.

The more than sixty different indigenous communities of Mexico represent approximately 15 percent of the national population (which is approximately 131 million).[18] The second article of the Mexican constitution as approved in 2013 recognizes the self-determination of the nation's indigenous communities, as long as such is consistent with "national unity." Clause 6 of this second article contains language to explain that the government must support communication rights and channel administration by constituent communities.[19] Consistent with the UN Declaration on the Rights of Indigenous Peoples, this means that media can be legally created to support languages and cultures of communities within Mexico, a signatory to this declaration. Dating back to the San Andres accords, which were signed in 1996 between the national government and the Zapatista National Liberation Army (EZLN) but never implemented, indigenous

communities within the country have at least been symbolically acknowledged as worthy of some forms of self-governance, including ownership of
communication channels and networks.

In the state of Oaxaca, which has a population of about 4 million, more
than half of its residents live in villages of less than 2,500 people. The state
also has at least sixteen distinct indigenous populations, though some estimates conclude that number to be far higher, given smaller cultural and
linguistic differences between communities that come, for example, from
Zapotec ancestry. Oaxaca represents the greatest linguistic, cultural, and,
interestingly, biological diversity in all of Mexico. This observation about
the state in which TIC "lives" is consistent with the perspectives of the cultural geographer David Turnbull, who has noted that biodiversity (of flora
and fauna) is always tied to human cultural diversity.[20]

Santa Maria Yaviche

Driving out of the Oaxaca valley, accompanied by a Mexican student who
kindly assisted me with my field research, I was excited to reach the town
of Guelatao, known with great local pride as the birthplace of Benito Juarez,
the first and only indigenous president in Mexican history. Yet as we continued on into the next town in the mountains, Ixtlan—the final place
where TelCel service could be accessed—nearby villagers warned us that our
destination, the Zapotec community of Santa Maria Yaviche, was another
three hours away, reachable only by dirt roads with hairpin turns leading
deep into a cloud forest.

Left with no choice, we moved forward. The biodiversity of the Oaxacan
Sierra Juárez range developed like a Polaroid, transforming from a parched
tangle into a landscape of evergreens. But none of this came close to what
lay farther ahead in the depths of the forest en route to Santa Maria Yaviche. As
we reached an altitude of over 7,000 feet, it seemed as if we had gone through
a magic portal: an impossible-seeming level of forest density, replete with
lush waterfalls, massive high-altitude trees, and washed-out roads unfolded
itself around us. Finally, just as our bumpy ride began to reveal small indigenous villages littered across the mountainside, we passed a sign:

In this community private property does not exist.
The buying and selling of communal lands is PROHIBITED.
Signed the Comisariat of Common Resources of Ixtlan de Juárez[21]

This was my second visit to Yaviche. During my first, in 2016, I interviewed Oswaldo, the local project coordinator, who was responsible for administering and localizing the TIC system within his community of approximately six hundred people.

As we met again, Oswaldo told me with pride that autonomy means everything to his community.[22] Autonomy is not simply about land, he explained, it is also about culture, language, and tradition. The fight for autonomy is also a fight to keep alive his indigenous Zapotec language, values, politics, and ways of knowing. He and his fellow villagers had named the TIC system Red Xhidza—"red" is the Spanish word for "network" and "Xhidza" the local name that his Zapotec community uses to refer to themselves—in the hope that TIC would indeed foster relationships and connections within Yaviche.

In a documentary about TIC that I viewed before my first visit, Oswaldo explained that the network represented a means of claiming territory. Frequencies in the air could be occupied, just like land, reinforcing the community's claim to a mobile phone network of its own. The TIC system offered a similar sense of empowerment since the communities could speak Zapotec to each other via their phones. Supporting these languages is especially valuable because they were rarely preserved in written form during a centuries-long period dating back the Spanish colonization of Mexico. Oswaldo stated:

> We are interested in the service here, with little money, and in being left alone. If they don't come to bother us we can live peacefully. So everything we are doing from an intellectual point of view challenges the system from above, one that is against us.[23]

In every visit I made to Yaviche I noted that many in the community considered the network to be tied to Oswaldo's family. So, while TIC was being used more widely, it was evident that the project's governance and administration lay in the hands of just a few people. Although that did not raise any complaints from those I spoke to about TIC, it does show that often a community technology is developed, organized, and administered by just a few individuals or families. Nonetheless, Oswaldo's focus clearly was less on himself than on supporting the goals, needs, and values of the community.

Before TIC, Yaviche had been able to operate and maintain a community radio station. As an inexpensive, oral, and easy to repair technology, radio has had great success as a vehicle of community expression across Oaxaca. Oswaldo's vision was for TIC to build on and extend the power of radio. It was also part of the larger struggle to achieve linguistic, cultural, political, and educational autonomy for the community. Schools, cultural institutions, technologies—for Oswaldo all of these represented opportunities for decentralization and indigenous autonomy.

Yet in 2017, as we interviewed dozens of community members on my second trip to Yaviche, a different story revealed itself. There had been a significant drop-off in TIC system users. Community members told us about numerous technical problems with long-distance calls, raising the issue that TIC may always have: being forced to depend on ISPs to allow calls from the local network to daisy-chain to the internet. We were also told that many users wished to switch to the corporate provider Movistar, which had just recently begun to provide service to Yaviche, though only indirectly through a nearby antenna located a few kilometers away in the hills behind the community. Many in Yaviche expressed concerns about the costs of administering TIC, and whether it would become the work of a single family. Though the entire community had decided on whether the project would be implemented, the question of how and if the system would be used would depend on each individual's choice to pay the subscription fee (or not). So I was learning that a community network in name might not be the same in practice.

Yaviche first signed onto TIC in September 2013. What I saw in 2017 was a system in transition. Threatened by influences like Movistar's entry into the area, TIC faced the intensive process of negotiation and adaptation that affects the growth and sustainability of any technology's application. Its ability to support the goals of the indigenous community was still in question, rather than fully established.

Santiago Nuyoo

The mountains of the Tlaxiaco district and the Mixteca region, which cover parts of Oaxaca and the neighboring Guerrero and Puebla provinces, lie to the west of the northern Sierra Juárez, where the Zapotec live. The TIC organizers suggested I visit Nuyoo, an indigenous Mixtec community of approximately 1,500 people located about five or six hours outside the city.

Nuyoo was a recent adopter of the TIC system (Talea, the first community to join, had done so in February 2013).

Like the Zapotecs, Mixtec communities are notable for their diversity. They speak a range of languages and dialects, have ancestors in the region dating back several thousand years, and have migrated across the nation and world (particularly within the United States).

As my research assistant and I entered Nuyoo, covered in rock formations and cacti unlike the dense forest of Yaviche, we noticed an excessive number of antennas on top of the central administration buildings set alongside the main square and church. This is a common sight in the Oaxaca region, a palimpsest of infrastructure that speaks to the attempts and failures of "developmental technology" implementation (see figure 22.1). We heard about a failed technology project, called Telecomm Nuyoo, administered by the federal government some years ago. Despite being lauded as a major breakthrough, this top-down "last mile" effort failed for the same reason so many others fall short: lack of buy-in from the community.

The community members with whom we spoke explained that Telecomm Nuyoo had been unreliable and paternalistic. The network had failed to function effectively in the rugged, cloudy, and wet terrain of the region. But even worse, a lack of communication by the governmental ministry responsible for the project had guaranteed a bad first impression in the eyes of the community. Unlike TIC, which is constantly in contact with different user communities, we were told that the agency implementing Telecom Nuyoo had shown no interest in working with the community to develop a sustainable solution to their concerns.

In contrast, when we visited the TIC administrative office in the center of Nuyoo, users spoke with great pride about the system; several described it as an extension of their own sovereign cultural voices and values. Indeed, murals depicting characters from Mixtec myths appeared throughout the building that housed the technology.

Despite being only in its second year, Nuyoo stood apart from the other two TIC communities, Yaviche and Talea. Absent was the corporate competition of Movistar, and present was a system that had few technical problems and a great deal of popular support.

Figure 22.1
Antennas and dead infrastructures of Nuyoo.

Talea de Castro

The first TIC system was built in Villa Talea de Castro, a town nestled among cloud forests amid numerous Zapotec and Mixtec indigenous community settlements (figure 22.2). With approximately 3,000 inhabitants, Talea is distinct from the other two villages I visited—and not only because of its size. The town is considered to be less indigenous and more "cosmopolitan" due to its reputation for attracting migrants.

The Talea environment posed different challenges than Yaviche and Nuyoo. At one point Talea users numbered close to eight hundred, the

Figure 22.2
A rooftop view from Talea, the first TIC member community.

maximum the TIC system could support. Yet as we arrived and began to conduct interviews, noting Movistar signs everywhere we looked, we were told that the number of users had now dropped to less than thirty, threatening the survival of the project in the internationally known location where it had first taken hold. The majority of TIC users had migrated to the external, corporate-managed system, we were told, after the community decided to pay 300,000 pesos for Movistar antenna infrastructure, approximately three times the cost for the TIC antennas. Why would they do this? Movistar marketers had falsely told them that the TIC system was not only illegal but also was a far poorer service with less reliability. The allure of Movistar, for many of Talea's community members, was *brand image*: access to the same "high quality" service that Movistar ostensibly provides to city dwellers in Oaxaca or Mexico City.

Is the near-death of TIC in Talea due to corporate influence and branding? Or, if we buy into the notion of Talea as a more "cosmopolitan" community, does it mean that members now see TIC as a technology of the past, and therefore less appealing? How is the presence, or vitality, of the *asamblea* a factor in ensuring that the project evolves to support community participation, autonomy, *comunalidad*, or sovereignty?

Sovereignty, Autonomy, and Ownership

My community is my root, my whole self. It is to take care of what I have so I can pass it to future generations. It is a feeling. Community and autonomy is to value who I am.

In our visits to the three indigenous villages, advocates of TIC consistently echoed sentiments like these as they described the meaning and importance of community. How, then, did the community express the meaning of TIC itself? The words most frequently used to describe the network's importance were "sovereignty," "autonomy," and "ownership."

"Autonomy," the most common word I heard, has a deep history in Latin American (and Oaxacan) political philosophy. *Autonomia*, the Spanish term for autonomy, describes the attempt by communities across the region to maintain their cultural, political, and economic vision despite a long history of colonization. As I learned from various community leaders in the Mixtec village Nuyoo and the Zapotec village Yaviche, *autonomia* does not describe a final state that can be achieved by any community, but instead a process that guides the practices, choices, and, at times, struggles of its members.

"Sovereignty" is a term I heard far less often, but it seems to best capture the political and cultural insights at the heart of my interviews. A term derived primarily from political science, sovereignty most commonly refers to the ability for a nation's governing body to have full authority over itself, without outside influence.[24] TIC represents a struggle for community sovereignty to overcome mere inclusion in someone else's system. If traditionally "last mile" user communities can become "first mile" designers, owners, and adopters of new technology—in their regions and languages, and through their community's deliberation processes—they will have sovereignty. Ownership and decision-making around the network through the *asambleas* is an important start.

There was a very real tension in each of the three communities between the desire for local ownership of the system on the one hand, and the desire to use technology that is global and modern (cosmopolitan) on the other.

TIC does not necessarily have to be at odds with modern technology—*if* modern technology is a vehicle by which a community wants to pursue its autonomy. Many people in the communities TIC wants to serve viewed the project that way; twelve out of the fifteen Talea members we interviewed, and several users we met in Yaviche who had switched to Movistar, perceived TIC (and criticized it) as being "anti-cosmopolitan."

Nonetheless, indigenous communities using TIC view autonomy, sovereignty, and ownership as values to pursue, not just by establishing the network itself but also by exploring how it could be used and adopted. For instance, in Yaviche, Oswaldo has established a local university called Xhidza, which teaches digital literacy and design from the perspectives, values, and cultures of Zapotec peoples. There are also attempts to take power over the technology infrastructures through solar, wind, and other forms of renewable energy. Thus autonomy, in the eyes of five TIC users we spoke to in Yaviche, does not just start and end with TIC. Their aspiration for autonomy extends to other its impact on other parts of their lives that they believe technology could help improve.

One way in which the network could better support autonomy would be to make the system accessible, open, and even modifiable by its user communities. As a project administrator in Yaviche described to us: "It is necessary to form technical teams who are knowledgeable about how the equipment and infrastructure works. And to offer those capacities to our youth or community members. The other thing is to bring together all the different phone technologies—to establish an intranet, have digital libraries, and for all to be in one single machine, so we can access everything in a more classified way according to our communities' interests."

Access to the internet or a mobile phone network is not magic bullet, for the question will always remain: Access to what? By connecting TIC to a number of valuable initiatives, such as a community-based intranet (for digital communications, texting, and even data-sharing with the community), a digital library through which to share cultural and educational resources, and the training of community members, the community can tap into its greatest resource: its people, their knowledge and traditions, and their ability to pursue a future that supports their collective dreams.

Autonomy or Cosmopolitanism?

Let's take a closer look at corporate competition, the major challenge the network faces. The influx of Movistar, TelCel, and other private telephony providers threatens TIC's sustainability not only because they offer different services at different price points, but also because global technology corporations offer a seductive product to these indigenous communities: the sense of cosmopolitanism that makes these rural users "feel" in step with the rest of the world. As TIC competes with Movistar in communities throughout Mexico, and around the world, a clash of values is in play: autonomy versus cosmopolitanism; tradition versus modernity; corporations versus the collective *asamblea*.

Price, quality of service, and the brand—all these factors influence what network villagers in Yaviche and Talea will use. To understand these choices, let's see how effective the brand advertising has been in each community. With Talea, as I mentioned earlier, the presence of Movistar has overwhelmed TIC, to the point where there are now about thirty users in a network that once had eight hundred. One convenience store owner told us, "Movistar came and washed our brains like coconuts. They offered us promotions and good deals, and at the end we ended up paying the same or even more."

Based on that assessment alone, TIC users who switched to the corporate Movistar network felt both demeaned and duped. This shift occurred mostly because of advertising and the competitive introductory offers Movistar used to gain entry into the Talea market. But the users paid a price.[25] Movistar forced them to join far more expensive plans, use phones that were incompatible with the TIC network (a strategy known as "network locking"), and settle for less service and talk-time for the same amount of money. The invasion strategy seems to work like this: enter the community at similar price points to the TIC network, initiate various forms of "guerilla" street advertising, and then shift subscriptions away from TIC to the point where it may no longer be financially viable. In each community, however, more than two-thirds of those we interviewed struggled to articulate exactly what it was that TIC lacked relative to the Movistar network. What, then, might account for the decision to switch away from TIC in the first place? Does the power of corporate branding have greater appeal than the idea of an indigenous-owned network?

Half of the Talea community members with whom we spoke, and every Yaviche Movistar user we met, described how they perceived Movistar service as more reliable due to its connection to a larger company that was functioning in other parts of the country and world. Perhaps, then, Movistar's appeal can be chalked up to its brand identity, not just among younger community members, but also among villagers who see themselves as cosmopolitan migrants. As an internet cafe–owner explained to us, it was not surprising that Movistar would be more popular within Talea, a village known as a destination for immigrants. But, she pointed out, those within the community more connected to traditional culture continued to be a part of TIC.

In an example of grassroots ingenuity, we learned that users have found creative ways to negotiate between TIC and Movistar services by using both—switching back and forth based on their varying user credit, quality of service, and other factors. For instance, in Yaviche some users of the community network have a Movistar phone, which they use to receive calls because the dialing codes for the inexpensive TIC network are so complicated. People want technology to be easy to access and use. TIC will have to find ways to leverage its symbolic value of autonomy for the indigenous and rural communities it serves, and overcome its reputation for being backwardly traditional and at a technical disadvantage to corporate-owned networks. The TIC leadership team will need to clarify how owning the network actually benefits them as a whole.

The Power of Our Visions and Voices

If the goal of TIC is to support community autonomy and sovereignty, it must be governed and administered locally, respecting the cultures and decision-making processes in place within local communities. As Oswaldo put it, "As long as we are not part of the technology, of that knowledge, we continue to be exploited by technology."[26]

By highlighting the ways that diverse sets of communities adapt, utilize, and govern technologies in the image of their histories, values, and needs, we can glimpse what might be possible when we approach the spread of technology to the global South as beyond mere "inclusion." In this way, technologies may reflect the many languages, worldviews, and lifestyles that exist throughout the world. As Peter Bloom points out:

> I have the feeling that technology wants to homogenize the world in some way.... And what we have to do is not let that happen. We must...appropriate technology and adapt it to us.[27]

This diversity-first ethic illustrates the experiment that TIC and Rhizomática are carrying out. Designing, owning, managing, localizing, and taking power over mobile technology are complicated tasks, but dozens of indigenous communities, in tandem with their hacker and activist partners, have already taken the leap. In the process they embark upon a journey of exploration outside of the producer-consumer relationship, which sets up a false either/or sense of opposition. These stories from southern Mexico show the power of *community agency*, the capacity to innovate and create technologies in line with the aspirations and values of people, not for-profit corporations.

As we continue to think about the future of digital technology, it is important that we take time to close our eyes and *listen*. Let's attune our ears to the *caracoles* like the Mayans who, as Zapatista leader Marcos writes, used the spiraling conch shell both as "a way to summon the community" and "as an aid to hear the most distant worlds."[28] Only by being awake to the lessons of other worlds can we ensure that efforts like TIC can overcome obstructions, continue to stir the desires of user communities, and produce new possibilities that support users and their values.

The TIC model in southern Mexico represents an opportunity for all of us to say, *yes, we as communities and users, regardless of where and who we are, can design technology systems in our image*. In chapter 17, I mentioned other examples: from the United States (Brooklyn's Red Hook neighborhood and inner-city Detroit), from Catalonia in Europe, and from other parts of the world where communities are building and managing their own collectively owned internet and intranet networks. We also see great interest in Mastodon, an open source software project that allows for any community to maintain its own federated communications via its own server. In that spirit of revolutionary technology, I next turn to a look at blockchain technology and cryptocurrencies, with the goal of unpacking the hype versus the potential.

V   Looking Toward Tomorrow: Our Path Is Not Locked In

23 Blockchain: A Crazy Free-for-All, and Maybe More?

with Adam Reese

> We are the Birth of a New Virtual Nation
> We are a Future for Our World and Humanity
> We are Sentinels, Universal and Inalienable
> We are Creativity and Visionary
> We are Rights and Freedoms
> We are Tolerant and Accepting
> We are Polity and Entity
> We are Privacy and Security
> We are Openness and Transparency
> We are a Dream and a Reality
> We are Bitnation

So ends the white paper describing Pangea, an app created by a blockchain startup called Bitnation.[1] These eleven declarations convey ideas that are fundamental to Bitnation's vision for a brighter tomorrow. Although some of the concepts expressed in the white paper may sound unusual to a mainstream audience, many of them are quite common in the "blockchain space" (we use this term throughout the chapter to refer to the overall environment and set of communities associated with the blockchain and cryptocurrency world).

Many digital systems today are structured so that some actors have the power to alter them in drastic ways. Blockchain technology, on the other hand, drives a type of network that is designed not to operate under the control of any single party. In this chapter, which I co-authored with Adam Reese, a Los Angeles–based writer who covers the blockchain space, we explain the basic concepts associated with this new tool and discuss how it relates to the central theme of this book: grounding digital technologies in the image of users' values.

But before we delve into the technology itself, and into the digital currencies it supports, let's examine what Bitnation proposes for "our world and humanity."

Virtual Nations

The authors of the Bitnation white paper describe a future in which people establish their own virtual "nations" on a blockchain platform created by the company. Different nations offer different services, and a user who desires what a particular nation has to offer can choose to become its "citizen." The authors suggest that such digital communities could eventually replace nation-states if they first wean citizens from nation-state dependence by providing "better, more secure, faster, cheaper...peer-to-peer alternatives for [state] services." When functioning virtual nations emerge in Bitnation's competitive, free-market environment, the authors claim, nation-states will by default "become increasingly irrelevant to our everyday lives." Their waning influence will liberate humankind from "the xenophobia and violence that is nurtured by the Nation State" because when "every one of us is a potential customer," there will be little need for violence.[2]

The Bitnation white paper doesn't include many examples of what virtual nations might actually do. But it does mention their potential to provide physical security via a complementary app, which alerts people when a friend is in trouble, for instance, and directs them to that friend's location. Such an app could be useful for operating a peaceful neighborhood watch program. Or it could just as well become an effective tool for assembling a vigilante mob or managing a mercenary police force. The paper also explains that virtual nations could help users manage land deeds, and even tie these ownership records to other important documents like marriage certificates. That way, a deceased person's spouse can automatically receive land ownership rights on the blockchain. Bitnation imagines that this feature will help gay and lesbian couples to assert property inheritance and other rights in jurisdictions like Uganda, Iran, and Chechnya, where "being homosexual results in prosecution by the government." But it's unclear how these blockchain records would hold up in any country that has its own land registry, never mind in a country that refuses to recognize a surviving same-sex partner's claim.

On these points and others, Bitnation has failed to consider an important reality: nation-states do not exist simply out of blind faith, nor will they simply disappear once we put our trust in some other kind of institution. It will take more than a "killer app" to dismantle the system by which the world's people are governed. The Bitnation proposal appears to have another major hole: Once nation-states have been scrapped, who will take on the task of providing essential services—like clean water, for example—to poor and far-flung populations when doing so is a money-losing venture? How could such communities entice service providers to come to them? Would another entity be willing to pay for their services? Would people have to move to where the services are?

We're not criticizing Bitnation in order to lampoon them, but rather because we believe they've encountered several pitfalls that are quite common in the blockchain space. First off, their interest in blockchain technology is closely tied to preexisting ideological convictions, in this case, libertarian notions that governments take more than they give. Ideology itself is not inherently a bad thing, but when it drives the development of a technology, the result may look good from some angles and half-baked from others. Bitnation's shortcomings in this regard are part of a wider trend in which people project their fantasies onto blockchain technology. We also take issue with the assumption that any technology *in and of itself* can be a path to desirable social, economic, or political outcomes: technologies matter, but outcomes depend on what people do with them.

Blockchain's Promise

Let's look at why people have become so excited about this technology. In short, the blockchain can serve as the basis for digital networks that don't have a single centerpoint. That is, each blockchain network comprises many *nodes*—in the blockchain space this term refers to *clients* (i.e., programs) on individual computers—and no single node is indispensable to the functioning of the network. Any one of these nodes could go offline and the network would be fine. Theoretically, a network that depends on no particular server is significantly harder for a single party to control than a network that depends on specific servers. Therefore, security is often counted among the technology's core promises, as is *decentralization* (roughly speaking, the ability to function in the absence of a central hierarchy).

Most of these networks are organized around a type of digital money commonly known as a *cryptocurrency*. (The Bitcoin network's cryptocurrency is called bitcoin; Ethereum's is called ether.) Blockchain networks are supposed to be un-gameable, so that people can use them to transact with parties they don't trust while having complete confidence that their money will be protected.[3] In other words, they trust the network to keep their money safe, even in dealings with sketchy people. There is no central authority to appeal to, and the need to make such an appeal is never supposed to arise.

Some blockchain enthusiasts say that once this technology reaches a certain stage of maturity, an explosion of innovation will follow. Others claim that if we can create a digital network that is genuinely decentralized, the concept of decentralization will inevitably permeate essential institutions, such as our political and financial systems, and radically improve them in the process.[4] One effect of this transformation, it's often said, will be to rid us of middlemen who find ways to take their cuts without creating any value.[5]

Politics Are Not Checked at the Door

The blockchain space reflects a range of political viewpoints. Toward the left of the spectrum are theories about how the technology could help support more equal economic outcomes.[6] Near what is normally described as the political center are projects that mix social liberalism with economic conservatism, such as the Plastic Bank. In low-income areas, this organization opens recycling centers where rubbish collectors receive cryptocurrency in exchange for plastic.[7]

Libertarianism has dominated political philosophy in the space from its early days and remains a strong presence today. It makes sense that people who distrust governments in general—and their currency-issuing apparati in particular—would take an interest in money that's not state-controlled. Indeed, many blockchain enthusiasts express glee at the idea of building something that governments can't regulate or destroy.[8]

In *The Politics of Bitcoin*, David Golumbia traces the ideological connections between free-market-loving libertarians in the blockchain space and "cyberlibertarianism," a school of thought described by the sociologist Langdon Winner in 1997. Cyberlibertarianism places an "emphasis upon radical individualism, enthusiasm for free-market economy, disdain for the role of government, and enthusiasm for the power of business firms" (all basic libertarian tenets), and couples them with the belief that "the dynamism of digital technology is our true destiny." As Winner saw it then, "Those able to rise to the challenge [of the digital age] are the champions of the coming millennium."[9] In the space today, people voice (self-congratulatory) praise for assembling a technology that will birth a shining new future, a future that few are farsighted enough to anticipate or bold enough to chase. For some thought leaders, this faith in individual potential comes coupled with a wariness of social safety net programs, especially welfare initiatives.[10]

We agree as well with Golumbia's assessment that although some people "who 'believe' in [cryptocurrency] think they do not subscribe to [extreme] theories, it frequently turns out that they rely on assumptions and concepts that do emerge from the far right."[11] Ardent advocates of right-wing philosophies are numerous in the space. Regular visits to "Crypto Twitter," the nickname for the social media platform's blockchain-focused corner, reveal memes and statements that express disdain for anyone identified as leftist, communist, socialist, or a "social justice warrior."[12] On the charge of extremism, we'll go a step further than Golumbia, who focused on the intellectual histories of libertarian ideas, and say that it's not unusual to catch white supremacist content buzzing around Crypto Twitter. Adam routinely encounters posts there, usually in the form of retweets, containing lines of reasoning that non-crypto social media accounts use to push claims of "white genocide," "demographic replacement," and other extreme-identitarian talking points.[13] This messaging only resonates with some participants in the space, however. Our overall impression is that while many see themselves as being on the bleeding edge of technology, few identify as ideological extremists.

If this all sounds abstract, perhaps a pop culture reference will help reveal the many strands of right-wing thought in the space.[14] The 3D-printed firearm inventor Cody Wilson was involved in cryptocurrency-related activities such as co-founding an anonymity-focused project and trying to dismantle the Bitcoin Foundation. Before he was arrested on charges of sexually assaulting a minor, many blockchain enthusiasts admired him for his apparent dedication to his convictions.[15] His defiant disregard for authority and his libertarian credentials—exemplified when the US Department of

Justice settled the lawsuit he filed for the right to distribute gun blueprints online—made Wilson an obvious choice as an early icon for the space.[16]

For the Love of Tech

The rush to adopt blockchain technology is evident around the world. Off the record a Chinese tech executive told Ramesh that her government is interested in seeing domestic firms become leaders in the space. Some of the country's largest tech companies, including Alibaba, Baidu, Tencent, and Huawei, have launched blockchain projects.[17]

In the West, experts have advised that blockchains make clunky databases and should only be used when their trust-enhancing features are needed. Nevertheless, many major companies have eagerly jumped into the field to launch projects that sometimes appear to miss the point of the technology.[18] Walmart is testing a blockchain-based system for tracking spinach from farm to shelf so that outbreaks of foodborne illness can be traced almost immediately. Clean spinach from disease-free farms, which currently gets thrown away as a safety measure, could then remain on the shelves.[19] Projects like this one have led to accusations that the blockchain is a solution in search of a problem: although the idea of improved tracking is a good one, blockchains are not the best tools for recording and storing all that data.[20] Moreover, by running the system privately, Walmart misses out on the trust that blockchain makes possible, which is arguably its most revolutionary feature. Critics describe such corporate experiments as PR stunts that aim to cash in on a trendy technology.[21]

While the Walmarts of the world are busying themselves trying to figure out how to use the blockchain as it stands today, developers in dozens of countries are working with startups and as free agents, pushing the technology forward with the hope that it might eventually be worthy of the hype. Many of these developers, for example those who want to use blockchain to create positive social impact, earnestly feel that their labor can contribute something valuable to humanity.[22] For instance, Taylor Gerring, a member of the original core Ethereum team, told me that the technology represents no less than a "social f—ing revolution," because he believes it can overcome the problem of online surveillance that is rampant today.

Many developers in the blockchain space believe strongly in keeping their projects "open source," meaning that all the code is a matter of public record. By committing to an open source standard, they allow others to build upon previous efforts.

For the Love of Money

The narrative of "blockchain as solution," however, is not equally exciting to everyone. Plenty of people are drawn in by the fantasy that cryptocurrencies are a magic moneymaking machine. Others aim to profit from the people chasing that fantasy (scams abound in the blockchain space), and some just relish good old-fashioned gambling.[23]

The "gamblers" include people who drifted in from Wall Street and the finance sectors of Europe and Asia to offer and trade cryptocurrency-linked derivatives.[24] These financial products have included leveraged CFDs (*contracts for difference*) with leverage as high as twenty times the stake; these are considered risky in normal financial markets where heads turn when a stock price goes up or down by a few percentage points in a day.[25] In the crypto space, where single-day, double-digit price swings are a fact of life, these derivatives can expose investors to stunning levels of risk.

In the second half of 2017, surging cryptocurrency values pushed the technology into the mainstream media spotlight, and a massive wave of investors poured in. At first it appeared that these newcomers would dilute the presence of long-standing groups in the space, such as cyberlibertarians. However, as we write in the midst of a long slump that has followed the highs of that winter, lower prices and the slightly less fervent tone of online chatter suggest that many of these newcomers have sold their holdings and lost interest in the technology.[26]

The influx of cash also coincided with an explosion in the number of initial coin offerings (ICOs), the fundraising events in which blockchain companies sell cryptocurrency to investors.[27] Well over a thousand companies have raked in many billions of dollars in total through this process, with some raising millions of dollars in minutes. ICOs have been described using the analogy of a company selling off stock. For reasons that are beyond the scope of this chapter, this is an apt comparison for many but not all of them.

When used correctly, this fundraising mechanism can empower teams to finance important work without relinquishing control over their projects to investors. At their worst, however, ICOs are little more than shameless cash grabs. Teams have skipped the step of starting to build anything before raising money, and have sometimes courted investments using nothing more than a vague white paper. "Celebrity ICOs" have raised eyebrows: The boxer Floyd Mayweather and the music producer DJ Khaled settled with the Securities and Exchange Commission (SEC) over "failing to disclose payments they received" in exchange for their roles promoting an ICO.[28] Mayweather's rival Manny Pacquiao apparently has one in the works as well.[29] The market has also seen ICOs for coins that made no sense at all, such as Jesus Coin, an ICO ostensibly held by Jesus himself that offers investors "cryptocurrency, healing, [and] forgiveness."[30]

Other cryptocurrency schemes appear remarkably cynical as well. For instance, a number of thought leaders who hold large amounts of a certain cryptocurrency try to control the public narrative around that coin as a way to boost its value.[31] Companies that have nothing to do with the space have pivoted and rebranded as blockchain firms. The worst offender may be the Long Island Iced Tea Corp., which watched its stock price jump 500 percent after it announced plans to exit the beverage game and changed its name to Long Blockchain Corp. in December 2017. The SEC eventually investigated the firm because of this move.[32]

And in keeping with mainstream tech culture, which increasingly pervades the blockchain space, several startups have branded themselves as humanitarian ventures and used images of African children in their marketing materials and public announcements.[33] Other blockchain players have found different ways to combine the fantasy of solving poor people's problems through technology with a focus on making money.

For example, after Hurricane Maria killed more than three thousand people in Puerto Rico and left the island devastated, a slew of blockchain entrepreneurs and investors began arriving there. They came with the idea of establishing a crypto-business hub, which they said would help the rebuilding effort. The journalist Naomi Klein accused these newcomers of choosing the island because they want to evade taxes, not help people.[34] (A cryptocurrency team that visited Puerto Rico in March 2018 wrote a blog post explaining that two laws on the books, Act 20 and Act 22, establish "no federal taxes in Puerto Rico" and "no taxes on cryptocurrencies profit made in Puerto Rico," respectively.[35] The memo also states that in its zeal to court blockchain investors, the government has "formed an advisory council focused on growing the local blockchain ecosystem.")

Klein criticized the child actor turned crypto entrepreneur Brock Pierce, who has been visiting the protectorate on business since 2013, for acting as a ringleader of the migration, but Pierce defends the techies' intentions.[36] For instance, he told a Puerto Rican YouTuber: "It's in these moments where we experience our greatest loss that we have our biggest opportunity to…upgrade.…And so Puerto Rico can become anything it wants.…We're here to help Puerto Rico, the people and the place, restart as well as it can."[37] This platitude-filled overture stands in contrast to another video, released in August 2018 by the *Guardian*, which was heavily edited and has since been taken down. It showed Pierce and a cohort of mostly white, mostly male, seemingly non-Spanish-speaking crypto people holding a town hall–style event in Puerto Rico.[38] During the meeting Pierce apparently got into a shouting match with a Puerto Rican woman and lectured other local attendees.

Pierce's eccentricity as well as the presumptuous position he appeared to represent helped the story garner media attention. But it's not the only tale out of the crypto space with whiffs of neocolonialism. About the same time that Klein voiced her criticisms, events were taking place on the other side of the world that appeared to contradict the cyberlibertarian mantra that blockchain technology is inherently a force of freedom.

The Chinese Canadian billionaire Changpeng Zhao made his fortune as the founder of Binance, an online cryptocurrency exchange that became one of the world's largest a mere six months after it opened for business.[39] In April 2018, he met with two dictators: Togo's Faure Gnassingbé, who has held power since 2005 (or 1967, if we count from the start of his father's presidency), and Yoweri Museveni, Uganda's president since 1986.[40] A full-blown movement is afoot in Togo to end the Gnassingbé family's grip on the country's political system. On a 2018 trip to Uganda, Adam and I met people who openly expressed their disapproval over Museveni's refusal to leave power and many others who hinted at such disapproval more subtly.

Zhao told Museveni that he hopes to set up Africa's largest cryptocurrency exchange in Uganda. He indicated on social media that a proposed partnership between Binance and the Gnassingbé government could bring "billions" of investment dollars to Togo.[41] These autocratic regimes would almost certainly receive a substantial share of any money brought in through such deals, and the funds would help these presidents hold onto power. In other words, the projects would represent a setback to Togolese and Ugandan efforts to win free and fair elections. Zhao's apparent openness to these partnerships is a clear demonstration that blockchain technology and cryptocurrency in and of themselves are no silver bullet for freedom.

And finally, on the subject of government interest, Adam spoke with a Ghanaian official who expressed great enthusiasm over the idea of a state-issued cryptocurrency. His vision is to use the currency to capture tax revenue from the informal economy—the gray zone of the economy that is sparsely monitored—whose participants include Ghana's poorest people.[42]

Taken together, the examples we've mentioned indicate that the blockchain has emerged as a kind of a fetish idol onto which people have projected their dreams, whether those dreams involve helping refugees,[43] building an Ayn Randian utopia where the pursuit of self-interest can be considered truly virtuous, or simply getting rich.[44] Next we discuss our take on the technology, setting aside the goals that other people have brought to the table.

Does Any of This Actually Matter?

If you ask us, it does. Not necessarily because of anything that has been built so far, but because today's blockchain networks may contain the seed for something that might benefit us all one day. If these systems become reliable and robust enough to support billions of users, they could form the basis of a more "trustworthy" online experience. Our online worlds today are so full of predatory forces that it can be hard to imagine what this shift could look like. What might we do with an internet that allows us to really trust the systems we use, and even trust our fellow users to a certain extent? Among other things, people would be able to enjoy many of their favorite features from Google, Facebook, and Amazon without any company capturing and storing their data, or taking a cut of a transaction between them and another private party.

Equally important, without a central authority mediating our digital experiences, we users could have more of a say about how platforms operate. In theory, we could have whatever kind of platform we like, as long as enough of us organize and get it built. So far, though, this is nothing more than a theory. Decentralized *networks* have been built, but it's a matter of opinion whether any viable models for decentralized *organizing* have emerged. Likewise, the common claim that decentralization will squeeze out people who make their money without creating value, although intriguing, is unproven. Like any tool, the blockchain must be coupled with specific strategies that address specific problems in all their complexity before it can truly become an effective instrument of positive social, political, or economic change.

Moreover, we're skeptical about this narrative because it feels similar to one that appears throughout this book. We were once told that the internet would liberate us into a more egalitarian and decentralized future. Instead, today's online world is full of power players, like internet service providers (ISPs) and major tech firms, who make important decisions about our digital experiences without consulting us. Once a space of imagination and freedom, the internet has also become a space of surveillance and data capture. It's also been widely argued that instead of expanding our social worlds, social media thrusts us into cultural echo chambers where we're insulated from opinions we don't share.

All the same, a new digital era may be headed our way—one that is based on the transfer of value in the same way that today's internet is defined by the transfer of data. Although Adam and I believe that a blockchain future could be just as frustrating and dangerous to users as the current age, we don't think it has to be. With the right guidance, these technologies could lead to some positive outcomes, which we touch on below. We can all make ourselves a part of this process right now by trying to understand how this technology relates to topics we know about and by looking critically at the solutions that are being proposed.

The Blockchain Itself

Before getting into specifics about what people might do with the blockchain, we want to introduce a few essential terms and concepts. Blockchain technology is a subset of what's called *distributed ledger technology* (DLT).[45] The blockchain itself is literally a ledger—a record of cryptocurrency transactions (and frequently other data as well) that is distributed among many computers around the world via the nodes we mentioned earlier." This list stretches all the way back to the beginning of the network.

When new transactions are added to the ledger, nodes receive word and update their copies accordingly. This way, rather than a central actor like a bank keeping the official tally of everyone's balances, thousands of nodes host that information. For an attacker to successfully tamper with it, they'd have to attack a huge number of nodes (or bribe the *node operators*) all at once. This makes it very difficult for any party, no matter how well funded, to interfere with the network in a significant way.

Bitcoin, Miners, Addresses, and Privacy

Bitcoin launched in early 2009 as the first blockchain network.[46] A few months earlier, its creator(s), whose true identity is still a matter of speculation, had published a white paper under the name Satoshi Nakamoto.[47]

The document describes a set of interconnected mechanisms that allow a coin-centric network to exist in perpetuity. One of these features is the need for *miners*—parties who are many in number and who compete for the privilege of making periodic updates to the blockchain.[48] These updates are bundles of transactions called *blocks* (hence blockchain). Creating these blocks is a privilege in the sense that a successful miner gets paid with newly minted bitcoin *and* the fees attached to each transaction. By including these transaction fees, senders essentially pay miners to compete, and as more miners compete independently, various types of attacks become more difficult to execute.

When a person "owns" cryptocurrency, they do not literally have those assets on their computer. Rather, they control an *address* that is listed on the ledger as containing a certain number of those assets.[49] The Bitcoin addresses that send and receive transactions are represented on the blockchain by a *public key*, a long string of numerals and letters. This public key is derived from a *private key*, a 64-character string made up of all 10 numerals and the letters A to F. The cryptographic relationship between these two keys entails that a public key can be inferred from a private key but never the reverse. Anyone who has a private key controls all the funds associated with the corresponding address, so it is of the utmost importance that blockchain users protect their private keys. By contrast, they *must* share their public keys in order to receive money.

Critics outside the space often allege that bitcoin is an effective tool for laundering money or making illegal purchases.[50] But this analysis is flawed in the sense that every bitcoin transaction literally becomes part of the blockchain. If a person's identity ever gets connected to an address they've used, then they're linked to every transaction that address has ever been involved in, and to the counterparties to those transactions. If one of those counterparties slips up, it can be enough to expose them.

This confusion may stem from the first major use case involving bitcoin. Between 2011 and 2013 people bought narcotics with the cryptocurrency on a darknet marketplace called Silk Road, falsely believing the transactions to be anonymous. In the end, not only did the FBI confiscate millions of dollars in bitcoin from Silk Road's alleged owner, Ross Ulbricht, but also two federal agents who had worked on the case were convicted of crimes relating to bitcoin they had illegally acquired over the course of the investigation.[51]

Other networks, most notably Monero and ZCash, do offer more robust transaction privacy. ZCash creator Zooko Wilcox has argued that concerns about criminals benefitting from privacy-enhancing software are short sighted, and that *privacy coins*, as they're known, can protect all users from "crime and fraud."[52]

Ethereum and Smart Contracts

Several Bitcoin features—blocks, miners, and key pair addresses—can be found on the Ethereum network as well. The overlap makes sense: Ethereum began as a project driven by Bitcoin users who were interested in building a broad range of applications on top of a blockchain but felt that Bitcoin was not sufficiently friendly to this goal.[53]

One major difference they devised was to use addresses that can be augmented with additional code, turning them into *smart contracts*. A smart contract is not "smart" in the same way as an AI system, which can learn from experience. On the contrary, it's a computer program that lives on a blockchain network and functions like any other algorithm: by reacting in a predefined way to incoming information.

As a simple example, imagine two friends who bet on a baseball game between the Los Angeles Dodgers and the Oakland Athletics. They both send their wagers to a smart contract that receives score data from a trustworthy source like an ESPN data feed. When the game is over, this betting contract sends all the money to the winner's address.

If designed well, these decentralized computer programs can prevent cheating. Notice how in the example above, the contract pays the winner without giving the loser a chance to renege. By contrast, normal contracts deter parties from violating the deal through the threat of consequences.

In the future, smart contracts may be able to handle more advanced tasks, such as helping communities that languish under inept governments to take control of their own road maintenance programs.[54] A contract might do this by automatically collecting fractions of a penny from each driver and storing that money. When repairs are needed, those funds could be used to pay for the work piece by piece, releasing money as crews provide evidence that each step of the job has been performed properly.[55] Or, for a more tangible use case, we can think of a smart contract that receives data from moisture sensors in a farm's soil. When a patch of earth gets too dry, the contract could trigger a payment to the water supplier for the exact amount of water needed and send that water to the thirsty area via an automated irrigation system.

The Beauty (and Terror) of Network Splits

No introduction to blockchain technology is complete without mentioning the *network split*, an event that can occur when different versions of a blockchain's history become available and nodes disagree on which one to store.[56] The split results in two networks that together have the same number of nodes as the original one had. People fear these splits because blockchain networks get less secure as fewer nodes support them. When both "sides" carry a significant number of nodes, it can be a serious problem.[57]

At the same time, we don't want to suggest that the ever-present threat of network splits is a bad thing. On the contrary, it's at the core of what makes this technology interesting. Anyone can alter the history of a blockchain network, or launch a new blockchain network using the same software that governs an existing network.[58] If that person can convince node operators to support their version of the network, then their efforts can be considered at least somewhat successful.[59] On the other hand, making a change that few node operators support basically amounts to launching a (tiny) new network that almost nobody cares about. This is a check on the threat of a single party taking control of these networks.

The Payment Superhighway

Now, with our quick-and-dirty technical summary out of the way, let's get back to why this stuff is worth understanding. Earlier, we mentioned that a new digital era is predicted, one in which blockchain networks function as an "internet of value" and payments are nearly free, nearly instant, and effortless to send and receive. This new internet would support the transfer of massive sums, but also very small ones, known as *micropayments*, which some tech enthusiasts speculate could become the basis for a whole new economy.

What would be the purpose of transferring 10 or 20 cents at time, or perhaps even fractions of a penny? For one thing, if internet users made micropayments to the websites they visited, more sites could turn a profit without collecting and selling user data. This funding model, which the virtual reality pioneer and technology critic Jaron Lanier has advocated, would not be an equally comfortable fit for every website, but it could help resolve concerns that our data are generating value for companies who neither ask our permission nor offer to compensate us.[60]

In addition to plugging into existing systems like the modern internet, micropayments can also open the door to new ways of interacting. Let's take an in-depth look at one use case that researchers are already building: peer-to-peer energy sales.[61]

Imagine a neighborhood where several houses have electrical meters that interact with a smart contract. Some of these houses generate energy from a renewable source (e.g., solar) and some don't. Those who generate their own power draw on it for personal use and also sell it to neighbors. (With or without a blockchain component, this kind of network is called a *microgrid*.) An energy sale would begin when Alice's meter alerts the smart contract that her reading is low. The contract would then recognize that her neighbor Bob has some energy to spare and manage the process of swapping Alice's money for Bob's electricity in a completely automated way.

Microgrids have been proposed as a more ecofriendly and resilient alternative to the massive grids that dominate today's energy landscape.[62] These systems are often structured so that if a central power provider goes down, huge sections of the grid go dark. The providers that operate these facilities sometimes struggle to restore power in a timely manner after natural disasters and occasionally find themselves at the center of scandals, accused of neglecting the public's well-being.[63] They're also reportedly vulnerable to sabotage.[64]

While it would take an enormous collaborative effort to achieve widespread microgrid adoption, cryptocurrencies and smart contracts could theoretically remove one barrier to this goal by automating the sales process. This is one example of how micropayments can help open the doors to innovation when combined with other strategies and technologies.

A Fleet of Computers, Always at the Ready

If we could send money effortlessly, that would make it easier to take existing resources that are sitting idle and put them to work. For instance, cryptocurrency payments can support decentralized storage platforms, where customers can store files in much the same way that they do on Dropbox, but with one crucial difference: instead of paying a company to hold data on its servers, customers can pay random people to store files in the spare space on their personal computers.[65] A project called the Interplanetary File System is one attempt to build such a network.

Golem, another company in the blockchain space, is building a marketplace where people buy and sell surplus processing power. Users are encouraged to treat the platform as a "decentralized supercomputer" and use it when they need a *lot* of computing power.[66]

As it turns out, the idea of harnessing several computers' processing power for a single purpose predates the blockchain. For instance, the Folding@Home project, launched in 2000 by the Stanford professor Vijay Pande, uses processing power from volunteers' home computers to simulate protein folding, a process that could reveal new information about Alzheimer's and other diseases.[67] By assembling his own network of computers for this purpose, Pande demonstrated that spare computing power could be used toward novel and ingenious ends. Imagine if vast amounts of computing power were constantly available to everyone, rather than just researchers and other money-backed actors. We could begin to develop all kinds of services—geared toward smartphone users, for example—that simply wouldn't have been conceivable before.

A Tool to "Decentralize the Internet"?

We mentioned earlier that the internet has fallen short on two of its greatest promises: to break our dependence on centralized systems run by people who are indifferent to our needs, and to make us freer. As it turns out, blockchain technology might be able to help make the internet less centralized, if only in some limited ways. To understand how, it's useful to begin with why some technologists think "decentralizing the internet" is important.

Ramon Roca, who co-founded the Guifi.net organization discussed in chapter 17, told us that every instance of centralization in the internet's architecture represents a possible point of failure. The most important one to address, he said, is the fact that most users connect to the internet through privately owned ISPs. These companies often act against their customers' best interests, according to Guifi.net's Roger Baig Viñas. They may arbitrarily increase rates, offer better service to higher-paying customers, or sell off vital network infrastructure that customers depend on if there is a profit to be made in it.[68] Roca explained that the easiest way to help internet users bypass ISPs is to build "community networks" that connect users to the "backbone of the internet," the actual fiber optic cables that constantly transmit vast amounts of data around the globe.[69] While blockchain technology has no obvious role to play in this effort, Roca believes that it might support the decentralization of another layer of the internet.

The domain name system (DNS) is a crucial component of our internet experience today. It connects the numerical addresses that actually help us find websites with addresses that people can read (e.g., google.com). DNS is quite effective at managing immense amounts of resolution requests and can thus support high volumes of online activity. But the system also relies on *root servers*, another possible point of failure.[70] Attackers can take these servers offline or manipulate them into sending users to the wrong site, where they may unwittingly reveal sensitive information, such as a password. Currently, several teams are working to create fully distributed blockchain-based alternatives that could replace various components of the DNS with systems in which these root servers have no role.

Additionally, the decentralized storage platforms discussed above might help keep user data out the hands of private, profit-seeking giants, among other benefits.[71] It's also important to point out, as obvious as it may be, that money is a driving force behind much of the activity on the internet.

By decentralizing the process of moving digital money, this technology offers to decentralize the financial infrastructure attached to the web.

Finally, a number of decentralized applications are coming online. To borrow an observation from the tech journalist Tom Simonite, most are basically more-private versions of services that already exist: imagine a You-Tube or Google Docs that doesn't collect user data.[72] While privacy is their main selling point today, these apps are only the first generation. Other benefits may reveal themselves as the technology develops.

Ultimately, there's no clear path to completely decentralizing every layer of the internet, and anyone engaged in turning that dream into reality must make certain pragmatic concessions. For instance, Blockstack, a public benefit corporation working to build "a new internet for decentralized applications" also relies on cloud storage services from the likes of Dropbox, Amazon, and Google.[73] In other words, the "race to decentralize everything" will be more of a marathon than a sprint.[74]

A Tool to Build Better Systems?

This technology is still in its infancy, but we believe that under the right circumstances, it could become a useful tool for achieving positive social and economic outcomes. For example, Adam has argued that smart contracts could support many of the core functions of labor unions, including voting, the payment of dues, notarizing audit data, and reporting employer malfeasance.[75] A decentralized union could allow workers across countries and continents to negotiate as a group. These laborers might be drivers for an international ridesharing firm.[76] They might be garment workers spread out across long, complex supply chains. The same model could also be used to launch less conventional unions, such as consumer unions that encourage food producers to adopt certain quality standards and reward compliance.

As we saw with the microgrid example, smart contracts may also have a role to play in making our large-scale systems more efficient. For instance, a startup called Hyperloop TT has proposed a transportation system that discourages wasteful behavior by pricing rides according to their relative environmental impact.[77] It would rely on a smart contract to analyze passenger preferences, lining up cheaper rides for people willing to carpool and offering more expensive, direct options to those who value speed. Other smart contracts would help passengers pay fees to toll booths, hyperloop owners, vehicle owners, and the vendors who will handle a vehicle's eventual ecofriendly disposal.

These two use cases, unions and mass transit, show how large groups might use this technology to achieve socially desirable outcomes. We believe that it may also enable individuals to contribute to the public good. For instance, the practice of using digital tokens to represent physical things, called *tokenization*, could allow several people to co-own things like a car or a storefront.[78] This way, three drivers could share a taxi and drive it in different shifts, or four different restaurateurs could use one space to service the breakfast, lunch, dinner, and late-night crowds, while smart contracts automate any financial dealings between the co-owners.[79] These types of schemes could help economize precious space (and other goods like vehicles) in crowded cities and potentially create new avenues for people with fewer resources to participate in the economy.

Roadblocks

There's a reason we keep talking about what blockchain might do in the future: many of the technology's "killer apps" simply don't exist yet, perhaps because the technology isn't ready to support them at this stage. There is a limited amount of time to remedy the situation; Ameen Soleimani, the CEO of SpankChain, a webcam model site that uses the Ethereum network for payments, warned in late 2018 that the public's attention won't be focused on blockchain forever.[80] If it fails to deliver compelling results, he predicted, people will forget about it and move on, and the opportunity to build something will be lost.

In our estimation, the biggest problems facing blockchain networks today is the limit on how many transactions they can support and the fact that cryptocurrencies can be challenging and risky to use. Let's break these down one by one.

Blockchain networks may be capable of processing over a million transactions a day, but this would be nowhere near enough to support the billions of transactions and smart contract calls that people would need to make if this technology were adopted globally.[81] Each block holds a limited amount of data, so if too many transactions are attempted, it takes longer to add them to the blockchain.[82] As transaction volumes mount, senders also have to attach bigger fees to get their transactions onto the block-chain.[83] This means the network temporarily costs more to use.

Another critical shortcoming apparent in this technology's first decade is usability. The blockchain may look sexy to believers and investors, but outsiders often find blockchain apps perplexing and have generally not adopted them in impressive numbers. For the most part, these apps can only shield users from a portion of the blockchain's complexity. As the SpankChain co-founder Wills de Vogelaere told Adam, nontechnical people will only adopt a technology en masse when they can use it safely and easily, and do so in spite of the fact that most of them are not familiar with what's happening under the hood.[84]

On the subject of safety, it's quite possible to lose large sums of money when handling cryptocurrency, even for people who have some idea of what they're doing. It happens when users mistakenly send funds to the wrong address because, for instance, they've transposed characters in the recipient's public key. (Bitcoin has safeguards against such an error, but on other blockchain networks including Ethereum, the danger is real.) People may also lose funds when they're victimized by scammers or malicious hackers and through a wide array of other pitfalls. For example, on Christmas Day 2017, one unfortunate user of the Verge blockchain, intending to make a $73,786.05 transaction, absentmindedly keyed in that amount as the *miners' fee*, so that they paid that staggering sum to a lucky group of miners instead of sending it to the intended recipient.

Unlike these hazards, price volatility presents a risk factor that appears almost impossible to guard against. Cryptocurrency markets are highly erratic, and holders are basically at their mercy. This volatility has real implications on adoptability: if a cryptocurrency's value can go down 10 percent in a day, people with financial challenges are unlikely to leave money in it. Blockchain developers have attempted to neutralize this threat by creating so-called *stablecoins*, tokens that are supposed to remain steady in value relative to some national currency. As we write this chapter, however, we're not convinced that any stablecoin on the market is viable for widespread, long-term use. [85]

These problems may not hold the space back forever, though, because they are widely known, and many developers are working to solve them. Some efforts to improve transaction times and costs (e.g., by using something called *plasma* blockchains) look promising at the moment, but it's not clear if or when they will be implemented.[86] We believe that if this technology ever truly takes off, it will be thanks to multiple scaling strategies, and the process may even involve more than one network.

Tackling usability will likely involve building a secure identity system of some sort—an easy and foolproof way for users to prove who they are. While partial solutions are already being developed, it will be awhile before a completely secure and truly convenient identity system is up and running.[87] As for the problem of price volatility, there is simply no clear path to a solution yet.

Long-Term Challenges

These are formidable obstacles, to be sure, but we're more concerned about what might happen if blockchain technology actually takes off on a grand scale. It promises to give us something incorruptible that we can hold onto as we navigate our messy, predatory, corrupting world, but will it protect us from forces of tyranny and inequality, as advertised, or be co-opted it into their service? We believe that the real challenge will not be building this technology but rather making it work on the ground. Our example of the meeting between the CEO of a leading cryptocurrency exchange and the kleptocratic dictator of Togo proves that this technology is not guaranteed to bring freedom to unfree places. When introduced in a careless or irresponsible way, it can even have the opposite effect. Or, as the Ethereum Foundation director Aya Miyaguchi said, it's entirely possible that this space could produce more "centralization [in] the name of decentralization."[88]

Why is this a threat? Partly because the question of how to actually decentralize the process of governing these networks remains unanswered despite all the fanfare around decentralization.[89] To find a solution, people in the space are exploring important questions, including: Who has the legitimate right to define these systems' rules? And by what process should users get to influence these systems' evolution? If we're serious about achieving better outcomes with blockchain-based systems than the ones we have with today's internet, then it's important for us to address these kinds of questions from the outset and revisit them often.

A 2018 conflict in the Ethereum community brought the governance question to the surface.[90] Following a series of events too long to describe here, and amid a hailstorm of online bickering, an Ethereum insider named

Yoichi Hirai resigned from what was essentially a software-editing role. We mentioned earlier that nodes are computer programs. Although node operators in many networks have several clients to choose from—and these clients may have slightly different ways of implementing the software that defines the network's rules—each of these clients is ultimately supposed to store identical copies of the blockchain. When Ethereum's so-called core developers deploy changes to this software, it's up to the teams behind these clients to update their own client accordingly.[91]

Hirai had been one of six people allowed to edit the "official" copy of this software. Just before resigning he argued that the system in place—core developers deploying changes that get implemented by a relatively small number of client providers—tended to result in node operators accepting changes without fully reviewing them. To combat centralization, he recommended getting the wider Ethereum community more involved in the process of creating these changes. "Ideally," he said, well-crafted changes would bubble up out of a "chaotic communication pattern without a centrepoint," rather than come from a core team. Hirai also questioned the wisdom in allowing people to change decentralized systems on the basis of any sort of "authority," especially people who (like him) had not been elected to their positions.[92]

Indeed, the existence of de facto decision-making bodies could be a problem if many people come to rely on blockchain networks. Furthermore, we worry that decentralization will be constantly under threat as these networks scale.

There could be other kinds of growing pains if cryptocurrencies take off as well. Let's consider a common crypto fantasy: a country's currency is steadily losing value, so residents decide to cut their losses and convert their savings from local currency into crypto. The dream is that cryptocurrency would save the day and the story would end there. Another possibility is that this process could accelerate the devaluation of the national currency and spur panic selling. If this happened, the last people to get the chance to dump out would inevitably include some of that country's most vulnerable people.

Finally, we'd be remiss if we did not comment on the trend of techie insider-ism in the blockchain space, where the fastest way to earn respect is usually to demonstrate technological expertise. Such an environment generally encourages people to seek technical solutions to all sorts of problems, whether or not they are actually technical in origin. What if outsiders' input is dismissed because they can't code, or because they haven't yet learned about certain aspects of DLT systems?[93] Then the space will lose out on their potentially valuable perspectives, and developers may get stuck using a narrow and overly specific toolkit to address complex problems.

Cause for Optimism

One confusing thing about the blockchain space: it's never quite clear where the cutting edge is. Dozens of teams are working on what sound like plausible strategies for making blockchains more useable. Regardless of whether any of these pan out, we believe that this collective experiment has already set a useful example for future endeavors in several ways.

First of all, this movement has seen people from all over the world apply their brainpower to what is basically a single problem. What's more, thousands of people—perhaps tens of thousands overall—have come together to create a modest-sized industry out of thin air, one that is far less dominated by corporate influence than most other niche tech industries. Regardless of what the movement achieves, it has already proven to be an impressive grassroots attempt at creating a technology that is not cast in the Silicon Valley mold.

We're also struck by how the fragile complexity of these systems sometimes frustrates efforts to think about the technology in a vacuum. For instance, a lot of things have to go right in order for a cryptocurrency to be widely usable: miners must never stop competing with each other, which requires that certain incentives be maintained;[94] market forces (including speculators and price manipulators) must be prevented from causing erratic price changes;[95] average Joes and Janes must have a safe and easy way to send, receive, and store funds; and so on. In this way, these systems push us to appraise technologies in terms of the outcomes they deliver, rather than to think of them as nothing more than a set of tools. Even if DLT doesn't ultimately teach us to evaluate technologies more holistically, it seems significant that so many members of this self-assembled army of technologists are having conversations about outcomes.

The blockchain's promise, in the words of the Aragon co-founder Luis Iván Cuende, is to give people "the power to create systems that better align incentives and distribute resources."[96] Perhaps, somewhere down the line, this technology can enable people to collaborate in creative and effective ways around shared challenges, like weaning humanity's dependence on oil. Here's the main point we want to convey: these kinds of positive outcomes will only be possible if a lot of people engage critically with both the tech and the challenges it faces in the real world. Or, as Cuende put it, technology can help improve our lot, but "without a global, conscious, and concerted effort, the outlook is incredibly bleak."[97]

Closing Thoughts

We have no illusions about blockchain saving the world. All the same, these networks' decentralized architectures make them potentially valuable tools for building grassroots solutions that support the needs of ordinary users. The technology may also offer major breakthroughs in areas where conventional computing experiences are lacking, like security and privacy.[98]

It is precisely because blockchain technology raises the possibility of digital services untainted by corporate ties that we believe it's worth keeping an eye on and trying to engage with. Like it or not, we may all end up blockchain users eventually. If you want to see this technology used to advance goals that matter to you, we encourage you to help steer it in that direction by joining the dialogue. For our part, we'll keep our ears open for emerging voices and continue to listen to those who talk past the hype.

In his 2018 book *Enlightenment Now*, the psychologist and linguist Steven Pinker sets out to debunk our pessimism about the present day. Yes, we must contend with problems like environmental destruction, income inequality, and political polarization. But Pinker argues that the real story of humanity is one of progress and achievement. Life expectancy has more than doubled since the 1800s, and extreme poverty has decreased. Workplace, automobile, and public safety have improved. Rates of violent crimes and deaths from war have declined. Literacy is on the rise. We humans are far better off now than we have ever been, says Pinker—we've just lost perspective.[1]

The psychologist's perspective, however, problematically treats "quality of life" as objectively quantifiable. Built into all of the measures he uses to map our progress are very typical Western assumptions about how the world works and what constitutes "the good life." Take automobile safety. It's great that better airbags, brakes, and backup cameras protect those of us who commute to work from injury or even death. Safety precautions like airbags, however, don't come out of the benevolence of the auto industry, they come when people stand up for what is right and reasonable.

On top of this automobiles themselves are not so great for the environment. We are in the depths of climate change so profoundly disruptive that many scientists say we've entered a new geologic era, the *anthropocene*, expresses the irreparable impact of the human race on the environment.

I don't mean to say that nothing has improved since the 1800s. But I do mean to poke a hole in the part of Pinker's argument that ignores what we have disregarded, forsaken, or damaged as we've pursued these "improvements" to human life. Attending to these trade-offs allows us to see a more *full* range of possibilities for future innovation, including ones our predecessors did not pursue, and to expand our understanding of what we can aspire toward.

The technologies of tomorrow are yet to be created, and those of today can be made better: the path is not "locked in." To refine how we design systems, regulate them, and embed them in our lives, we need to involve diverse stakeholders (especially those most vulnerable) in positions of real authority, in roles that come with a real "say." The ethic of *active collaboration* can overcome negative effects, like climate change, that occur when a relentless drive for productivity and profit leave other values, like clean air and water, by the wayside.

Many of the tech world players I've discussed in this book—such as the Silicon Valley billionaire inventor and investor Peter Thiel, or those in the blockchain and cryptocurrency world—promote a myth: by empowering individuals and freeing corporations from needless constraints, they claim, the internet acts "naturally" as a force for good. There's a problem, though. With everyone let off the leash, it's back to the law of the jungle, where corporations are usually bigger, stronger, and more ruthlessly focused than the motley collection of humans called "society." The vulnerability of our basic cultural and economic needs becomes scary when faced by the stolid forces of capitalism.

My argument is not with engineers or with technology writ large. Economic inequality, threats to democracy, cultural misunderstandings—these all exist independent of the internet, and their causes can't be simply pinned on technologies or their libertarian evangelizers. The largest tech companies have done exactly what they were supposed to do: grow and increase their valuation and bottom line. As they maximized self-interest, the rest of us began to encounter problems. But it didn't have to be that way, nor does it have to continue that way moving forward. The fact that we can't pin our societal challenges solely on tech only reinforces the notion that it's our responsibility, together as users and citizens, to force the changes we want to see.

We can't assume that governments and politicians are going to come up with solutions to ensure that online platforms are accountable, transparent, and return value to us all. Too many politicians around the world are forced to answer to their wealthiest funders and supporters. And some, including liberals, are directly supported by the wealthiest 1 percent in their campaigns, whose interests naturally differ from the rest of ours.

On some occasions, though, politicians *can* be our allies. One who is running in the 2020 US presidential election promises to make big tech accountability a part of her campaign platform: Senator Elizabeth Warren has called for a "cop on the beat" to regulate and break up the unfair advantages held by Google, Amazon, and Facebook. It's fine for Amazon Marketplace to host the buying and selling of products, for instance. But it's not okay for Amazon to use the vast troves of data it collects on our buying habits in order to compete with, manipulate, and dominate small- and medium-sized businesses and innovators on the platform. Rather than have us continue to be puppets controlled from behind the stage, Warren boldly asks us to get back to a digital world where everyone has a chance.

Since May 2018 a measure that sinks its teeth into tech company transparency has been in effect in the European Union. The General Data Protection Regulation (GDPR) has set up legal repercussions and fines as a means to ensure that users are given opportunities to understand how their data may be used, and the choice to consent to these uses or not. The European Commission invited a high-level group of experts (called the HLEG) to advise on policy initiatives to address fake news and online disinformation (defined as false, inaccurate or misleading information that causes public harm).[2] The resulting HLEG report underscores the importance of making sources of online news transparent, improving user understanding of news content, empowering local journalists, and ensuring that the diversity of the news ecosystem is maintained, rather than invisibly filtered by online platforms.

Technologies can be designed with an eye toward their shorter- and longer-term impacts on equity, justice, and diversity—*if* those are values truly held by those doing the designing. How about "move slowly and don't break things" as a new mantra instead of the de facto "move fast and break things" offered by Mark Zuckerberg?

To embark upon that path we first have to recognize the designer/user dynamic as contingent—that is to say, *recognize ourselves, the users, as potential designers*. And then we have to answer an important question: Do *we* want an internet that extracts data from all of us to amplify the power and wealth of those who build automated and gig economy technology for their personal profit. Many of us care about climate change, for example. The battle for an equal internet may sound more abstract than this clear global crisis. But, given how technology influences our decisions and shapes what we think and know, it is as essential to our society as air or water is to the natural environment. The internet's intimate connection in our lives should not merely give us pause but also light a fire under us. What infrastructure, what design and dissemination processes, and most importantly, what sorts of social roles and relationships do we want to promote in our world? "No opinion" is no longer an option.

As we come toward the end this book, I'd like to highlight some of the important initiatives underway that demonstrate how technologies can serve us all, not just a few of us at the cost of everyone else. Even within some of the large tech companies that I have critiqued in these pages there exist efforts to rein in the negative effects their platforms have caused.

The Algorithmic Justice League

Joy Buolamwini calls herself a "poet of code" and a "daughter of art and science." These epithets reflect the influence of her artist-writer parents and her journey as a Ghanaian American woman whose childhood fascination with robot technology led to a promising career as a computer scientist. In fact, her 2016 TED talk, "How I'm Fighting Bias in Algorithms," has elevated her visibility.[3] I met Buolamwini when she was busily preparing for an interview the following day on ABC TV's *Matter of Fact*, a weekly public affairs program hosted by Soledad O'Brien.

As a graduate student at the MIT Media Laboratory, Buolamwini developed an art/science project that depended on the use of a webcam with facial recognition software. But because the software did not read her black skin, she had to don a white mask to get the camera to "recognize" her. Her encounters with other systems revealed serious biases as well. She was both unnerved and galled by how often they saw her, as a black woman, in an unflattering light (if they saw her at all). Now her work aims to raise awareness of the troubling fact that the digital systems most of us use perpetuate the biases and inequalities of everyday life. No longer willing to modify herself and her identity to work with these biased technologies, she asks a simple question: Why can't technology serve all of us?

Buolamwini hopes to stir up a social movement. She formed the comics-inspired Algorithmic Justice League to fight the "coded gaze," a term she coined to describe the harmful biases in artificial intelligence. Fusing her interests in computer science and the arts, she created a visual poem, "AI,

Ain't I A Woman?," which shows how AI systems frequently fail to correctly identify the gender of important women of color, including Oprah Winfrey, Serena Williams, and Michelle Obama.[4] In that poem Joy shares how Face++, a Chinese government–funded facial recognition AI program, consistently identifies Williams, the African American professional tennis player deemed by many to be the greatest athlete of all time, as male—despite having access to pictures spanning more than ten years.[5]

It doesn't matter whose facial recognition system we are talking about: Microsoft, IBM, Google, Facebook, and Amazon have all have created technologies that regularly fail to identify famous black women. Why? Likely because the data sets used to train AI software are mostly composed of male and white faces. In one prominent case, reported the *New York Times*, a facial recognition data set was estimated to be more than 75 percent male and more than 80 percent white.[6] It may seem obvious that data sets should include all kinds of faces if the software is going to work for all kinds of people. But the gender and race of the AI engineers who build these systems (overwhelmingly white and male) likely played a role in allowing this point to fall through the cracks. As AI becomes ever more embedded into the institutions we all depend on—including the criminal justice system, hiring, and even lending—the creation of fair and just AI systems becomes an important civil rights issue of our time.

Ensuring fairness will be tough, though, if there continues to be little transparency around AI. The media artist and AI-developer Parag Mital expressed his concern in a TechCrunch article about the dangers of allowing a few private companies to own most of our digitized data. "Much of what is researched with AI may not be public knowledge," he wrote, "and is likely internal research that's closely held by just a few very wealthy corporations. How can the public make informed decisions about something that is kept secret?"[7]

We are on the brink of engineering a digital future that perpetuates bias. Recognizing that women, indigenous peoples, racial minorities, and the poor are in vulnerable positions around the world is a great first step, but it isn't enough; we need to take action by actually building this recognition, and the ability to audit it, into our technology. To treat technologies as neutral because we know so little about the ways in which they are engineered is to stick our heads in the sand. The biases that pervade our world won't go away if we ignore them. In fact, they may become more pervasive than ever, spreading far and wide as people use technology without recognizing whose voices and values drove its design—and whose didn't.

Aspirational AI

Perhaps there is reason to hope that Silicon Valley will address the issue of algorithmic bias. Technology magnates and representatives of the biggest technology companies, including Sam Altman, the founder of Y Combinator and co-founder of OpenAI, and the Tesla founder and former CEO Elon Musk, have begun to express their concerns about bias in AI.

Recognizing the problem is the first step. Altman told me that "the field of AI researchers is by far the least diverse field [he's] ever seen." He backed this up with a statistic he'd recently learned, that "99 percent of machine-minded PhD degrees were men. Quite frankly," Altman commented, they "have the biggest impact in the field."[8]

So what then has OpenAI done about this? In its advertising the nonprofit research organization adopted this famous saying—*The best way to predict the future is to invent it*—from Alan Kay, a pioneer in computer science who made major contributions to object-oriented programming languages and graphical user interfaces.[9] In that spirit, the group's stated goal is to advance AI to benefit humanity as a whole. The organization has pledged to support intelligent systems that are an "extension of individual human wills and in the spirit of liberty, as broadly and evenly distributed as possible." Though this goal sounds noble, it points to a tension that exists between the generalized approach of creating for us all (in a way that optimizes "individual liberty") and an equity-specific approach (that directly addresses bias and inequity in AI systems).

Consider how the insurance industry is shifting as we speak. John Hancock, one of the oldest and largest life insurers in the United States, has now stopped underwriting traditional life insurance policies and instead only sells "interactive" ones that track fitness and health data through wearable devices.[10] Responding to this, Kate Crawford, one of the most important researchers studying the effects of AI, tweeted: "We saw this coming, and here it is. Endless trapdoors ahead: data inaccuracies, intentional gaming, surveillance 24/7, data breaches that will be infinitely worse."[11]

The John Hancock case might soon become the norm rather than the exception. Although the Academy Award–winning filmmaker Michael

Moore focused his documentary *Fahrenheit 11/9* on the rise of Donald Trump and the threats his presidency poses to American democracy, he also highlighted the massive West Virginia teachers' strike of early 2018.[12] The strike began as a protest when the state "upgraded" its employees' health-care in lieu of a pay raise. In a dystopian twist, the new provider forced teachers to wear Fitbit-like devices to track their physical activity. If they did not earn enough points per year to meet predetermined fitness goals, they were liable for an additional $500 deductible and a monthly surcharge.[13] The teachers eventually negotiated to secure a pay raise *and* overturned the invasive tracking policy.

Although we shouldn't dismiss the concerns of some technologists who see AI systems as proliferators of a superintelligent species (think *The Terminator*), the most pressing issues we face today transcend the tropes found in science-fiction films. The digital future we are inventing is likely to reflect, normalize, and advance the discriminatory world of today. Neither a space colony on Mars nor hand wringing about super robots addresses these concerns. In order to ensure a more just future, we must engage with what is actually in front of us, in the here and now.

What do those inside the AI world think about its dangers, and how do they imagine what a balanced and just world with AI may look? I recently put this question to Terah Lyons, the director of the Partnership on AI, a group that has gained wide-ranging support from the consulting group McKinsey, prominent technology companies (including Amazon, Apple, Google, and Facebook), and activist and human rights organizations like the American Civil Liberties Union (ACLU) and the Electronic Frontier Foundation.[14] The Partnership brings together more than seventy organizations spanning nine countries and three continents in an effort to educate the public on AI and foster AI projects actively designed for social benefit.[15]

That the Partnership group even exists can be considered a coup in the tech world, and the range of its member organizations can be seen as a good sign. Leading this organization is not easy though, Lyons tells me, nor does it mean that positive changes are inevitable.[16] As an advisory group, the Partnership is limited because tech companies pursuing AI might not implement any conclusions or policy proposals the group makes. That's why convincing companies to adopt recommendations, whether they are Partnership members or not, is of the utmost importance.

Lyons is optimistic, nonetheless, that there are people within the technology industry who have the "presence of mind and goodwill" to make decisions that will benefit society as a whole.[17] These people are crucial, she told me, because they have great power, and can influence how systems are designed, audited, and regulated within the companies themselves.

Two Partnership board members, Amazon and the ACLU, are positioned to show the feasibility for tech industry actors to take a common-ground approach. Amazon, through its facial Rekognition technology, has been assisting the US Immigration and Customs Enforcement (ICE) agency in tracking, detaining, and deporting immigrants. I discussed in chapter 4, for instance, how the COMPAS and PredPol systems, designed for policing and the criminal justice system respectively, tend to make mistakes when it comes to people of color. In the case of ICE, the risk applies to Latinx people who might be misidentified as matches for deportation orders and thus forced to leave the country, simply for being undocumented. According to Mijente, a nonprofit immigrant rights group, "It's time to hold Amazon accountable for its outsized share in building the deportation machine, and demand that they stop. It won't be easy—Amazon makes billions from these contracts."[18] We've seen that Amazon's facial recognition systems are currently being sold to the police and military contractors, which has spurred the ACLU's public records requests to help make the workings of technology transparent. The question remains whether these two Partnership board members can use this opportunity to engage in frank and open dialogue, and whether Amazon will create systems that acknowledge the concerns of those who question the humanity of their business choices.

For now, there are more questions than answers. We should address them before we release AI systems far and wide in ignorance about how they will be governed, regulated, designed, or understood by their users. We must not rush forward with initiatives that give power and legitimacy to AI systems owned by private corporations. Otherwise we may end up at the place where we treat robotic systems as persons under the law, an option being discussed as I write. Taking this further, will we someday live in a society in which we allow robots the right to vote or have free speech?[19]

The Google, the Bad, and the Ugly

Of the companies I most often discuss in this book, Google was more willing to talk to me than the rest. That's important because the search engine pioneer created the most visited website in the world, and its presence has only grown given how many applications and services it has developed.

I've written about Google (so far) with a heavy dose of concern. Why? For at least three reasons: First, the company remains secretive about what personal data it collects and, like most other internet companies, has an unreadable privacy policy. Second, Google might be a monopoly already and is the biggest lobbyist in Washington, DC. It secretively assisted the US military with its potentially illegal and lethal drone project Maven. Google also might have had a hand in the firing of the anti-monopoly researcher Barry Lynn, from the influential New America foundation.[20] Third, Google is able to manipulate search results and suggestions to sway voting behavior. We might assume that these are solely left-wing worries, but in reality the concerns about how information finds us (and how we find information) are bipartisan.[21]

But Google is also refreshingly different from the other big tech companies in its approach to transparency and sharing. It not only releases a great deal of open source software code, but it also created a way to share open source code related to machine learning applications through its TensorFlow library, the most popular of its kind available online today. In addition, Google hosts talks that are available for anyone to view online. Scientists, politicians, artists, scholars, Zen masters, and activists have all used this platform to share their knowledge with the world for free. Does any of this serve to counteract Google's unrivaled and largely unregulated power in tech?

I decided to ask the company directly. When I reached out for an interview and was paired with Reena Jana, Google's head of Product Inclusion, it seemed like I had found the ideal person to answer my questions. Jana, a South Asian woman of color, has worked for many years as a journalist, editor, and storyteller, skills that supplement her role at Google to oversee the company's effort to increase employee, user, and supplier diversity.[22]

Our conversation centered on Google's intentions and the question of its responsibilities to its users and our world. I was surprised to find out how inspired Jana was by Bryan Stevenson, a social justice advocate and founder of the Equal Justice Initiative. Stevenson, whose 2012 TED talk has been viewed nearly six million times, has dedicated his life to fighting injustice in the US prison industrial complex, which locks up far more people—disproportionately black and brown—than every other economically powerful country in the world put together.[23] The Equal Justice Initiative offers legal representation to those who cannot afford it or who have been abused in prison—particularly children, the mentally ill, and death row inmates.

Jana chimed in with valuable insights about Google's ambitions and intentions. Unlike most other major tech companies, she explained, Google has a large percentage (nearly 45 percent) of women in senior management.[24] Its tools have driven growth for small businesses across the United States and much of the world. And its services like Gmail, Maps, Search, and Drive are all starting to work with many languages. Google Maps is now available in more than forty languages, spoken together by 1.25 of the world's 7 billion people.[25] While that is less than 20 percent of the overall global population, and just 0.5% of the number of languages in a world where linguistic diversity is being threatened, it reflects an attempt to expand and customize services to a more diverse group of people.

Like other multinational companies, Google now operates in at least fifty countries worldwide. Jana explained to me that this global expansion has allowed the company to develop services that are tailored to the concerns and constraints of users in those locations.[26] By increasingly calling itself an AI company, Google signals that machine learning and intelligent systems are going to be a central point of emphasis. It reinforces that awareness by publishing its AI principles and practices for fairness, and by hosting an active AI blog that shares its latest research projects.[27] Google claims to recognize the importance of training its AI systems using data sets inclusive of its increasingly diverse user base—although it remains unclear how exactly these systems would be designed for (or learn from) "minority" users. If Google wishes to overcome stereotypes, or even physical constraints (like poor or unreliable infrastructure), the company will need to go deeper, maybe even give up some power and control over the design process. The belief systems, values, and protocols of ownership and community roles that make up a community's knowledge systems, or *ontologies*, are extremely important, yet they rarely considered in technology "access" projects.

True to form, the company's Next Billion Users (NBU) team is determined to expand its footprint, recognizing some of the constraints and realities of different places and societies. NBU has offered free Wifi in train stations in India, Mexico, Nigeria, and Indonesia.[28] One product, Files Go (renamed Files by Google in 2018), helps users free up space on their mobile phones and share files offline. Its designers created it for use in nations where users cannot afford laptops or desktops, and instead primarily access the internet via phones. The Files by Google services are customized to combine the team's observations about how people actually use their phones with educated guesses about what they may need.

As Jana described these various efforts to me, I found her perspective to be intelligent and refreshing. She acknowledged the shortcomings of Silicon Valley and how unrepresentative it is of the diversity within United States and the world. She spoke with me openly about the race, gender and class biases that pervade technological systems, including Google's. Technologies are indeed biased, she admitted, because their creators reflect the inequalities in our world. She did qualify her remark, however, by saying that "these biases have existed in the world for centuries and millennia.... We [at Google] have to be innovative, and we are experimenting with different ways of creating equality of opportunity around our algorithms."[29]

It's hard to know whether promises of increased equity and inclusiveness will every truly come to be. By acknowledging that disconnect between engineers and the users of their systems is real, Google has taken a heartening first step. But if Google continues to hide the inner workings of their systems from us, can we really trust that they will support equality and inclusion? One option is to just wait and see. A better one would be to take steps now to ensure transparency and accountability, so we can *know* that the digital world will serve us all in the future.

It's not surprising at Google, which employs over 88,000 people, to come across perspectives steeped in authoritarian thought. For instance, let's look at the alarming views in a video titled *Selfish Ledger*, which have been almost universally received by the public and press as "creepy" and "dystopian." *Selfish Ledger* describes a future in which Google gathers personal data about its users via its digital services and stores that data in a "ledger," or virtual account book. From there it relies on AI to determine what gaps exist in Google's knowledge about those users, and to fill those gaps by suggesting other data-harvesting products it thinks users will find appealing. The real goal, however, is to get users to serve the missing data to the ledger on a silver platter. (For example, it could suggest a bathroom scale that not only tracks weight and body mass index, but is also connected by Bluetooth to the user's fitness app, which would be accessible to the ledger.) Although Google says it only intended *Selfish Ledger* to circulate internally, the video was leaked to the press in 2018.[30]

Nick Foster, whose vision drove the project, tells us: "'By thinking of user data as multigenerational it becomes possible for emerging users to benefit from the preceding generation's behaviors and decisions.' Foster imagines mining the database of human behavior for patterns, 'sequencing' it like the human genome, and making 'increasingly accurate predictions about decisions and future behaviors.'"[31] The company swears that the ledger is only a thought experiment or "speculative design" rather than a blueprint "related to any current or future products." Nonetheless, the *Selfish Ledger* could be the logical next step on the path Google has laid out for itself and its billions of users.

This represents a shift from seeing data as something that serves human users, to instead seeing users as resources to serve the technology. From this perspective we, the humans of today, are custodians: mere carriers of information, rather than private owners of our own personal data.

The *Selfish Ledger* video shows no interest in exploring the messy ethical questions of whether such data should be collected, how it may be analyzed, who would have access to it, how it might be regulated or audited, or whether it should be privately owned. There's no room for privacy and self-determination in this scary but increasingly realistic scenario, where corporate machines take complete control.

I recognize that this futurist manifesto is not the mainstream view at Google, nor is the anti-diversity manifesto I described in chapter 5 that questioned women's capacity to be effective engineers.[32] Nonetheless, the fact that these radical perspectives exist within a private company that is so embedded in all of our lives speaks volumes.

This is why it's simply not good enough for us to put the onus for solving these problems on the companies who, let's face it, played a large role in creating the problems in the first place. Elites can't be expected to solve the problems of a hierarchical dynamic that they have profited from, though they can be part of a process that includes us all and shifts the entire system.

Rally for a Living Wage at Amazon

Today people can push back against big tech with significant force. Bernie Sanders, a longtime progressive US politician and now candidate for the 2020 presidential election, decided to take on Amazon and its CEO Jeff Bezos in the middle of 2018. He criticized the company for not paying its workers a living wage, and demanded that a company run by the wealthiest person in the history of the world do better. "Count to ten," he tweeted in August 2018. "In those ten seconds, Jeff Bezos, the owner and founder of Amazon, just made more money than the median employee of Amazon makes in an entire year."[33] In fact, some of Amazon's employees, who make an average of $28,000 per year, have been forced to use food stamps, Medicaid, and subsidized housing just to survive.[34] Meanwhile Bezos makes approximately $275 million per day.[35]

It may be surprising to know that the movement worked. Rallies led by Sanders and his team, along with protests from employees, forced a decision by Amazon, made in October 2018, to guarantee each of its workers $15 per hour. But it's important to note that this success isn't the be-all and end-all when it comes to a living wage. At that hourly wage the average American still can't afford the rent for a modest one-bedroom apartment. Indeed, with the current national minimum wage ($7.25), one would have to work 2.5 full-time jobs to afford a one-bedroom apartment in most of the country.[36]

Additionally, the Amazon wage change doesn't account for the possibility that these employees may lose their jobs to automation or that American workers may soon have their jobs outsourced to nations where labor laws are nonexistent and wages and conditions are far worse. It also doesn't deal with the stunner that Amazon paid no taxes on its $11.2 billion profits last year. While Amazon's current wage increase may not combat globalized inequality or the concerns I've raised regarding automation, it does show that change can happen. And it raises hope that small victories could lead to greater ones, where worker security is protected or, proactively, new jobs of the future are identified and made available to those at risk.

Movements like this are important to pay attention to because they show that as workers, consumers, and users we all have power. We can also appeal to politicians, businesses, designers, activists, journalists, and public-interest organizations to join us in advocating for fairness, equity, and diversity.

Are doctors and engineers more similar than different? Natasha Singer connected these two professions in a 2018 *New York Times* story stating: "The medical profession has an ethic: First, do no harm. [But] Silicon Valley has an contrasting ethos: Build it first and ask for forgiveness later." Her story, "Tech's Ethical Dark Side," focused on how top engineering universities in the United States, such as MIT, Harvard, Stanford, and New York University (NYU), are now trying to bring a "medicine-like morality to computer science."[1] Singer goes on to discuss a wave of new class offerings at universities around the country bringing social, political, and cultural issues into a much needed dialogue with engineering and design. Her point is that a new "human-centered" approach toward technology can and should start with what we learn in our schools and universities.

Building on this, it's time to consider the relationship between technology and our planet given that the "planned" shelf life of many of the devices we use is a fraction as long as they will stay on our planet. For example, our iPhones, which we usually keep for only a couple of years, might vanish from our sights and minds when we trade them in for a new one, but in reality they might be just at the beginning of their lives on Earth. They might travel to landfills and cause environmental harm. Or they might move to other parts of our cities, countries, or world—and once there, be repaired and distributed to those who lack the resources to buy into the corporate policy of "planned obsolescence," a strategy that has made Apple so incredibly wealthy.[2] This is why legislation that would provide anyone with the "right to repair" is an important topic now debated in many parts of the world.

Currently in the United States, though, this right does not exist. Consider the story of Eric Lundgren, an entrepreneur from the Los Angeles area who is obsessed with recycling. According to a report from the *Los Angeles Times*, he built an electric car out of recycled parts that drove further than a Tesla, and has created a recycling facility for electronics that processes 41 million pounds of e-waste each year.[3] He counts IBM, Motorola, and Sprint among his clients. He has done pro bono work to clean up the e-waste that has accumulated in Ghana and China, and donated recycled cell phones to US soldiers overseas.

Lundgren is an iconic figure of selflessness and civic engagement, right? Nope. For his recycling efforts, he is considered a criminal and is going to jail. He should be a hero for helping our society and world, lessening the environmental impact of the electronic devices we throw away while creating a sustainable business that provides jobs. He thought he could get more affordable computers into people's hands while contributing to a second-hand industry in digital technology devices. But because doing good for the community sometimes clashes with the business interests of tech companies, Lundgren has instead earned himself a powerful enemy: Microsoft. Why has the company targeted him for retribution? Because in some cases, turning the junk he recycled into workable technology that supports its users has required him to install Microsoft Windows on the personal computers he has salvaged. In fact, Microsoft and the government worked together to put him in prison for fifteen months. The assistant US attorney on the case told him explicitly, "Microsoft wants your head on a platter and I'm going to give it to them."[4]

Lundgren should be feted as a role model for leveraging the recycler's ethic—recycle, reduce, reuse—to contribute to the common good. His story parallels the examples I've shared from Africa, where repair and recycle are necessities, not luxuries. Entrepreneurs like Lundgren can transform the "build to die" model that's made computer and phone manufacturers so wealthy into a different kind of business focused on employing people, helping the environment, and providing consumers in the second-hand market an alternative, one that will likely complement rather than compete with the current consumer market for new technology.[5]

Repair is just one of the many social or environmental themes that have been overlooked by a traditional education in engineering or computer science. Schools and universities will need to open up and revise their curricula, incorporating new themes and subjects of study if they wish to educate students for a digital future that is inclusive, sustainable, and collaborative.

Another unfortunate legacy is that most education systems treat the science and engineering fields as separate from the humanities and social sciences. This is why we rarely see courses in which code writing or software design taught along with materials that "understands" the places where the software would "work."

As we begin to see cultural or social topics being taught in conjunction with engineering, I suspect that we will also see engineers who are better equipped to think deeply about the world they are transforming with the systems they design. New jobs can be created for those who can translate across technical and ethical domains as technologies are innovated and rolled out. In a smattering of new offerings and initiatives in the United States, this process is just beginning: the prestigious Association for Computing Machinery has released a code of ethics, although it's uncertain how it will be interpreted and taught across the world.[6] A newly released list of computer science ethics classes taught at dozens of universities around the world reveals new course titles, such as "Race and Gender in Silicon Valley" or "Ethics in Video Games."

Sure, our education is supposed to prepare us to work, to enter the job market. But it is also supposed to prepare us to be creative, reflective, deliberative humans. Education that discourages reflection and criticism treats people as tools, not as humans with social, ethical, and creative needs. Bringing different disciplines together in our schools, for example by marrying the sciences and the arts or by pairing engineering with social sciences, will not only prepare us for the jobs of the future but will also be ethically and intellectually enriching to us as human beings. Science, technology, engineering, and mathematics (STEM) education need not stand on its own. It's unrealistic to think of science or technology as a given—as some sort of airtight study of "what is"—without recognizing how deeply philosophy, ethics, human behavior, politics, and the arts influence each of these fields.

Mitchell Baker, the executive chairwoman of the Mozilla foundation, echoed these concerns recently. She believes that "we are intentionally building the next generation of technologists who have not even the framework or the education or vocabulary to think about the relationship of STEM...to society or humans or life."[7] She warns that as users (who number in the billions) become complacent, blindly following what technologies tell us to do, we lose our ability to ask fundamental questions like

"Who does this serve?" or "How might we apply this technical knowledge in different manners?"[8]

What about design, which is often only seen as a way to make something look pleasing or to make it "usable"? From this limited perspective, we give designers all the power, leaving us none. But this is not the only way we can think about design or engineering. Despite the "lone genius" myths we tend to circulate, great scientists (like Newton) or split-brain artist-engineers (like Leonardo da Vinci) didn't work in a vacuum; their technical and artistic expertise evolved in response (and shaped) to societal visions of their times. Design is also a process that can be imaginative and speculative. For example, what if supporting user autonomy were a design principle itself?

These values—design as process, design as communication, even design as humility—can be guiding lights for how we build technology for the future. Tim Wu, a legal scholar and well-known author, recently applied the term *toxic design* to describe the most popular internet technologies and social media sites (to Facebook in particular).[9] Toxic design encourages unhealthy behavior, giving us incentives to act in ways that are counter to our best interests as human beings; it exploits our weaknesses and plays to our instinctive selves, subject to the twin forces of punishment and reward. These forces, which are at work in our bodies at all times, come from the ways our normal, central nervous system functions. The dopamine, adrenaline, and other neurotransmitter rushes that get us geared-up to pay attention or respond can be hijacked. Toxic design (and toxic technology) works the same way.[10]

Online recommendation systems are an obvious example of the problem: they can either expose us to extreme content to keep our attention, or they can enhance our addiction to affirmation, training us to obsessively log onto Instagram or Facebook in search of more likes, comments, or shares. It's not that different from "rubbernecking" when there's a car accident on the highway, or from bingeing on junk food. Both acts release dopamine because the brain believes that information related to danger and food is important to our survival.[11] In the long term though, such obsessions, attractions, and addictions can do us harm.

Good design can also be a source of empowerment, a way of delivering value to everyone. What we need to do, argues Wu, is escape from the "false loops" that make our experiences online feel perpetually incomplete and drive our endless need to check in "just one more time," whether when scrolling through our Facebook feeds or clicking on a YouTube "Up next" auto-recommendation. Google, meanwhile, follows us all over the internet, only to lure us back to its site through targeted advertising so we can sign on to additional Google products and services. Instead, Wu asks, "can social media be like, here's the stuff you check up on, and then it ends?"[12]

Related to design ethics is the potential of *digital literacy*.[13] The term might seem self-explanatory, something like "ensuring that everyone knows how to use the existing technology." But, in reality, it is far subtler and more important. Literacy, in its traditional definition, isn't just the ability to read or write—it's actually about the capacity to reflect, analyze, and create.[14] It's about taking a newspaper or magazine article, book, even a fictional story and reflecting what's behind it: who wrote it, what their assumptions were, what world they were a part of, what other information there might be on a similar subject. After all, children who can sound out the words in a book still aren't quite "reading" if they can't understand what the characters are doing or why. It's even more powerful to grasp the meaning of the story— why someone would tell it, or what cultural significance it has.

What about digital literacy? It should be about how to use a technology and reflect upon it. But we've gotten to see how easy it is to blindly trust the information that finds us rather than critically reflect on what we see. For example, assuming that Google search results are neutral, trustworthy, and "all we need to know" has become a popular strategy for trying to stay afloat in a tidal wave of digital information.

This is a huge problem for two interrelated reasons: first because the information pushed onto our phones, social media feeds, or search results does not reflect some universal truth; and second because the information it does give us is based on computational choices that support private corporate goals rather than our own. In other words, it's not as if Google might return the wrong answer to our search by mistake; rather, the answers suit whichever corporation is feeding us this information. Truth-value, social value, or our individual preferences aren't being attacked; they simply don't matter much. Our experience of the "open universe of information"—the way the internet was supposed to be—has been clouded by algorithmic goggles that filter the near-infinite possibilities available on the internet with results that masquerade as truth or knowledge.

In our secondary schools and universities, it will be important to teach building blocks for digital literacy: how different platforms are built, the basic concepts of computing even if we don't wade into code, and real glimpses into what is (or is not) happening behind the scenes when we use a system. Digital literacy is just the doorway, then, to other literacies that we must wrap our heads around to ensure that technology serves all our best interest. As we step through that door we can develop algorithmic literacy (understanding bias in AI systems, or how a search engine system works), data literacy (how/when/where data is collected, how is it aggregated and retained, by whom, and with what effects), and political and economic literacy (what technologies are owned by whom, what industries are shaped by technology in what manners, how technologies shape public and political life, and the relationships between corporate and public/political interests).

Economic Security and Our Options

What about economic security within a world of profound inequality, where it was recently estimated that the world's eight wealthiest people had equivalent wealth to the bottom half (about 3.7) of the global population?[15] It's time for our digital economy to counter the shifts we see today, this profound transfer of money and power. In part III of this book I discussed a few alternative efforts: the universal basic income (UBI), micropayments for user data, portable benefits packages, worker re-skilling, cooperatives, and new forms of collective bargaining (that follow the examples of guilds and worker councils that have long existed in various forms across the world).

We should seriously consider each of these alternatives as we convince tech moguls and government policymakers that our current path will likely lead to massive unemployment and underemployment and to increasing stratification. If we stay on this path, even those who are able to find jobs will not be secure. They will likely work harder than ever, yet make less, and many employed people will still struggle to meet the basic costs of housing, health, and food.

A tech industry infused with this compassionate ethos—*that we are all in this together*—can bridge these divides and create new jobs. In chapter 3 I discussed the problematic PredPol (predictive policing) project that uses AI machine-learning techniques to process data from crime reports. The AI decides whether a crime will be classified as "gang related" based on a collection of facts about the case, for instance the weapon used, the location of the crime, or the number of suspects. I also mentioned the blasé response of PredPol's engineers when the project was met with widespread disapproval, including the story of Hau Chan, a computer scientist based at Harvard, who was presenting PredPol research at the Artificial Intelligence Ethics and Society conference. When he encountered a firestorm of criticism he responded, "I'm just an engineer."[16] Then, according to *Science* magazine, Black Lemoine, a Google software engineer in the audience, stormed out after saying, "Once the rockets are up, who cares where they come down?"[17]

Lemoine's overall point about accountability is well taken. But should we condemn Chan? Engineers have rarely had ethics training or exposure to the critical social sciences, and they are not usually responsible for the downstream, often unanticipated, effects of the systems they create. Their attention is focused on optimizing, debugging, designing, and creating, and it's almost always been that way.

But instead of seeing this is as crisis, what if we saw it as an opportunity, a clear and open space in the digital economy where we can train and hire people to do socially conscious, dignified work? What if we saw it as a way to transform our educational systems, and an opportunity to bring professional training onto the job for existing engineers to bridge these gaps? We need to learn from examples like PredPol, not simply criticize them. It is important that we have a frank and inclusive dialogue—with engineers front and center—about whether we should be building these kinds of systems. We can use this broad conversation to identify a whole new suite of jobs, paying fair wages to evaluate, track, and otherwise oversee ethically "borderline" projects like PredPol and other applications of AI. Ethical translators, political translators, cultural translators—all of these can be jobs that allow society to connect its heart and spirit with the digital economy of tomorrow.

Consider that the EU's GDPR might create as many as 75,000 new jobs, which will have to be filled by humans, just to ensure that its requirements are being met.[18] Or consider another project, called Rising Frames, which supports minority and at-risk populations to analyze their online representation and share their voices to correct bias.[19] Ensuring that the information created through automation and algorithms is appropriate and accurate is a role that people around the world can be hired to do, and that they will be needed to do as data becomes ever more granular, local, and hyperspecific.

Conversations to identify the types of youth and inner-city work opportunities that may emerge in an AI and machine learning–shaped economy have surfaced thanks to a partnership between Van Jones, the CNN commentator and former Obama administration "green jobs czar," and Fei-Fei Li, head of AI for Google Cloud. Jones and Li started an AI summer camp for high school girls, which in turn spearheaded the organization AI4ALL to focus on jobs and training for women and minority working-class children.

Jones has led in these spaces before: he has not only advocated for green jobs, and fought for education and employment for underserved youth, but he has long stressed the importance of democratizing coding and digital literacy for the inner city. He points out, "We are going to have unintended consequences from Artificial Intelligence. But unintended consequences are one thing. Consequences based on ignorance or malice or just not caring enough is totally different."[20]

So, although AI and automation could knock out a large portion of existing jobs, it has the potential to provide quite a few new jobs. But we should move mindfully toward this future. We can't simply assume that any job coming out of AI or automation is a good one. We must make certain that new technologies serve us, as well as the firms that create them, and require that the private sector be accountable to the public it directly impacts.

We've seen that a low unemployment rate is a misleading indicator of economic well-being if someone has a job but must work harder and longer to afford the same or less. The idea that it's beneficial to give people jobs if those jobs are poorly compensated, uncreative, or even traumatizing is similarly wrongheaded. Take, for example, the job of content moderator who weeds out offensive or unacceptable material before it reaches a platform's users. Or consider call centers, another industry that has boomed across the world, though primarily in Asia and Africa. These centers require workers to adapt to the business-week hours of Western clients while earning wages far lower than their Western counterparts. The psychological and social trauma associated with these jobs should be weighed heavily when we think of jobs for the future.

It's time to look at the economic security proposals I discussed in chapters 14 and 15, such as portable benefits, worker councils, and the universal basic income (UBI) with both an open mind and critical eye. Take UBI. It's gotten a lot of attention, and I gave particular consideration to some of its proponents. But Douglas Rushkoff, one of the leading technology writers of the past thirty years, calls UBI nothing but a "booby prize," and "Silicon Valley's latest scam."[21] He characterizes it as a way of *supporting* the interests of tech moguls, not the rest of us. Why? First of all, it gives them an excuse not to pay a living wage. On top of that we need UBI, he says, because tech companies and other "1 percenters" have extracted so much wealth from the rest of us. UBI also supports big business interests because without it consumers become valueless. Rushkoff states that with UBI the ability to "create or exchange value is stripped from us, all we can do with every consumptive act is deliver more power to people who can finally, without any exaggeration, be called our corporate overlords."[22] It's a chilling vision, and it's worth taking seriously.

Why not instead consider proposals that give users power from the start? Rushkoff sees UBI as an attempt to put a bandage on the sucking chest wound of a broken system, one that will only make us ever more disenfranchised. Instead he asks us to think about approaches like *universal basic assets* being experimented upon in Denmark. In this program, people who are born poor will be provided with even greater access to shared public resources.[23]

Decentralization and Sovereignty

Throughout this book I've shared stories that shift the power dynamics associated with the internet as we know it. They are reminders that we need not follow the top-down pattern that Silicon Valley, or perhaps even the Chinese tech companies take for granted today when it comes to providing access to digital technologies. From Brooklyn to Detroit to Catalonia, we've seen successful and sustainable mesh networks created from the ground up, showing the potential for users everywhere to create and use technologies in ways that benefit them, without any hidden strings of exploitation attached.

Having power over technology's deployment can allow jobs, skills training, and educational opportunities to be dictated by communities that might otherwise might have had data and money *extracted* from them. I've shown how even indigenous peoples in the cloud forests and jungles of southern Mexico can create mobile phone networks on *their own terms*. These communities, thanks to their collective ownership, have been able to redistribute money made out from their projects to support families and local organizations. I've also shown how businesses can grow in ways that are inclusive, maintaining fidelity to and respect for the places that gave them their start. I provided a range of examples from Africa, from business conducted on street corners to perspectives from a Google-funded AI lab. The BRCK organization in particular, though birthed in Nairobi, has now become a force in changing how people worldwide think about connectivity in ways that are attuned to their realities, not just those of wealthy executives or investors sitting thousands of miles away.

These examples highlight the strengths of *decentralization*, but do not suggest their total autonomy or sovereignty. They all have dependencies, connections, and relationships with others. What's important is that decision-making power is placed in the hands of the communities themselves, which is a far cry from the standard, largely unintelligible terms of service that most of us are used to having imposed upon us. This is not only important when it comes to economic or political decisions, but is also about the support and respect of cultural diversity. We would be promoting a program of cultural homogenization if we were to move forward with a digital world that flattens its billions of users, forcing them to conform to systems created thousands of miles away by engineers who do not represent the gender, culture, class, or race of many of their users. In a world where diversity in all its forms is being threatened, we must also remember that the most vulnerable among us are at further risk. For example, networks and systems that provide their users with greater power and voice can allow indigenous languages, which may otherwise eventually go extinct, an opportunity to remain vibrant.

I've also discussed in chapter 23 the powerful ways in which blockchain and cryptocurrency technologies could be deployed to support peer-to-peer exchanges of energy and money. This technology opens up the possibility of supporting the decisions of individual users and communities in different parts of the world, allowing them to transact and exchange as they see fit. As of today, the blockchain industry is admittedly still in its infancy—it's a crowded environment flush with people looking to make money and, in some cases, erode the safety net of the state or government or pursue other uber-libertarian fantasies. But again, these examples reflect paths taken by those building and monetizing the technology *at present*, rather than where they could go in the future.

We can also look to interesting examples of how social networks might be decentralized using networks like Mastodon. This open source platform resembles Twitter but is completely different because it allows users to create their own community, based on their own rules, on their own servers. The civic media scholar and writer Ethan Zuckerman tells us that there are Mastodon servers that "have anywhere from a few hundred to a few tens of thousands of people...but they confederate: you can share information between those servers."[24] The server administrator and the members make decisions about sharing, unlike with Facebook, Google, or any other centralized platform that makes decisions from the top-down using the de facto corporate-surveillance model ubiquitous today.

Zuckerman and his colleagues at the MIT Media Laboratory are also working on a promising platform called Gobo. Its users will be able to tweak the algorithms on Facebook, Twitter, and other platforms to filter posts by gender, geography, timeliness, and other contexts. This feature could give users greater power to control what algorithms otherwise "automagically" make visible to us. We know, for example, that men dominate many tech spaces. But according to Zuckerman, Gobo could change that experience temporarily, or for as long as we wish, allowing us to say things like, "I'd like to hear from more women in my feed,' or even 'Mute all the men!'"[25]

The computer scientist Sir Timothy Berners-Lee, who created the World Wide Web, has also been displeased with the directions the internet has taken, particularly the rise of data-sniffing intermediaries that control our experiences online.[26] The inventor wants to get back to a decentralized internet, as does nearly every internet pioneer I spoke to while doing research for this book. Berners-Lee is launching Inrupt, a startup built on the decentralized web platform Solid that he and others at MIT have been working on for past few years. Inrupt will provide all of the applications we commonly use—calendar, chatting, video viewing, search, even a virtual assistant like Siri—in a package that works like Google products do, but with one huge difference: we, rather than the company behind the technology, will have control over our personal data. As for the big tech companies? "We are not talking to Facebook and Google about whether or not to introduce a complete change where all their business models are completely upended overnight." Berners-Lee says. "We are not asking their permission."[27]

Digital and Organizational Inclusion

In October 2018, I had the opportunity to present research I gathered for this book concerning algorithms, justice, and bias at a meeting where Eric Holder, the former US attorney general, gave the opening speech.

A number of business leaders who attended were interested in the increasingly popular topic of diversity and inclusion in the corporate world, including within the technology industry. At the end of his presentation I asked Holder this question: What do we do when *implicit bias* becomes the hidden blueprint by which AI systems function and technologies of all forms are designed?

All of us carry around a modicum of implicit bias. It's not foremost in our minds; in fact, it's usually subconscious and may even contradict the values we genuinely believe we hold. Compared to a consciously held belief, implicit bias is like a mental shortcut or habit running in the background when, without realizing it, we use stereotypes to make decisions and predictions about the world and the people in it. And yet we can catch the implicit bias in our behavior if we are attentive to it—for instance, when we interview prospective engineers for an open position and assume that the female applicant is less-experienced or -trained for the job. Research has shown that we can measure implicit bias, and that an increased level of racial bias predicts police shootings and infant health outcomes.[28]

By suggesting that implicit bias is "the blueprint by which AI systems function," I meant that most learning algorithms need to be able to categorize any bit of information fed into them, and that these categories, created by humans, will inevitably incorporate some amount of our culture's biases. ProPublica recently discovered that Facebook offers its advertisers more than 29,000 categories of people to target, many of which would have no meaning if they didn't invoke a whole world of culture and stereotypes.[29] The "mom" category, for instance, includes entries for "green moms," "corporate moms," "stay-at-home moms," "fit moms," "big city moms," and "trendy moms."[30] Why not working or professional moms?[31]

Holder seemed surprised but intrigued by my question about technology and bias, and nodded his head repeatedly while I spoke. He told me that the effort to get people to recognize implicit bias as a real issue of concern should be at the "top of the decision tree," that is, the first issue to consider.[32] Only from there, he said, can we overcome the biases in our society that get baked into technology.

There are several things we can do to get to the top of the decision tree. We can make sure to reinforce a consciousness of inequity and bias in our institutions, and include in our education models, be they schools or professional training sessions, the scientific work that proves such biases exist. But we can also overcome bias by placing people from underrepresented groups in positions of power right now. By shifting who designs and creates technology we can actually transform the very structures within which decisions are made.

In Silicon Valley, the inclusion of women and racial minorities is still a major issue. The state of California, however, has taken some steps to remedy this, including 2018 legislation that requires every company to have a woman on its board. This is smart; it builds on research showing how corporate boards that include women outperform those without a woman by at least 10 percent, in terms of net income and business performance.[33]

I spoke with Ed Jean-Louis, a partner at a venture capital firm called Tale Venture Partners. Jean-Louis is African American, of Haitian descent, and has had a successful career thus far, building on an MBA he received at UCLA's Anderson School of Business. But he is determined to do more than simply make money out of his business background. His firm focuses on funding founders of companies, including in the technology space, that are from minority backgrounds. He says that they represent "qualified candidates that have diverse perspectives and viewpoints of the world, which will enhance what they are trying to build. The data backs that up."[34]

And what does this evidence look like? First, teams seem to always outperform individual decision-makers, contributing to better business performance and the bottom line. In a study done by *Forbes* with more than six hundred companies, gender-diverse teams outperform individuals 73 percent of the time, and teams that include people of different ages and geographic locations have been found to make better decisions 87 percent of the time.[35] Perhaps the most conclusive study of all, by McKinsey & Company, shows a powerful correlation between the inclusion of diverse races and genders in the leadership of companies. Analyzing data gathered from hundreds of organizations and thousands of executives in the United Kingdom, Latin America, Canada, and the United States, the consultancy found statistically significant connections between a diverse leadership team and financial performance. Consider the top quartile of the companies in terms of diversity: financial returns outperformed the median in the industry by 15 percent and 35 percent when they elevated gender and race/ethnic diversity, respectively. As the study authors conclude: "More-diverse companies are better able to win top talent, and improve their customer orientation, employee satisfaction, and decision making, leading to a virtuous cycle of increasing returns."[36]

So it seems like a no-brainer to say that organizations, including technology companies, must be inclusive on every level. But changing human habits and practices is challenging, and a much slower process than simply demonstrating that the world would be a better place if they were to change. Increasing diversity in tech is a step that still hasn't been fully embraced by the industry, as the absence of women and minorities in many positions of leadership (and within Silicon Valley generally) attests. A recent study found that at Apple, 21 percent of employees are Asian, 9 percent are black, 13 percent are Hispanic, 3 percent are multiracial, and 54 percent are white.[37] Women make up only 23 percent of workers in tech roles and 32 percent of employees overall, according to Apple's own records. The numbers are similar at Google, which is 56 percent white.[38]

Some organizations, like Backstage Capital and its founder, Arlan Hamilton, have created venture capital funds to address diversity; Hamilton's has invested some $40 million in more than eighty companies founded by women, racial minorities, and LGBTQ people. Black women are the most marginalized when it comes to funding today, receiving only 0.2 percent of the overall total).[39]

But, as funders like Hamilton are profiled in business magazines like *Forbes* and invited to speak at glitzy conferences alongside figures like Howard Schultz from Starbucks, it is useful to return again to an important concern: Inclusion and diversity within an economic system that systemically harms racial minorities, women, and other vulnerable people might be like rearranging the deck chairs on the *Titanic*. Is putting a black or brown face on an extractive system, which leaves investors and CEOs wealthier than ever at the cost of everyone else, really the solution? Forget what the siding looks like: What *relationships* do we want to build into the structures that will define our future?

At the top of the food chain, though, are those who fund, manage, and build the technology ecosystem. Jean-Louis tells me that the transformation to include more diverse people in these positions of power must come. After all, shouldn't those in charge of our digital future be representative of our diversity? But there are powerful psychological, political, economic, and social reasons why those with great power want to hang onto it, raising a tantalizing question: Could diversity actually be a value that challenges the very systems in our world that have threatened it?

In a powerful piece written nearly twenty years ago, the sociologist Phil Agre laid out the distinction between *shallow diversity* and *deep diversity*, concepts he developed in relation to the information economy and the free market.[40] Shallow diversity, he points out, is all about passive inclusion, fitting a new face or figure into a system that is preset and not open to change or deep transformation. Deep diversity, on the other hand, involves the meeting of very different worldviews, experiences, languages, traditions, and perspectives.

As they practice deep diversity, organizations can ask the important basic questions about what they stand for, what they value, how they are similar to (or different from) the status quo, and even how they build and conceive of technology. As Agre tells us, "It is only through the encounter with difference that we are able to question our own assumptions, and it is only through the encounter with difference that we can tell the difference between our own heads and the radical strangeness and challenge of the real world."[41]

Conclusion

Science and technology march on. We are building nanotechnologies, constructed on the tiny scale of individual atoms and molecules, which are far faster than computers that once took up entire rooms. We are creating robots to emulate and exceed human capacity in various areas, from computer programs—like Google's AlphaGo, which can beat world champion Go players—to automated systems and machines that replace human labor. If we consider what these breakthroughs together might mean for our world, we can see ourselves approaching a place where the biological and technological are no longer separate: the cyborg fusion of man and machine. Depending on whom you ask, this is either an amazing feat for mankind or the beginning of a terrifying, species-wide nightmare.

But beyond a simplistic approach that judges whether a technology is good or evil, I believe we can ask more important questions, the kinds that will actually help us move past the malaise and get excited and mobilized about where we go now with technology: In whose image are these powerful technologies produced? In whose interest? What paths are left to conquer for those driving at breakneck speed thought the tech revolution? And for that matter, what are the dynamics built into "conquering"? Rather than produce and develop technologies without thinking much about what they may mean—socially, culturally, psychologically, even spiritually—we might want to take an honest look at where our world can heal, become more equal, dignified, and balanced.

In 2018, when the technology-writer Douglas Rushkoff gave a talk to a group of investment bankers on "the future of technology," he correctly surmised that the group would have less interest in the potential impacts of technologies than on the specific "binary choice of whether or not to invest in them."[1] He was surprised, however, to find that many of them spoke with ease about what they called "The Event," shorthand for "the environmental collapse, social unrest, nuclear explosion, unstoppable virus, or Mr. Robot hack that takes everything down."[2]

This scenario sounds like something straight out of a sci-fi novel. But maybe it's more real than it seems. What if billionaires like Peter Thiel and Elon Musk approach the digital future as a means to secure resources for themselves in the face of oncoming doom, rather than to improve humankind or protect the world from disasters caused by climate change? If anything, Rushkoff's story tells us that it is naïve to rely on the Thiels and Musks, whose allegiances are most likely to their bottom-lines. That's why *all of us* must step up to ensure that the digital world is more balanced.

As technology extends its reach, pulling in those at the margins of "development," we all have an opportunity to simply pause and listen to the voices on the other end of this newly installed line, far "beyond the valley." At the same time, the diversity in our world that can take leadership over the next era of technology is potentially larger, deeper, and more profound than ever: from Oaxaca to Kampala, London, and Beijing—the latter two are arguably the most surveilled cities in the world—we have a wider swath of experience with technology, and see its greater potential for the people it can serve, than ever before. As we look at the stories I've shared throughout this book, we can once and for all upend the myth that technology is neutral, even benevolent, regardless of who builds it and for what purposes.

In just the last three decades, a few folks sitting in Silicon Valley (and to a lesser extent in China) have managed to re-engineer our society, our governments, even how we feel, behave, and think about ourselves. We can see this in perfect, crystallized evolution as we survey the scene of the global tech industry: a picture of futuristic (maybe even misanthropic) dreaming for some, and dire straits for almost everyone else, as the industry continually builds and invests, and then dismembers and shifts its operations to automated systems or (when needed) new, cheaper, or more profitable locations around the world.

We are now waking from that spellbound slumber. What alternative visions ensure that we will all have voice and power over the technologies of our present and future? Can we ground them in values of cooperation rather than suspicion and needless competition? What models for human relationships, human thriving, and human sociality exist beyond top-down, extractive, and conquering paradigms that have now taken power over the most powerful technology industries? I've shared many in this book, but that's just a start. After all, this is the next design adventure, one that you and I will go on, this time as co-creators, if we dare.

Notes

Introduction

1. It's worth noting here that the common term "open internet" goes back to an idea from philosophers like Henri Bergson and Karl Popper, who famously developed the concept of an "open society" in opposition to the realities of authoritarian societies.

2. Sally French, "China Has 9 of the World's 20 Biggest Tech Companies," MarketWatch, May 31, 2018, https://www.marketwatch.com/story/china-has-9-of-the-worlds-20-biggest-tech-companies-2018-05-31.

3. Allen Cheng, "Alibaba vs. the World," *Institutional Investor*, July 25, 2017, https://www.institutionalinvestor.com/article/b1505pjf8xsy75/alibaba-vs-the-world.

4. Sissi Wang, "Want to See the Future of Shopping? Look to China," CBC, January 10, 2018, https://www.cbc.ca/radio/spark/379-integrated-shopping-leaving-silicon-valley-wifi-enabled-plastic-and-more-1.4474658/want-to-see-the-future-of-shopping-look-to-china-1.4474665.

5. Lily Kuo, "World's First AI News Anchor Unveiled in China," *The Guardian*, November 9, 2018, https://www.theguardian.com/world/2018/nov/09/worlds-first-ai-news-anchor-unveiled-in-china.

6. Ben Snyder, "Today in Shitty Machine Learning Startups, This Company Claims to Predict IQ, Personality, and Violent Tendencies by Applying Deep Learning to Facial Features and Bone Structure. That's Phrenology," Tweet, *@jbensnyder* (blog), November 19, 2018, https://twitter.com/jbensnyder/status/1064759174574280704?s=12.

7. E. F. Schumacher, *Small Is Beautiful: Economics as if People Mattered* (London: HarperCollins, 2010).

1   The Power of Data

1. Natalie Gagliordi, "Apple Touts New 'Town Square' Retail Store Concept," ZDNet, September 12, 2017, https://www.zdnet.com/article/apple-touts-new-town-square -retail-store-concept/.

2. Mark Zuckerberg, "Building Global Community," *Facebook* (blog), February 16, 2017, https://www.facebook.com/notes/mark-zuckerberg/building-global-community /10103508221158471/?pnref=story.

3. Cathy O'Neil, *Weapons of Math Destruction: How Big Data Increases Inequality and Threatens Democracy*, (New York: Crown, 2016).

4. Cathy O'Neil, "Transcript of 'The Era of Blind Faith in Big Data Must End,'" TED2017, April 2017, https://www.ted.com/talks/cathy_o_neil_the_era_of_blind_ faith_in_big_data_must_end/transcript.

5. Judith Butler, *The Psychic Life of Power: Theories in Subjection* (Stanford, CA: Stanford University Press, 1997).

6. Jodi Dean, *Blog Theory: Feedback and Capture in the Circuits of Drive* (Malden, MA: Polity Press, 2010).

7. Ibid., 13.

8. Joris Toonders, "Data Is the New Oil of the Digital Economy," *Wired*, July 2014, https://www.wired.com/insights/2014/07/data-new-oil-digital-economy/.

9. Richard Florida, "The Extreme Geographic Inequality of High-Tech Venture Capital," CityLab, March 27, 2018, https://www.citylab.com/life/2018/03/the-extreme -geographic-inequality-of-high-tech-venture-capital/552026/.

10. Pamela Newenham, "Supporting Start-Ups with Accelerator Programmes," *Irish Times*, August 11, 2014, https://www.irishtimes.com/business/supporting-start-ups -with-accelerator-programmes-1.1891367.

11. Jacob Poushter, "Smartphone Ownership and Internet Usage Continues to Climb in Emerging Economies," Pew Research Center's Global Attitudes Project, February 22, 2016, http://www.pewglobal.org/2016/02/22/smartphone-ownership -and-internet-usage-continues-to-climb-in-emerging-economies/.

12. Tim Adams, "Jaron Lanier: 'The Solution Is to Double down on Being Human,'" *The Guardian*, November 12, 2017, http://www.theguardian.com/technology/2017 /nov/12/jaron-lanier-book-dawn-new-everything-interview-virtual-reality.

13. Jaron Lanier, *Who Owns the Future?* (New York: Simon and Schuster, 2014), xxviii.

14. Jaron Lanier, *How We Need to Remake the Internet*, TED2018 (TED, 2018), https:// www.ted.com/talks/jaron_lanier_how_we_need_to_remake_the_internet.

15. Tim O'Reilly, *WTF? What's the Future and Why It's Up to Us* (New York: Harper-Business, 2017).

16. Edward S. Herman and Noam Chomsky, *Manufacturing Consent: The Political Economy of the Mass Media* (New York: Pantheon, 2002).

17. Noam Chomsky, interview by author, March 20, 2018.

18. Ibid.

19. Noam Chomsky, "The Most Effective Way to Restrict Democracy Is to Transfer Decision-Making from the Public Arena to Unaccountable Institutions: Kings and Princes, Priestly Castes, Military Juntas, Party Dictatorships, or Modern Corporations. Noam Chomsky," Tweet, *@noamchomskyT* (blog), April 13, 2018, https://twitter.com /noamchomskyT/status/984799440832794624.

20. Jacob Hamburger, "(Still) Manufacturing Consent: An Interview with Matt Taibbi," *Jacobin*, October 2018, https://www.jacobinmag.com/2018/10/matt-taibbi -interview-fairway-manufacturing-consent.

21. Robert M. Bond et al., "A 61-Million-Person Experiment in Social Influence and Political Mobilization," *Nature* 489, no. 7415 (September 2012): 295–298, https://doi .org/10.1038/nature11421; Robert Epstein and Ronald E. Robertson, "The Search Engine Manipulation Effect (SEME) and Its Possible Impact on the Outcomes of Elections," *Proceedings of the National Academy of Sciences* 112, no. 33 (August 18, 2015): E4512–4521, https://doi.org/10.1073/pnas.1419828112.

22. Hugo Juárez Olguín et al., "The Role of Dopamine and Its Dysfunction as a Consequence of Oxidative Stress," *Oxidative Medicine and Cellular Longevity* 2016 (2016), https://doi.org/10.1155/2016/9730467.

23. Bill Davidow, "Exploiting the Neuroscience of Internet Addiction," *The Atlantic*, July 18, 2012, https://www.theatlantic.com/health/archive/2012/07/exploiting-the -neuroscience-of-internet-addiction/259820/.

24. Yann Truong, Rod McColl, and Philip Kitchen, "Practitioners' Perceptions of Advertising Strategies for Digital Media," *International Journal of Advertising* 29, no. 5 (January 1, 2010): 709–725, https://doi.org/10.2501/S0265048710201439.

25. This is what the social theorist Michel Foucault referred to as *discourse* throughout his many books.

26. Paul R. La Monica, "Tech's Top Five Now Worth More than $3 Trillion," CNNMoney, October 31, 2017, http://money.cnn.com/2017/10/31/investing/apple -google-alphabet-microsoft-amazon-facebook-tech/index.html.

27. Paul Oliver, "According to the Data: How Social Media Matters for Protest and Activism," *NYU Center for Data Science* (blog), March 9, 2018, https://medium.com

/center-for-data-science/according-to-the-data-how-social-media-matters-for-protest
-and-activism-f88caa05b333.

28. Siva Vaidhyanathan, interview with author, February 23, 2018.

29. Laura Stevens, "Amazon's Fiscal Discipline Pays Off as It More Than Doubles
Profit," *Wall Street Journal*, April 26, 2018, https://www.wsj.com/articles/amazons
-fiscal-discipline-pays-off-as-results-beat-expectations-1524774174.

30. "3 Reasons Why AWS Is Dominating the Cloud Marketplace," Privo, May
26, 2016, https://www.privoit.com/3-reasons-why-aws-is-dominating-the-cloud
-marketplace/.

31. Ibid.

32. Prachi Bhardwaj, "Jeff Bezos Says He Liquidates a Whopping $1 Billion of
Amazon Stock Every Year to Pay for His Rocket Company, Blue Origin," Business
Insider, April 25, 2018, http://www.businessinsider.com/amazon-ceo-jeff-bezos
-liquidates-billions-to-fund-blue-origin-2018-4.

33. *Sanders. Amazon Has Gotten Too Big—CNN Video* (CNN, 2018), https://www.cnn
.com/videos/politics/2018/04/01/bernie-sanders-amazon-is-too-big-sotu.cnn.

34. Bernie Sanders, "Count to Ten. In Those Ten Seconds, Jeff Bezos, the Owner and
Founder of Amazon, Just Made More Money than the Median Employee of Amazon
Makes in an Entire Year…," Tweet, *@BernieSanders* (blog), August 21, 2018, https://
twitter.com/BernieSanders/status/1032069466052538368.

35. Franklin Foer, "Amazon Must Be Stopped," *New Republic*, October 9, 2014,
https://newrepublic.com/article/119769/amazons-monopoly-must-be-broken
-radical-plan-tech-giant.

36. Ethan Zuckerman, interview by author, February 13, 2018.

37. Jillian D'Onfro, "Google and Facebook Digital Ad Marketshare Growth: Pivotal,"
CNBC, December 20, 2017, https://www.cnbc.com/2017/12/20/google-facebook
-digital-ad-marketshare-growth-pivotal.html.

38. "Google Search Statistics—Internet Live Stats," accessed May 1, 2018, http://
www.internetlivestats.com/google-search-statistics/.

39. Elisa Shearer and Jeffrey Gottfried, "News Use Across Social Media Platforms
2017," Pew Research Center's Journalism Project, September 7, 2017, http://www
.journalism.org/2017/09/07/news-use-across-social-media-platforms-2017/.

40. "Apple Reports First Quarter Results," Apple Newsroom, February 1, 2018,
https://www.apple.com/newsroom/2018/02/apple-reports-first-quarter-results/.

41. Todd Haselton, "Facebook New Privacy Settings Don't Address Instagram,
WhatsApp," CNBC, March 28, 2018, https://www.cnbc.com/2018/03/28/facebook
-new-privacy-settings-dont-address-instagram-whatsapp.html.

42. Gregorio Zanon, "No, End-to-End Encryption Does Not Prevent Facebook from Accessing WhatsApp Chats," *Medium* (blog), April 12, 2018, https://medium.com /@gzanon/no-end-to-end-encryption-does-not-prevent-facebook-from-accessing -whatsapp-chats-d7c6508731b2.

43. "The Techlash against Amazon, Facebook and Google—and What They Can Do," *Economist*, January 20, 2018, https://www.economist.com/printedition/2018 -01-20.

44. Farhad Manjoo, "Tech's 'Frightful 5' Will Dominate Digital Life for Foreseeable Future," *New York Times*, December 21, 2017, https://www.nytimes.com/2016/01/21 /technology/techs-frightful-5-will-dominate-digital-life-for-foreseeable-future.html.

2 The Social Contract

1. Travis M. Andrews, "Hawaii Governor Didn't Correct False Missile Alert Sooner Because He Didn't Know His Twitter Password," *Washington Post*, January 23, 2018, https://www.washingtonpost.com/news/morning-mix/wp/2018/01/23/hawaii -governor-didnt-correct-false-missile-alert-sooner-because-he-didnt-know-his-twitter -password/.

2. Maureen Dowd, "The Doyenne of DNA Says: Just Chillax With Your Ex," *New York Times*, November 20, 2017, https://www.nytimes.com/2017/11/18/style/anne -wojcicki-23andme-genetics.html.

3. Similarly, we should not assume that our retinal scans are incorporated into Apple's databases, but should Apple really have such power to collect and own intimate data about our bodies?

4. Ray Kurzweil, "The Coming Merging of Mind and Machine," *Scientific American*, March 23, 2009, https://www.scientificamerican.com/article/merging-of-mind-and -machine/.

5. Ray Kurzweil, *The Singularity Is Near: When Humans Transcend Biology* (New York: Viking, 2005).

6. Andy Gregory, "Adam Curtis: Why We Need a New Kind of Radical Politics," iNews, November 7, 2017, https://inews.co.uk/culture/film/adam-curtis-politics/.

7. Friedrich Wilhelm Nietzsche, *Beyond Good and Evil: Prelude to a Philosophy of the Future* (Edinburgh: T. N. Foulis, 1907), 146.

8. Andy Gregory, "Adam Curtis."

9. "Our Company," Google, accessed June 23, 2018, //www.google.com/about /our-company/; John Durham Peters, "Google Wants to Be God's Mind: The Secret Theology of 'I'm Feeling Lucky,'" (note that the former motto not to be evil was formally discarded as Google incorporated in 2015 into its parent holding company Alphabet.), *Salon*, July 19, 2015, https://www.salon.com/2015/07/19

/google_wants_to_be_gods_mind_the_secret_theology_of_im_feeling_lucky/; Rob Siltanen, "The Real Story Behind Apple's 'Think Different' Campaign," *Forbes*, July 19, 2015, https://www.forbes.com/sites/onmarketing/2011/12/14/the-real-story-behind-apples-think-different-campaign/#15e2b76c62ab; Sheryl Hampton, "Amazon: Earth's Most Customer-Centric Company," Delight, November 13, 2013, http://delight.us/earths-most-customer-centric-company/; Craig Timberg and Jia Lynn Yang, "Jeff Bezos, The Post's Incoming Owner, Known for a Demanding Management Style at Amazon—The Washington Post," *Washington Post*, August 7, 2013, https://www.washingtonpost.com/; Lindsay Stein, "Microsoft's New 'Make What's Next' Ad Shows Girls How to Pursue STEM Careers," *AdAge*, March 7, 2017, http://adage.com/article/cmo-strategy/microsoft-s-make-ad-shows-pursue-stem/308189/.

10. In one example, recent calls for Twitter to block one of Trump's online accounts initiated public debate about whether Twitter was a platform for free speech or an arbiter of inflammatory content.

11. William Saletan, "Goon Control," *Slate*, January 11, 2013, http://www.slate.com/articles/health_and_science/human_nature/2013/01/guns_don_t_kill_people_people_kill_people_so_keep_dangerous_people_away.html.

12. Marc Lallanilla, "The Dark Side of the Nobel Prizes," *Live Science*, October 4, 2013, https://www.livescience.com/40188-dark-history-alfred-nobel-prizes.html.

13. Christina Zhao, "Is the iPhone X's Facial Recognition Racist?," *Newsweek*, December 18, 2017, http://www.newsweek.com/iphone-x-racist-apple-refunds-device-cant-tell-chinese-people-apart-woman-751263; Natasha Lomas, "FaceApp Apologizes for Building a Racist AI," *TechCrunch*, April 25, 2017, http://social.techcrunch.com/2017/04/25/faceapp-apologises-for-building-a-racist-ai/.

14. Amy Kraft, "Microsoft Shuts down AI Chatbot after It Turned into a Nazi," March 25, 2016, https://www.cbsnews.com/news/microsoft-shuts-down-ai-chatbot-after-it-turned-into-racist-nazi/.

15. Marshall McLuhan, *Understanding Media: The Extensions of Man*, Reprint edition (Cambridge, MA: The MIT Press, 1994).

3   Foreclosing the Future

1. "Monopoly," Investopedia, November 24, 2003, https://www.investopedia.com/terms/m/monopoly.asp.

2. "Citizens United v. Federal Election Commission," Oyez, accessed May 16, 2018, https://www.oyez.org/cases/2008/08-205.

3. Stacy Mitchell, "Amazon Doesn't Just Want to Dominate the Market—It Wants to Become the Market," *The Nation*, February 15, 2018, https://www.thenation.com/article/amazon-doesnt-just-want-to-dominate-the-market-it-wants-to-become-the

-market/; "One-Stop Shop: Jeff Bezos Wants You to Buy 'Everything' on Amazon," NPR.org, October 14, 2013, https://www.npr.org/2013/10/14/232204962/one-stop -shop-jeff-bezos-wants-you-to-buy-everything-on-amazon.

4. Mitchell, "Amazon Doesn't Just Want to Dominate."

5. Will Knight, "China and the US Are Bracing for an AI Showdown—in the Cloud," *MIT Technology Review*, January 31, 2018, https://www.technologyreview.com/s /610140/china-and-the-us-are-bracing-for-an-ai-showdownin-the-cloud/.

6. Sally French, "All the Companies in Jeff Bezos's Empire, in One (Large) Chart," MarketWatch, January 30, 2018, https://www.marketwatch.com/story/its-not-just -amazon-and-whole-foods-heres-jeff-bezos-enormous-empire-in-one-chart-2017 -06-21.

7. Roger McNamee, interview by author, January 16, 2018.

8. Kenneth P. Vogel, "Google Critic Ousted from Think Tank Funded by the Tech Giant," *New York Times*, August 30, 2017, https://www.nytimes.com/2017/08/30/us /politics/eric-schmidt-google-new-america.html.

9. Gideon Resnick and Sam Stein, "How the Democratic Party Is Learning to Love Being Anti-Monopoly," *The Daily Beast*, July 25, 2017, https://www.thedailybeast .com/how-the-democratic-party-is-learning-to-love-being-anti-monopoly.

10. Vogel, "Google Critic Ousted From Think Tank."

11. Ibid.

12. Barry Lynn, "I Criticized Google. It Got Me Fired. That's How Corporate Power Works," *Washington Post*, August 31, 2017, https://www.washingtonpost.com /news/posteverything/wp/2017/08/31/i-criticized-google-it-got-me-fired-thats-how -corporate-power-works/.

13. Julia Angwin and Jeff Larson, "Machine Bias," ProPublica, May 23, 2016, https:// www.propublica.org/article/machine-bias-risk-assessments-in-criminal-sentencing.

14. Ibid.

15. Ibid.

16. Becky Pettit and Bryan Sykes, "Incarceration," Stanford Center on Poverty and Inequality, 2017, https://inequality.stanford.edu/sites/default/files/Pathways _SOTU_2017_incarceration.pdf; Leah J. Floyd et al., "Adolescent Drug Dealing and Race/Ethnicity: A Population-Based Study of the Differential Impact of Substance Use on Involvement in Drug Trade," *The American Journal of Drug and Alcohol Abuse* 36, no. 2 (March 2010): 87–91, https://doi.org/10.3109/00952991003587469; Emmanuel Rudatsikira, Adamson S Muula, and Seter Siziya, "Variables Associated with Physical Fighting among US High-School Students," *Clinical Practice and*

Epidemiology in Mental Health: CP & EMH 4 (May 29, 2008): 16, https://doi.org/10.1186/1745-0179-4-16.

17. Lauren Nichol Gase et al., "Understanding Racial and Ethnic Disparities in Arrest: The Role of Individual, Home, School, and Community Characteristics," *Race and Social Problems* 8, no. 4 (December 2016): 296–312, https://doi.org/10.1007/s12552-016-9183-8.

18. Ali Winston, "A Pioneer in Predictive Policing Is Starting a Troubling New Project," The Verge, April 26, 2018, https://www.theverge.com/2018/4/26/17285058/predictive-policing-predpol-pentagon-ai-racial-bias.

19. Sungyong Seo et al., "Partially Generative Neural Networks for Gang Crime Classification with Partial Information," *Proceedings of the 2018 AAAI/ACM Conference on AI, Ethics, and Society*, 2018, https://dl.acm.org/citation.cfm?doid=3278721.3278758.

20. Jackie Wang, "'This Is a Story About Nerds and Cops': PredPol and Algorithmic Policing," *e-flux* 87 (December 2017), https://www.e-flux.com/journal/87/169043/this-is-a-story-about-nerds-and-cops-predpol-and-algorithmic-policing/.

21. Jackie Wang, *Carceral Capitalism*, Semiotext(e) Intervention Series 21 (Cambridge, MA: MIT Press, 2018), 231–232.

22. Wang, "'This Is a Story About Nerds and Cops.'"

23. Richard Winton, "California Gang Database Plagued with Errors, Unsubstantiated Entries, State Auditor Finds," *Los Angeles Times*, August 11, 2016, http://www.latimes.com/local/lanow/la-me-ln-calgangs-audit-20160811-snap-story.html.

24. Winston, "A Pioneer in Predictive Policing."

25. Will Nicol, "What Is Reddit: A Beginner's Guide to the Front Page of the Internet," July 19, 2018, https://www.digitaltrends.com/social-media/what-is-reddit/.

26. Steve Huffman, interview by author, May 8, 2018.

27. Nicol, "What Is Reddit."

28. Jay Caspian Kang, "Should Reddit Be Blamed for the Spreading of a Smear?," *New York Times*, October 19, 2018, https://www.nytimes.com/2013/07/28/magazine/should-reddit-be-blamed-for-the-spreading-of-a-smear.html.

29. Sarah Phillips, "A Brief History of Facebook," *The Guardian*, July 25, 2007, http://www.theguardian.com/technology/2007/jul/25/media.newmedia.

30. Julia Kollewe, "Google: A 10 Year Timeline," *The Guardian*, September 5, 2008, http://www.theguardian.com/business/2008/sep/05/google.google.

31. Lawrence Lessig, interview by author, March 13, 2018, and April 3, 2018.

4 Disconnection and Connection

1. Paul Virilio, *The Information Bomb*, Radical Thinkers 10 (London; New York: Verso, 2005).

2. Henry A. Kissinger, "How the Enlightenment Ends," *The Atlantic*, May 15, 2018, https://www.theatlantic.com/magazine/archive/2018/06/henry-kissinger-ai-could -mean-the-end-of-human-history/559124/.

3. Michal Kosinski, David Stillwell, and Thore Graepel, "Private Traits and Attributes Are Predictable from Digital Records of Human Behavior," *Proceedings of the National Academy of Sciences* 110, no. 15 (April 9, 2013): 5802–5805, https://doi.org/10.1073 /pnas.1218772110.

4. Wu Youyou, Michal Kosinski, and David Stillwell, "Computer-Based Personality Judgments Are More Accurate than Those Made by Humans," *Proceedings of the National Academy of Sciences* 112, no. 4 (January 27, 2015): 1036–1040, https://doi .org/10.1073/pnas.1418680112.

5. Roberto Baldwin, "Study: Facebook Likes Can Be Used to Determine Intelligence, Sexuality," *Wired*, March 12, 2013, https://www.wired.com/2013/03/facebook-like -research/.

6. Cal Parker, "Steve Jobs Never Wanted Us to Use Our Phones Like This," *New York Times*, January 25, 2019, https://www.nytimes.com/2019/01/25/opinion/sunday /steve-jobs-never-wanted-us-to-use-our-iphones-like-this.html.

7. Siva Vaidhyanathan, *Antisocial Media: How Facebook Disconnects Us and Undermines Democracy* (New York: Oxford University Press, 2018).

8. Siva Vaidhyanathan, interview by author, February 23, 2018.

9. Ibid, my emphasis.

10. Cady Lang, "Newspaper Accidentally Mistakes Alec Baldwin for Donald Trump," *Time*, February 13, 2017, http://time.com/4669465/newspaper-mistakes-alec -baldwin-for-trump/.

11. Jean Baudrillard, *Simulacra and Simulation: The Body, in Theory* (Ann Arbor: University of Michigan Press, 1994).

12. Baudrillard; Jorge Luis Borges, and Andrew Hurley, *A Universal History of Iniquity*, Penguin Classics (New York: Penguin, 2004).

13. Baudrillard, *Simulacra and Simulation*, 1.

14. McNamee, interview with author, January 17, 2018.

15. Ibid.

16. Ibid.

17. Ibid, my emphasis.

18. Ibid.

19. Julia Carrie Wong, "Former Facebook Executive: Social Media Is Ripping Society Apart," *The Guardian*, December 12, 2017, http://www.theguardian.com/technology /2017/dec/11/facebook-former-executive-ripping-society-apart; Antonio Garcia-Martinez, "I'm an Ex-Facebook Exec: Don't Believe What They Tell You about Ads," *The Guardian*, May 2, 2017, http://www.theguardian.com/technology/2017/may/02 /facebook-executive-advertising-data-comment; Nick Bilton, "'What Have I Done': Early Facebook Employees Regret the Monster They Created," *Vanity Fair*, October 12, 2017, https://www.vanityfair.com/news/2017/10/early-facebook-employees-regret -the-monster-they-created.

20. Justin Brown, "Former Facebook Executive: 'You Don't Realize It, but You Are Being Programmed,'" Ideapod, December 13, 2017, https://ideapod.com/former -facebook-executive-dont-realize-programmed/.

21. Mike Allen, "Sean Parker Unloads on Facebook: 'God Only Knows What It's Doing to Our Children's Brains,'" Axios, November 9, 2017, https://www.axios.com /sean-parker-unloads-on-facebook-god-only-knows-what-its-doing-to-our-childrens -brains-1513306792-f855e7b4-4e99-4d60-8d51-2775559c2671.html.

22. McNamee, interview, January 19, 2019.

23. Lucas Matney, "YouTube Has 1.5 Billion Logged-in Monthly Users Watching a Ton of Mobile Video," *TechCrunch*, June 22, 2017, http://social.techcrunch.com /2017/06/22/youtube-has-1-5-billion-logged-in-monthly-users-watching-a-ton-of -mobile-video/.

24. AI systems commonly use personalized data to make their recommendations, but also take into account numerous other factors and data points.

25. Paul Lewis, "'Fiction Is Outperforming Reality': How YouTube's Algorithm Distorts Truth," *The Guardian*, February 2, 2018, http://www.theguardian.com /technology/2018/feb/02/how-youtubes-algorithm-distorts-truth.

26. Steve Grove, "The CNN/YouTube Republican Debate," *Official Google Blog* (blog), November 28, 2007, https://googleblog.blogspot.com/2007/11/cnnyoutube -republican-debate.html.

27. Daria J Kuss and Olatz Lopez-Fernandez, "Internet Addiction and Problematic Internet Use: A Systematic Review of Clinical Research," *World Journal of Psychiatry* 6, no. 1 (March 22, 2016): 143–176, https://doi.org/10.5498/wjp.v6.i1.143.

28. McNamee, interview, January 19, 2018.

29. Ibid.

30. Ibid.

31. Ibid.

32. Ibid.

33. "France Investigates IPhone Slowdown," *BBC News*, January 8, 2018, http://www.bbc.com/news/world-europe-42615378.

34. Melanie Ehrenkranz, "Apple Details Alarming Labor Issues in Latest Supply Chain Report," Gizmodo, March 8, 2018, https://gizmodo.com/apple-details-alarming-labor-issues-in-latest-supply-ch-1823611397; Brian Merchant, "Life and Death in Apple's Forbidden City," *The Guardian*, June 18, 2017, http://www.theguardian.com/technology/2017/jun/18/foxconn-life-death-forbidden-city-longhua-suicide-apple-iphone-brian-merchant-one-device-extract.

35. Avery Hartmans, "Apple Makes 23 Different Dongles—and It Would Cost You $857 to Buy Them All," Business Insider, August 12, 2017, http://www.businessinsider.com/apple-dongles-2017-8; David Morgenstern, "Apple: Why the Big-Brother Attitude with Peripherals?," ZDNet, July 28, 2014, https://www.zdnet.com/article/apple-why-the-big-brother-attitude-with-peripherals/.

36. Peter Kafka, "Elizabeth Warren Says Apple, Amazon and Google Are Trying to 'Lock out' the Competition," Recode, June 29, 2016, https://www.recode.net/2016/6/29/12060804/elizabeth-warren-apple-google-amazon-competition.

37. Carole Cadwalladr and Emma Graham-Harrison, "Facebook Accused of Conducting Mass Surveillance through Its Apps," *The Guardian*, May 24, 2018, http://www.theguardian.com/technology/2018/may/24/facebook-accused-of-conducting-mass-surveillance-through-its-apps.

38. Chris Smith, "Google Hoards Way More of Your Personal Data than Facebook," *New York Post*, April 23, 2018, https://nypost.com/2018/04/23/google-hoards-way-more-of-your-personal-data-than-facebook/.

39. The idea is to follow the approaches taken in DNA and genomic sequencing to then create a "selfish" universal ledger of planetary and human data. Quote from Vlad Savov, "Google's Selfish Ledger Is an Unsettling Vision of Silicon Valley Social Engineering," The Verge, May 17, 2018, https://www.theverge.com/2018/5/17/17344250/google-x-selfish-ledger-video-data-privacy.

40. Jason Ward, "Dystopian Abuse of Microsoft's Powerful AI Camera Tech Is All but Inevitable," Windows Central, May 18, 2017, https://www.windowscentral.com/microsofts-ai-driven-camera-technology-will-almost-certainly-lead-abuse; Matt Kapko and Matthew Finnegan, "What Is Windows Hello? Microsoft's Biometrics Security System Explained," *Computerworld*, April 18, 2018, https://www

.computerworld.com/article/3244347/microsoft-windows/what-is-windows-hello
-microsofts-biometrics-security-system-explained.html; James Titcomb, "Should You
Be Worried about the IPhone X's Facial Recognition?," *Telegraph*, September 14,
2017, https://www.telegraph.co.uk/technology/2017/09/14/should-worried-iphone
-xs-facial-recognition/; Jefferson Graham and Edward C. Baig, "Facial Recognition:
IPhone Today, Tomorrow the Airport?," *USA Today*, accessed May 29, 2018, https://
www.usatoday.com/story/tech/talkingtech/2017/09/14/facial-recognition-iphone
-today-tomorrow-airport/664000001/; Thuy Ong, "Apple Is Reportedly Removing
Apps That Share Your Location Data with Third Parties," The Verge, May 9, 2018,
https://www.theverge.com/2018/5/9/17334602/apple-targeting-apps-location-data
-sharing-third-parties.

41. Matt Cagle, "Amazon Teams Up with Law Enforcement to Deploy Dangerous
New Face Recognition Technology," ACLU of Northern CA, May 22, 2018, https://
www.aclunc.org/blog/amazon-teams-law-enforcement-deploy-dangerous-new-face
-recognition-technology.

42. Natasha Singer, "Amazon Is Pushing a Technology That a Study Says Could
Be Biased," *New York Times*, January 24, 2019, https://www.nytimes.com/2019/01
/24/technology/amazon-facial-technology-study.html?action=click&module=Well
&pgtype=Homepage&section=Technology.

43. Russell Brandom, "Amazon Is Selling Police Departments a Real-Time Facial Recognition System," The Verge, May 22, 2018, https://www.theverge.com/2018/5/22
/17379968/amazon-rekognition-facial-recognition-surveillance-aclu.

44. Singer, "Amazon Is Pushing."

5   Blind Solutions

1. Evgeny Morozov, *To Save Everything, Click Here: The Folly of Technological Solutionism* (New York: PublicAffairs, 2013), xiv.

2. Ian Tucker, "Evgeny Morozov: 'We Are Abandoning All the Checks and Balances,'" *The Observer*, March 9, 2013, http://www.theguardian.com/technology
/2013/mar/09/evgeny-morozov-technology-solutionism-interview.

3. Ibid.

4. Tim Wu, *The Master Switch: The Rise and Fall of Information Empires* (New York:
Vintage, 2010), 6.

5. Tim Wu, *The Attention Merchants: The Epic Scramble to Get Inside Our Heads* (New
York: Vintage, 2017).

6. Tim Wu, interview by author, February 1, 2018.

7. Ibid.

8. George Soros, "Remarks Delivered at the World Economic Forum," George Soros, January 25, 2018, https://www.georgesoros.com/2018/01/25/remarks-delivered-at -the-world-economic-forum/.

9. Victor Tangermann, "A Timeline for Humanity's Colonization of Space," Futurism, October 30, 2017, https://futurism.com/a-timeline-for-humanitys-colonization -of-space/; Kristen V. Brown, "The 7 Most Insane Ways That Tech Enhanced the Human Body in 2015," Splinter, December 31, 2015, https://splinternews.com/the -7-most-insane-ways-that-tech-enhanced-the-human-bod-1793853808; Clay Dillow, "Will People Alive Today Have the Opportunity to Upload Their Consciousness to a New Robotic Body?," *Popular Science*, March 2, 2012, https://www.popsci.com /technology/article/2012-03/achieving-immortality-russian-mogul-wants-begin -putting-human-brains-robots-and-soon.

10. James Surowiecki, "This Woman's Viral Post on Diversity in Tech Inspired a Cultural Shift in the Industry," *MIT Technology Review*, 2017, https://www .technologyreview.com/lists/innovators-under-35/2017/visionary/tracy-chou/.

11. Tracy Chou, "Where Are the Numbers?," Medium, October 11, 2013, https:// medium.com/@triketora/where-are-the-numbers-cb997a57252.

12. Louise Matsakis, "Google Employee's Anti-Diversity Manifesto Goes 'Internally Viral,'" Motherboard, August 4, 2017, https://motherboard.vice.com/en_us/article /kzbm4a/employees-anti-diversity-manifesto-goes-internally-viral-at-google.

13. Joe Nocera, "Google Has a Diversity Problem. And a Lawsuit Problem," Bloomberg, March 7, 2018, https://www.bloomberg.com/view/articles/2018-03 -07/discriminating-against-white-men-isn-t-google-s-big-diversity-problem; Cleve R. Wootson Jr., "A Google Engineer Wrote That Women May Be Unsuited for Tech Jobs. Women Wrote Back," *Washington Post*, August 6, 2017, https://www.washingtonpost .com/news/the-switch/wp/2017/08/06/a-google-engineer-wrote-that-women-may -be-genetically-unsuited-for-tech-jobs-women-wrote-back/.

14. "Google Systemically Underpays Women, Regulator Says," *Los Angeles Times*, April 7, 2017, http://www.latimes.com/business/technology/la-fi-tn-google-women -20170407-story.html.

15. Daniel J. Hemel, "Summers' Comments on Women and Science Draw Ire," The Harvard Crimson, January 14, 2005, https://www.thecrimson.com/article/2005/1 /14/summers-comments-on-women-and-science/.

16. "Population, Female (% of Total)," The World Bank, 2017, https://data.worldbank .org/indicator/SP.POP.TOTL.FE.ZS.

17. Laura Paddison, "Workplace Automation Is Happening, And Women Will Be Hit The Hardest," *Huffington Post*, January 23, 2018, sec. Impact, https://www.huffingtonpost.com/entry/automation-displaces-women-jobs _us_5a671ba8e4b0e5630073361e.

18. Chou, "Where Are the Numbers?"

19. Tracy Chou, interview by author, December 7, 2017.

20. Chou, interview.

6   Brave New Digital World

1. Aldous Huxley, *Brave New World* (London: Chatto & Windus, 1966).

2. Yilun Wang and Michal Kosinski, "Deep Neural Networks Are More Accurate than Humans at Detecting Sexual Orientation from Facial Images," *Journal of Personality and Social Psychology* 114, no. 2 (2018): 246–257, https://doi.org/10.1037/pspa0000098.

3. Colin Marshall, "Aldous Huxley Tells Mike Wallace What Will Destroy Democracy: Overpopulation, Drugs & Insidious Technology (1958)," *Open Culture* (blog), April 4, 2018, http://www.openculture.com/2018/04/aldous-huxley-tells-mike -wallace-what-will-destroy-democracy-overpopulation-drugs-insidious-technology -1958.html.

4. Lee Rainie and Maeve Duggan, "Privacy and Information Sharing," Pew Research Center: Internet, Science & Tech, January 14, 2016, http://www.pewinternet.org /2016/01/14/privacy-and-information-sharing/; "Data Privacy: What the Consumer Really Thinks" (UK: The Direct Marketing Association, February 23, 2018), 12, https://dma.org.uk/uploads/misc/5a857c4fdf846-data-privacy—what-the-consumer -really-thinks-final_5a857c4fdf799.pdf.

5. Marshall, "Aldous Huxley."

6. Micah L. Sifry, "Facebook Wants You to Vote on Tuesday. Here's How It Messed with Your Feed in 2012," *Mother Jones*, October 31, 2014, https://www.motherjones .com/politics/2014/10/can-voting-facebook-button-improve-voter-turnout/.

7. David Axelrod, interview by author, June 13, 2018.

8. Nicole Hemner, "The Conservative War on Liberal Medial Has a Long History," *Atlantic*, January 17, 2014, https://www.theatlantic.com/politics/archive/2014/01 /the-conservative-war-on-liberal-media-has-a-long-history/283149/.

9. Axelrod, interview.

10. "Research: Media Coverage of the 2016 Election," Shorenstein Center on Media, Politics and Public Policy, September 7, 2016, https://shorensteincenter.org/research -media-coverage-2016-election/.

11. Axelrod, interview.

12. Daniel Kreiss, interview by author, December 14, 2017.

13. Michael Beschloss, "Eisenhower, an Unlikely Pioneer of TV Ads," *New York Times*, December 21, 2017, https://www.nytimes.com/2015/11/01/upshot/eisenhower-an -unlikely-pioneer-of-tv-ads.html.

14. Daniel Kreiss, *Prototype Politics: Technology-Intensive Campaigning and the Data of Democracy*, Oxford Studies in Digital Politics (New York: Oxford University Press, 2016).

15. "A History of Satellite TV," The Satellite Museum, 2013, http://www.satellite museum.com/a_history_of_satellite_tv.htm.

16. Robert W. McChesney, *Rich Media, Poor Democracy: Communication Politics in Dubious Times* (New York: The New Press, 2015); Robert W. McChesney, *Digital Disconnect: How Capitalism Is Turning the Internet Against Democracy* (New York: The New Press, 2013).

17. McChesney, *Rich Media*, 2.

18. McChesney, *Digital Disconnect*, 3.

7 Cambridge Analytica and Global Disinformation

1. "Cambridge Analytica: The Data Firm's Global Influence," BBC, March 22, 2018, http://www.bbc.com/news/world-43476762.

2. Ibid.

3. Andrew Orlowski, "Did Somebody Say Brexit? Cambridge Analytica Grilled: Brit MPs' Fake News Probe," *The Register*, February 27, 2018, https://www.theregister.co .uk/2018/02/27/dcms_fakenews_cambridge_analytica/.

4. Nicholas Confessore and Danny Hakim, "Data Firm Says 'Secret Sauce' Aided Trump; Many Scoff," *New York Times*, March 6, 2017, https://www.nytimes.com /2017/03/06/us/politics/cambridge-analytica.html.

5. Nanjala Nyabola, "Politics in the Digital Age: Cambridge Analytica in Kenya," Al Jazeera, March 22, 2018, https://www.aljazeera.com/indepth/opinion/politics -digital-age-cambridge-analytica-kenya-180322123648852.html.

6. Privacy International, "Voter Profiling in the 2017 Kenyan Election," *Privacy International* (blog), June 6, 2017, https://medium.com/@privacyint/voter-profiling -in-the-2017-kenyan-election-8d9ac1e52877.

7. "Major Cambridge Analytica Role in Kenyan Poll Denied," *Independent Uganda*, March 22, 2018, https://www.independent.co.ug/major-cambridge-analytica-role-in -kenyan-poll-denied/.

8. Matthew Rosenberg, Nicholas Confessore, and Carole Cadwalladr, "How Trump Consultants Exploited the Facebook Data of Millions," *New York Times*, March 17,

2018, https://www.nytimes.com/2018/03/17/us/politics/cambridge-analytica-trump
-campaign.html.

9. Olivia Solon and Oliver Laughland, "Cambridge Analytica Closing after Facebook
Data Harvesting Scandal," *The Guardian*, May 2, 2018, http://www.theguardian
.com/uk-news/2018/may/02/cambridge-analytica-closing-down-after-facebook-row
-reports-say; "Cambridge Analytica and SCL Elections Commence Insolvency Proceedings and Release Results of Independent Investigation into Recent Allegations,"
CA Commercial, May 2, 2018, http://ca-commercial.com/news/cambridge-analytica
-and-scl-elections-commence-insolvency-proceedings-and-release-results-3.

10. Tom Cheshire, "Behind the Scenes at Donald Trump's UK Digital War Room,"
Sky News, October 22, 2016, https://news.sky.com/story/behind-the-scenes-at-donald
-trumps-uk-digital-war-room-10626155.

11. Hannes Grassegger and Mikael Krogerus, "The Data That Turned the World
Upside Down," Motherboard, January 28, 2017, https://motherboard.vice.com/en_us
/article/mg9vvn/how-our-likes-helped-trump-win.

12. "How Trump Team Uses Social Media to Impact the Public," *Morning Joe* (MSNBC,
March 13, 2017), http://www.msnbc.com/morning-joe/watch/how-trump-team-uses
-social-media-to-impact-the-public-896561731555.

13. Jonathan Allen and Jason Abbruzzese, "Cambridge Analytica's Effectiveness
Called into Question despite Alleged Facebook Data Harvesting," NBC News, March
20, 2018, https://www.nbcnews.com/politics/politics-news/cambridge-analytica-s
-effectiveness-called-question-despite-alleged-facebook-data-n858256; Kris-Stella
Trump, "Four and a Half Reasons Not to Worry That Cambridge Analytica Skewed
the 2016 Election," *Washington Post*, March 23, 2018, https://www.washingtonpost
.com/news/monkey-cage/wp/2018/03/23/four-and-a-half-reasons-not-to-worry
-that-cambridge-analytica-skewed-the-2016-election/; Gabrielle Lynch, Justin Willis,
and Nic Cheeseman, "Cambridge Analytica's Role in African Elections Was Real but
Overstated," Quartz, March 31, 2018, https://qz.com/1242223/cambridge-analytica
-facebook-had-little-impact-on-kenya-nigeria-elections/.

14. Donie O'Sullivan, "Facebook Investigating Employee's Links to Cambridge
Analytica," CNN, March 18, 2018, http://money.cnn.com/2018/03/18/technology
/business/facebook-cambridge-analytica/index.html.

15. Richard Nieva, "Facebook Says Cambridge Analytica Had Data on 87 Million
People," CNET, April 4, 2018, https://www.cnet.com/news/facebook-says-cambridge
-analytica-had-data-on-87m-people/.

16. Lesley Stahl, "Aleksandr Kogan: The Link between Cambridge Analytica and Facebook," CBS News, September 2, 2018, https://www.cbsnews.com/news/aleksandr
-kogan-the-link-between-cambridge-analytica-and-facebook-60-minutes/.

17. "Graham on Facebook Hearing," United States Senator Lindsey Graham, April 10, 2018, https://www.lgraham.senate.gov/public/index.cfm/2018/4/graham-on -facebook-hearing.

18. Stahl, "Aleksandr Kogan."

19. Grassegger and Krogerus, "The Data That Turned the World Upside Down."

20. What's more, the incident followed Robert Mueller's indictment of thirteen Russians, stating they had used Facebook to cause "information warfare" against the United States.

21. "Suspending Cambridge Analytica and SCL Group from Facebook," *Facebook Newsroom* (blog), March 16, 2018, https://newsroom.fb.com/news/2018/03 /suspending-cambridge-analytica/.

22. James Vincent, "Academic Who Collected 50 Million Facebook Profiles: 'We Thought We Were Doing Something Normal,'" *The Verge*, March 21, 2018, https:// www.theverge.com/2018/3/21/17146342/facebook-data-scandal-cambridge -analytica-aleksandr-kogan-scapegoat.

23. David Streitfeld, Natasha Singer, and Steven Erlanger, "How Calls for Privacy May Upend Business for Facebook and Google," *New York Times*, March 24, 2018, https://www.nytimes.com/2018/03/24/technology/google-facebook-data-privacy .html.

24. CBS/AP April 11, 2018, 5:14 p.m., "Mark Zuckerberg's Testimony Reveals Congress' Confusion about Facebook," accessed April 12, 2018, https://www.cbsnews .com/news/mark-zuckerberg-testimony-reveals-congress-confusion-about-facebook/.

25. Emily Stewart, "Lawmakers Seem Confused about What Facebook Does—and How to Fix It," *Vox*, April 10, 2018, https://www.vox.com/policy-and-politics/2018 /4/10/17222062/mark-zuckerberg-testimony-graham-facebook-regulations.

26. Sheera Frenkel et al., "Delay, Deny and Deflect: How Facebook's Leaders Fought through Crisis," *New York Times*, November 16, 2018, https://www.nytimes.com /2018/11/14/technology/facebook-data-russia-election-racism.html.

27. Drew Harwell, "AI Will Solve Facebook's Most Vexing Problems, Mark Zuckerberg Says. Just Don't Ask When or How," *Washington Post*, April 11, 2018, https:// www.washingtonpost.com/news/the-switch/wp/2018/04/11/ai-will-solve-facebooks -most-vexing-problems-mark-zuckerberg-says-just-dont-ask-when-or-how/.

28. Camila Domonoske, "Mark Zuckerberg Tells Senate: Election Security Is An 'Arms Race,'" NPR.org, April 10, 2018, https://www.npr.org/sections/thetwo-way /2018/04/10/599808766/i-m-responsible-for-what-happens-at-facebook-mark -zuckerberg-will-tell-senate.

29. "Facebook CEO Mark Zuckerberg Senate Hearing on Data Protection," C-SPAN, April 10, 2018, https://www.c-span.org/video/?443543-1/facebook-ceo-mark -zuckerberg-testifies-data-protection.

30. The Honest Ads Act is a bill in the US Senate that promotes the regulation of online campaign advertisements used by companies such as Facebook and Google.

31. "Transcript of Mark Zuckerberg's Senate Hearing," *Washington Post*, April 10, 2018, https://www.washingtonpost.com/news/the-switch/wp/2018/04/10/transcript-of -mark-zuckerbergs-senate-hearing/?noredirect=on&utm_term=.dab79707060e.

32. Russell Brandom, "Facebook-Backed Lawmakers Are Pushing to Gut Privacy Law," *The Verge*, April 10, 2018, https://www.theverge.com/2018/4/10/17218756 /facebook-biometric-privacy-lobbying-bipa-illinois.

33. Natasha Lomas, "Facebook Starts Its Facial Recognition Push to Europeans," *TechCrunch* (blog), April 20, 2018, http://social.techcrunch.com/2018/04/20/just -say-no/.

34. Richard Chirgwin, "Facebook Admits It Does Track Non-Users, for Their Own Good," *The Register*, April 17, 2018, https://www.theregister.co.uk/2018/04/17 /facebook_admits_to_tracking_non_users/.

35. Caroline Kelly, "Cambridge Analytica Whistleblower: Facebook Data Could Have Come from More than 87 Million Users," CNN, April 8, 2018, https://edition .cnn.com/2018/04/08/politics/cambridge-analytica-data-millions/index.html.

36. Edward Moyer, "Cambridge Analytica Whistleblower's Startup Reportedly Had Data Access Too," CNET, March 28, 2018, https://www.cnet.com/news/facebook -cambridge-analytica-whistleblowers-startup-had-data-access-too/.

37. Noam Chomsky, interview by author, March 20, 2018.

38. For example, this was found to be the case in Sri Lanka in April 2018, when anti-Muslim rumors and false information spread like wildfire on Facebook, leading to a spate of killings and lynching (see Taub and Fisher, note 49).

39. Sarah Wildman, "The World's Fastest-Growing Refugee Crisis Is Taking Place in Myanmar. Here's Why," Vox, September 18, 2017, https://www.vox.com/world /2017/9/18/16312054/rohingya-muslims-myanmar-refugees-violence.

40. C. L. Kellow and H. L. Steeves, "The Role of Radio in the Rwandan Genocide," *Journal of Communication* 48, no. 3 (February 7, 2006): 107–128, https://doi.org/10 .1111/j.1460-2466.1998.tb02762.x.

41. Russell Smith, "The Impact of Hate Media in Rwanda," BBC News, December 3, 2003, http://news.bbc.co.uk/2/hi/africa/3257748.stm.

42. "Rwanda Radio Transcripts," Montreal Institute for Genocide and Human Rights Studies, accessed May 31, 2018, https://www.concordia.ca/content/concordia/en/research/migs/resources/rwanda-radio-transcripts.html?null.

43. Daniel Cooper, "Facebook Accused of Supporting 'Ethnic Cleansing' in Myanmar," Engadget, September 20, 2017, https://www.engadget.com/2017/09/20/facebook-accused-of-supporting-ethnic-cleansing-in-myanmar/.

44. Julia Carrie Wong, Michael Safi, and Shaikh Azizur Rahman, "Facebook Bans Rohingya Group's Posts as Minority Faces 'Ethnic Cleansing,'" *Guardian*, September 20, 2017, http://www.theguardian.com/technology/2017/sep/20/facebook-rohingya-muslims-myanmar.

45. "UN Human Rights Chief Points to 'Textbook Example of Ethnic Cleansing' in Myanmar," UN News, September 11, 2017, https://news.un.org/en/story/2017/09/564622-un-human-rights-chief-points-textbook-example-ethnic-cleansing-myanmar.

46. Wong, Safi, and Rahman, "Facebook Bans Rohingya.'"

47. Olivia Solon, "Civil Rights Groups: Facebook Should Protect, Not Censor, Human Rights Issues," *The Guardian*, October 31, 2016, http://www.theguardian.com/technology/2016/oct/31/facebook-human-rights-censorship-civil-rights-mark-zuckerberg-aclu.

48. Betsy Woodruff, "Exclusive: Facebook Silences Rohingya Reports of Ethnic Cleansing," *The Daily Beast*, September 18, 2017, https://www.thedailybeast.com/exclusive-rohingya-activists-say-facebook-silences-them.

49. Amanda Taub and Max Fisher, "Where Countries Are Tinderboxes and Facebook Is a Match," *New York Times*, April 21, 2018, https://www.nytimes.com/2018/04/21/world/asia/facebook-sri-lanka-riots.html.

50. Ezra Klein, "Mark Zuckerberg on Facebook's Hardest Year, and What Comes Next," Vox, April 2, 2018, https://www.vox.com/2018/4/2/17185052/mark-zuckerberg-facebook-interview-fake-news-bots-cambridge.

51. "Facebook Is Again Criticized for Failing to Prevent Religious Conflict in Myanmar," *TechCrunch* (blog), April 10, 2018, http://social.techcrunch.com/2018/04/10/facebook-is-again-criticized-for-failing-to-prevent-religious-conflict-in-myanmar/.

52. Samantha Bradshaw and Philip N. Howard, "Troops, Trolls and Troublemakers: A Global Inventory of Organized Social Media Manipulation," Computational Propaganda Research Project (University of Oxford, December 2017), 37.

53. Annie Gowen and Elizabeth Dwoskin, "Indians Are Wild about WhatsApp. But Some Worry It's Hurting Democracy," *Washington Post*, May 14, 2018, https://www.washingtonpost.com/world/asia_pacific/why-whatsapp-may-present-a-greater

-challenge-to-democracy-than-facebook/2018/05/14/11124fea-5630-11e8-a6d4
-ca1d035642ce_story.html.

54. David Nemer, "The Three Types of WhatsApp Users Getting Brazil's Jair Bolso-naro Elected," *The Guardian*, October 25, 2018, https://www.theguardian.com/world
/2018/oct/25/brazil-president-jair-bolsonaro-whatsapp-fake-news.

8   The Great Radicalizers

1. Michael Bonfiglio, "It's a Whole New Ball Game Now," *My Next Guest Needs No Introduction With David Letterman* (Netflix, January 12, 2018), https://www.netflix
.com/watch/80209188?trackId=13752289&tctx=0%2C3%2C83140fd0-43b5-41bd
-a7f0-9e22305fe43b-9633241%2C%2C.

2. Nyshka Chandran, "Former President Barack Obama Warns on Polarizing Media, US Electoral System," CNBC, January 12, 2018, https://www.cnbc.com/2018/01/12
/former-president-barack-obama-warns-on-polarizing-media-us-electoral-system
.html.

3. Bonfiglio, "It's a Whole New Ball Game Now."

4. Soumitra Dutta and Matthew Fraser, "Barack Obama and the Facebook Election," *US News & World Report*, November 19, 2008, https://www.usnews.com/opinion
/articles/2008/11/19/barack-obama-and-the-facebook-election.

5. Julie Bort, "Obama: Our Democracy and Economy Won't Survive If America Stays Divided," Business Insider, May 26, 2018, http://www.businessinsider.com/barack
-obama-democracy-economy-wont-survive-america-divided-2018-5.

6. Zeynep Tufekci, "YouTube, the Great Radicalizer," *New York Times*, March 11, 2018,   https://www.nytimes.com/2018/03/10/opinion/sunday/youtube-politics
-radical.html.

7. Jonathan Albright, "Untrue-Tube: Monetizing Misery and Disinformation," Medium, February 25, 2018, https://medium.com/@d1gi/untrue-tube-monetizing
-misery-and-disinformation-388c4786cc3d.

8. Ethan Zuckerman, "What Happens When You Normalize the Abnormal," CNN, November 23, 2016, https://www.cnn.com/2016/11/23/opinions/bannon-spencer
-white-supremacy-trump-zuckerman/index.html.

9. Zeynep Tufekci, *We're Building a Dystopia Just to Make People Click on Ads*, TEDGlobal>NYC, September 2017, https://www.ted.com/talks/zeynep_tufekci_we
_re_building_a_dystopia_just_to_make_people_click_on_ads.

10. Echo chambers (i.e., louder and more extreme rather than merely being echoes themselves) and filter bubbles (i.e., information being gathered about us is being used to filter our experiences of the content we see online).

11. MarketWatch, "Obama Tells Letterman How Algorithms Undermined Political Promise of Social Media," MarketWatch, January 18, 2018, https://www.marketwatch .com/story/obama-tells-letterman-how-algorithms-undermined-political-promise-of -social-media-2018-01-17.

12. Kreiss, *Prototype Politics*.

13. Christopher H Achen and Larry M. Bartels, *Democracy for Realists: Why Elections Do Not Produce Responsive Government*, Princeton Studies in Political Behavior (Princeton, NJ: Princeton University Press, 2017).

14. "Worldwide Desktop Market Share of Leading Search Engines from January 2010 to April 2018," Statista, accessed June 14, 2018, https://www.statista.com/statistics /216573/worldwide-market-share-of-search-engines/.

15. Robert Epstein and Ronald E. Robertson, "The Search Engine Manipulation Effect (SEME) and Its Possible Impact on the Outcomes of Elections," *Proceedings of the National Academy of Sciences*, August 18, 2015, https://www.pnas.org/content /112/33/E4512.

16. Robert Epstein, "How the Internet Flips Elections and Alters Our Thoughts," Aeon, February 18, 2016, https://aeon.co/essays/how-the-internet-flips-elections -and-alters-our-thoughts.

17. As they conducted their initial studies, the guess was that they would see a 2 to 3% effect based on how results were ordered. Stunningly, over a series of replicated experiments conducted within the United States, Australia, and India, they found that that voting percentages of undecided voters could shift 20% based on whether the results in this initial tier were skewed toward one candidate or party versus another, and that the top-ranked candidate could have their support increased by 48.4%.

18. Ed Sanora, *Robert Epstein Full Interview with Tucker Carlson (3/23/2018)*, accessed May 10, 2018, https://www.youtube.com/watch?v=FAfOv9VHnTs.

19. Epstein, "How the Internet Flips Elections."

20. "2016 Presidential Election Results," accessed May 10, 2018, http://www.cnn .com/election/2016/results/president. In these states his victory margin totaled to less than 1% of the vote.

21. Adam Rogers, "Google's Search Algorithm Could Steal the Presidency," *Wired*, August 6, 2015, https://www.wired.com/2015/08/googles-search-algorithm-steal -presidency/.

22. Bond et al., "A 61-Million-Person Experiment in Social Influence and Political Mobilization," *Nature* 489 (September 13, 2012), 295–298, https://www.nature.com /articles/nature11421.

23. Safiya Umoja Noble, *Algorithms of Oppression: How Search Engines Reinforce Racism* (New York: New York University Press, 2018).

24. Safiya Umoja Noble, "Google Has a Striking History of Bias against Black Girls," *Time*, March 26, 2018, http://time.com/5209144/google-search-engine-algorithm -bias-racism/.

25. Ibid.

26. Darran Simon, "Five Years after Trayvon Martin's Death, a Movement Lives On," CNN, February 26, 2017, https://www.cnn.com/2017/02/26/us/trayvon-martin -death-anniversary/index.html.

27. Aylin Caliskan, Joanna J. Bryson, and Arvind Narayanan, "Semantics Derived Automatically from Language Corpora Contain Human-like Biases," *Science* 356, no. 6334 (April 14, 2017): 183–186, https://doi.org/10.1126/science.aal4230.

28. Hannah Devlin, "AI Programs Exhibit Racial and Gender Biases, Research Reveals," *The Guardian*, April 13, 2017, http://www.theguardian.com/technology /2017/apr/13/ai-programs-exhibit-racist-and-sexist-biases-research-reveals.

29. Hannes Grassegger and Mikael Krogerus, "The Data That Turned the World Upside Down," Motherboard, January 18, 2017, https://motherboard.vice.com /en_us/article/mg9vvn/how-our-likes-helped-trump-win.

30. Michal Kosinski, interview by author, March 8, 2018.

31. Ibid. For a comprehensive account of the printing press and its affect from the Reformation onward, see: Elizabeth L. Eisenstein, *The Printing Press as an Agent of Change* (Cambridge: Cambridge University Press, 2009).

32. Kosinski, interview.

33. Mary J. DiMeglio, "Britney Spears, Mariah Carey Call for Gun Control Laws after Mass Shooting," iHeartRadio, November 6, 2017, https://www.iheart.com/content /2017-11-05-britney-spears-mariah-carey-call-for-gun-control-laws-after-mass -shooting/.

34. Kosinski, interview.

35. "What Is Personalization?—Definition from Techopedia," Techopedia .com, accessed May 30, 2018, https://www.techopedia.com/definition/14712 /personalization.

36. Edmund L. Andrews, "The Science Behind Cambridge Analytica: Does Psychological Profiling Work?," Stanford Graduate School of Business, April 12, 2018, https://www.gsb.stanford.edu/insights/science-behind-cambridge-analytica-does -psychological-profiling-work.

9   Bernie Is Born

1. "U.S. Census Bureau QuickFacts: Vermont," United States Census Bureau, accessed May 10, 2018, https://www.census.gov/quickfacts/VT.

2. Dan Merica, "Bernie Sanders Announces His Presidential Run," CNN, April 30, 2015, https://www.cnn.com/2015/04/29/politics/bernie-sanders-announces -presidential-run/index.html; Trail Guide, "Bernie Sanders Nets His First Endorse- ment from a Member of Congress," *Los Angeles Times*, October 8, 2015, http://www .latimes.com/nation/politics/trailguide/la-na-trailguide-10072015-htmlstory.html.

3. Philip Rucker, "Bernie Sanders Raises $1.5 Million in 24 Hours, Says His Campaign," *Washington Post*, May 1, 2015, https://www.washingtonpost.com/news/post-politics /wp/2015/05/01/bernie-sanders-raises-1-5-million-in-24-hours-says-his-campaign/.

4. Patrick Healy, "Hillary Clinton and Bernie Sanders Battle for Party's Future," *New York Times*, January 19, 2018, https://www.nytimes.com/2016/01/25/us/politics /hillary-clinton-bernie-sanders-democratic-party-iowa-new-hampshire.html; John Santucci, "Trump Calls Bernie Sanders 'Maniac' and 'Socialist-Slash-Communist,'" ABC News, October 15, 2015, https://abcnews.go.com/Politics/trump-calls-bernie -sanders-maniac-socialist-slash-communist/story?id=34484030.

5. Clare Foran, "Bernie Sanders's Big Money," *The Atlantic*, March 1, 2016, https:// www.theatlantic.com/politics/archive/2016/03/bernie-sanders-fundraising/471648/.

6. Eli Yokley, "America's Most and Least Popular Senators—October 2018," Morn- ing Consult, October 10, 2018, https://morningconsult.com/2018/10/10/americas -most-and-least-popular-senators-10-10/; Justin McCarthy, "Americans Maintain a Positive View of Bernie Sanders," Gallup, October 5, 2018, https://news.gallup.com /poll/243539/americans-maintain-positive-view-bernie-sanders.aspx.

7. Sam Stein and Jason Cherkis, "The Inside Story of How Bernie Sanders Became the Greatest Online Fundraiser In Political History," *Huffington Post*, June 28, 2017, https://www.huffingtonpost.com/entry/bernie-sanders-fundraising_us_59527587 e4b02734df2d92c1.

8. Adam Gabbatt, "Millennials 'Heart' Bernie Sanders: Why the Young and Hip Are #FeelingtheBern," *The Guardian*, August 20, 2015, sec. US news, http://www .theguardian.com/us-news/2015/aug/20/bernie-sanders-millennials-young-voters.

9. Many Americans reflected disenchantment with the vanishing of middle-class jobs that had come partly as a result of trade policies enacted over the past couple decades, which permitted US-headquartered corporations to solicit work overseas for far cheaper rates.

10. Dana Tims, "Bernie Sanders Brings Populist Message to Thousands at Moda Center," *The Oregonian*, March 25, 2016, http://www.oregonlive.com/politics/index .ssf/2016/03/bernie_sanders_brings_populist.html.

11. Charlotte Alter, "'Birdie Sanders' Gets a Standing Ovation at Rally," *Time*, March 25, 2016, http://time.com/4272885/bernie-sanders-bird-podium/.

12. Brad Kim, "Birdie Sanders," Know Your Meme, 2016, http://knowyourmeme .com/memes/birdie-sanders; Politico, "People Flip Out over Bird Joining Bernie at the Lectern.Pic.Twitter.Com/KKptU17mDF," Tweet, *@politico* (blog), March 25, 2016, https://twitter.com/politico/status/713474676765417477; Charlotte Harris, "Birdie Sanders: The Flight of a Meme Uniquely Positioned for Virality," *Medium* (blog), June 8, 2017, https://medium.com/@harrischarlotteb/birdie-sanders-the-flight-of-a-meme -uniquely-positioned-for-virality-d27547fb186f.

13. Rachel Dicker, "Why Is the Internet Still Obsessed With #BirdieSanders?," *US News & World Report*, March 28, 2016, https://www.usnews.com/news/articles/2016 -03-28/birdiesanders-hashtag-trends-after-bird-lands-on-bernie-sanders-podium-at-a -rally.

14. Nick Corasaniti, "Bernie Sanders Campaign Showed How to Turn Viral Moments into Money," *New York Times*, January 20, 2018, https://www.nytimes.com/2016/06 /25/us/politics/bernie-sanders-digital-strategy.html.

15. Michael D. Shear and Matthew Rosenberg, "Released Emails Suggest the D.N.C. Derided the Sanders Campaign," *New York Times*, December 21, 2017, https://www .nytimes.com/2016/07/23/us/politics/dnc-emails-sanders-clinton.html; "Bernie Sand-ers' Long and Winding Road to Backing Hillary Clinton," *PBS NewsHour*, July 12, 2016, https://www.pbs.org/newshour/show/bernie-sanders-long-winding-road-backing -hillary-clinton.

16. Keegan Goudiss, interview by author, March 9, 2018.

17. Ibid.

18. Nicole Gaudiano, "Revolution Messaging Helps Drive Sanders' 'Political Revolu-tion,'" *USA Today*, March 18, 2016, https://www.usatoday.com/story/news/politics /elections/2016/2016/03/18/revolution-messaging-helps-drive-sanders-political -revolution/81977160/.

19. These include: Justice for Flint, an advocacy group that fights for clean drinking water in a Michigan city where lead contamination has become epidemic; #NoDAPL, which partnered with the Standing Rock Sioux Native American tribe to attempt to stop the construction of the Dakota Access Oil Pipeline; and Daily Action, which gives citizens the opportunity to make one phone call a day to support one political cause. See Maeve Stier, "Bringing Justice to Flint with Donations via Text Message," Revolution Messaging, March 3, 2016, https://revolutionmessaging.com/news /revolution-messaging-helps-bring-justice-flint-donations-via-text-message; Morgan Hill, "Far from Afield: Digital Organizing at Standing Rock," Revolution Messaging, December 13, 2016, https://revolutionmessaging.com/news/digital-organizing-at -standing-rock; Carla Aronsohn, "Resisting Extremism in America, One Phone Call

at a Time," Revolution Messaging, January 30, 2017, https://revolutionmessaging
.com/news/daily-action.

20. #JusticeforFlint was second highest trending topic on Twitter after #Oscars on
the night of the awards ceremony, thanks in part to a partnership between Revolution and several well-known celebrities.

21. Chaudhary was also Obama's videographer from 2009 to 2011 and was New
Media Road director for the former president's 2008 campaign.

22. Arun Chaudhary, interview by author, April 17, 2018.

23. Ibid.

24. Ibid.

25. Ibid.

26. Gabbatt, "Millennials 'Heart' Bernie Sanders."

27. Chaudhary, interview.

28. Goudiss, interview.

29. Dae Levine, interview by author, April 18, 2018.

30. "Daily Action Case Study," Revolution Messaging, July 5, 2017, https://
revolutionmessaging.com/cases/daily-action.

31. Carla Aronsohn, interview by author, April 27, 2018.

32. Emily Witt, "How the Survivors of Parkland Began the Never Again Movement,"
New Yorker, February 19, 2018, https://www.newyorker.com/news/news-desk/how
-the-survivors-of-parkland-began-the-never-again-movement.

33. The NRA lobby continues to oppose any significant gun restrictions, even for
the types of military-grade weapons used in the shooting.

34. Brad Davis, "First National Bank of Omaha Ends Relationship with NRA,"
Omaha World-Herald, February 23, 2018, http://www.omaha.com/money/first
-national-bank-of-omaha-ends-relationship-with-nra/article_c55d1003-5c88-59a1
-83d5-3e7712863459.html.

35. Emma Sarran Webster, "Here's How to Call the President While the White House
Comment Line Is Closed," *Teen Vogue*, January 26, 2017, https://www.teenvogue
.com/story/how-to-call-white-house-comment-line-closed.

36. Daniel Victor Webster, "Here's Why You Should Call, Not Email, Your Legislators," *New York Times*, January 20, 2018, https://www.nytimes.com/2016/11/22/us
/politics/heres-why-you-should-call-not-email-your-legislators.html.

37. Aronsohn, interview.

38. Brian Hanley, "20 Reasons Bernie Sanders Is the One to Beat Trump in 2020," Medium, September 23, 2018, https://medium.com/@brianhanley_41165/20-reasons -bernie-sanders-is-the-one-to-beat-trump-in-2020-backed-by-data-593b0ad179e4 ?fbclid=IwAR1bapAyDkyIeIvJ00Fo-hRIBxnK_alxrAoacO6FPshB-_AFsgs69wIG0Ac.

39. Scott Goodstein, interview by author, March 9, 2018.

10 Digital War Games around the World

1. Noam Chomsky, interview by author, March 20, 2018.

2. Laura Rosenberger, interview by author, March 21, 2018.

3. Henry Farrell, "The Chinese Government Fakes Nearly 450 Million Social Media Comments a Year. This Is Why," *Washington Post*, May 19, 2016, https://www .washingtonpost.com/news/monkey-cage/wp/2016/05/19/the-chinese-government -fakes-nearly-450-million-social-media-comments-a-year-this-is-why/?utm_term= .ded2d92502ed.

4. Alexandra Ma, "China Has Started Ranking Citizens with a Creepy 'Social Credit' System—Here's What You Can Do Wrong, and the Embarrassing, Demeaning Ways They Can Punish You," Business Insider, April 8, 2018, http://www.businessinsider .com/china-social-credit-system-punishments-and-rewards-explained-2018-4.

5. Ibid.

6. Don Lee, "At This Chinese School, Big Brother Was Watching Students—and Charting Every Smile or Frown," *Los Angeles Times*, June 30, 2018, http://www .latimes.com/world/la-fg-china-face-surveillance-2018-story.html.

7. Anna Mitchell and Larry Diamond, "China's Surveillance State Should Scare Everyone," *The Atlantic*, February 2, 2018, https://www.theatlantic.com /international/archive/2018/02/china-surveillance/552203/.

8. Ibid.

9. Josh Rogin, "Can the Chinese Government Now Get Access to Your Grindr Profile?," *Washington Post*, January 12, 2018, https://www.washingtonpost.com/news /josh-rogin/wp/2018/01/12/can-the-chinese-government-now-get-access-to-your -grindr-profile/.

10. Tony Romm, "Facebook Granted Devices from Huawei, a Chinese Telecom Firm, Special Access to Social Data," *Washington Post*, June 5, 2018, https://www .washingtonpost.com/news/the-switch/wp/2018/06/05/facebook-granted-devices -from-huawei-a-chinese-telecom-firm-special-access-to-social-data/; Rogin, "Can the Chinese Government."

11. Stephen Nellis and Cate Cadell, "Apple Moves to Store ICloud Keys in China, Raising Human Rights Fears," Reuters, February 26, 2018, https://uk.reuters.com

/article/us-china-apple-icloud-insight/apple-moves-to-store-icloud-keys-in-china -raising-human-rights-fears-idUKKCN1G8060.

12. Brendan I. Koerner, "Inside the OPM Hack, the Cyberattack That Shocked the US Government," *Wired*, October 23, 2016, https://www.wired.com/2016/10/inside -cyberattack-shocked-us-government/.

13. Julianne Pepitone, "Clapper: China Is 'the Leading Suspect' in OPM Hacks," NBC News, June 25, 2015, https://www.nbcnews.com/tech/security/clapper-china -leading-suspect-opm-hack-n381881.

14. Rosenberger, interview.

15. Nick Harding, "'Deepfake' Videos Produced by Russian-Linked Trolls Are the Latest Weapon in Fake News War, Official Monitors Warn," *Telegraph*, May 26, 2018, https://www.telegraph.co.uk/news/2018/05/26/deepfake-videos-produced-russian -linked-trolls-latest-weapon/.

16. Glenn Greenwald and Ewen MacAskill, "NSA Prism Program Taps in to User Data of Apple, Google and Others," *The Guardian*, June 7, 2013, http://www .theguardian.com/world/2013/jun/06/us-tech-giants-nsa-data.

17. Ellen Nakashima and Joby Warrick, "Stuxnet Was Work of U.S. and Israeli Experts, Officials Say," *Washington Post*, June 2, 2012, https://www.washingtonpost .com/world/national-security/stuxnet-was-work-of-us-and-israeli-experts-officials -say/2012/06/01/gJQAlnEy6U_story.html.

11 Disrupting Jobs and Lives

1. Ginia Bellafante, "A Driver's Suicide Reveals the Dark Side of the Gig Economy," *New York Times*, February 11, 2018, https://www.nytimes.com/2018/02/06/nyregion /livery-driver-taxi-uber.html.

2. Doug Schifter, "Doug Schifter—To Those It May Concern, I Have Been Finan- cially…," *Facebook* (blog), February 5, 2018, https://www.facebook.com/permalink .php?story_fbid=1888367364808997&id=100009072541151.

3. Nikita Stewart and Luis Ferré-Sadurní, "Another Taxi Driver in Debt Takes His Life. That's 5 in 5 Months," *New York Times*, May 28, 2018, https://www.nytimes .com/2018/05/27/nyregion/taxi-driver-suicide-nyc.html.

4. Gigesh Thomas and Johan De Tavernier, "Farmer-Suicide in India: Debating the Role of Biotechnology," *Life Sciences, Society and Policy* 13 (May 11, 2017), https://doi .org/10.1186/s40504-017-0052-z.

5. "2016 TLC Factbook" (New York City Taxi and Limousine Commission, 2016), 1, http://www.nyc.gov/html/tlc/downloads/pdf/2016_tlc_factbook.pdf; Peter Szekely, "New York Taxi Drivers Call for Pay Guarantee, Limits on Cars for Hire," Reuters,

April 26, 2018, https://www.reuters.com/article/us-new-york-taxi/new-york-taxi
-drivers-call-for-pay-guarantee-limits-on-cars-for-hire-idUSKBN1HW37F.

6. Winnie Hu, "Taxi Medallions, Once a Safe Investment, Now Drag Owners into
Debt," *New York Times*, September 10, 2017, https://www.nytimes.com/2017/09/10
/nyregion/new-york-taxi-medallions-uber.html; "The History of New York's Taxi
Medallions," Taxi Intelligence, September 24, 2017, http://www.taxiintelligence
.com/the-history-of-new-yorks-taxi-medallions/.

7. Stephen Vita, "The Companies That Are Funding Uber and Lyft," Investopedia,
January 15, 2016, https://www.investopedia.com/articles/markets/011516/companies
-are-funding-uber-and-lyft.asp.

8. Heather Somerville, "True Price of an Uber Ride in Question as Investors Assess
Firm's...," Reuters, August 23, 2017, https://www.reuters.com/article/us-uber
-profitability/true-price-of-an-uber-ride-in-question-as-investors-assess-firms-value
-idUSKCN1B3103.

9. Leslie Hook, "Can Uber Ever Make Money?," *Financial Times*, June 23, 2017,
https://www.ft.com/content/09278d4e-579a-11e7-80b6-9bfa4c1f83d2.

10. Tim O'Reilly, *WTF: What's the Future and Why Is It Up to Us?* (New York: Harper-
Collins, 2018).

11. "Uber Driver-Partner Insurance—How It Works—What It Covers," Uber,
accessed August 20, 2018, https://www.uber.com/drive/insurance/.

12. Stephen Zoepf et al., "The Economics of Ride Hailing: Driver Revenue, Expenses
and Taxes," Working Paper Series (MIT Center for Energy and Environmental Policy
Research, February 2018), http://orfe.princeton.edu/~alaink/SmartDrivingCars
/PDFs/Zoepf_The%20Economics%20of%20RideHialing_OriginalPdfFeb2018.pdf;
James Doubek, "Uber, Lyft Drivers Earning a Median Profit of $3.37 per Hour, Study
Says," NPR.org, March 2, 2018, https://www.npr.org/sections/thetwo-way/2018/03
/02/590168381/uber-lyft-drivers-earning-a-median-profit-of-3-37-per-hour-study
-says.

13. Saba Waheed et al., "More Than a Gig: A Survey of Ride-Hailing Drivers in Los
Angeles" (UCLA Institute for Research on Labor and Employment, May 2018), 3–4.

14. Ibid., 4.

15. Carl Benedikt Frey and Michael A. Osborne, "The Future of Employment: How
Susceptible Are Jobs to Computerisation?," *Technological Forecasting and Social Change*
114 (January 2017): 254–280, https://doi.org/10.1016/j.techfore.2016.08.019.

16. Michael K. Spencer, "The Future of Work, Jobs in 2030 Will Be Different,"
Medium, November 30, 2017, https://medium.com/@Michael_Spencer/the-future
-of-work-jobs-in-2030-automated-f4de965d0953.

17. Jelle Visser, "Union Membership Statistics in 24 Countries," *Monthly Labor Review* 129 (January 2006): 38–49.

18. Anna Ratcliff, "Just 8 Men Own Same Wealth as Half the World," Oxfam International, January 16, 2017, https://www.oxfam.org/en/pressroom/pressreleases/2017 -01-16/just-8-men-own-same-wealth-half-world.

19. Noah Kirsch, "The 3 Richest Americans Hold More Wealth Than Bottom 50% of the Country, Study Finds," *Forbes*, November 9, 2017, https://www.forbes.com/sites /noahkirsch/2017/11/09/the-3-richest-americans-hold-more-wealth-than-bottom -50-of-country-study-finds/.

20. Hillary Hoffower and Andy Kiersz, "A Minimum-Wage Worker Needs 2.5 Full-Time Jobs to Afford a One-Bedroom Apartment in Most of the US," Business Insider, June 14, 2018, https://www.businessinsider.com/minimum-wage-worker-cant-afford -one-bedroom-rent-us-2018-6.

21. It is estimated that Bajaj exports more than 40% of its vehicles to Africa.

22. UberAfricaForum, "Uber Office in Dar es Salaam—Address & Contact," February 18, 2018, https://www.uberafricaforum.com/uber-office-dar-es-salaam/. There is one office for Tanzania in the capital city of Dar es Salaam.

23. Theodore Schleifer, "Uber's Latest Valuation: $72 Billion," Recode, February 9, 2018, https://www.recode.net/2018/2/9/16996834/uber-latest-valuation-72-billion -waymo-lawsuit-settlement.

24. Uber filed the IPO paperwork with the SEC at the end of 2018, but it plans to reach settlement with 160,000 drivers in nine states before going public. See "Uber IPO Lowered to $90B," January 8, 2019, https://www.pymnts.com/news/ipo/2019 /uber-ipo-lowered/.

25. Bellafante, "A Driver's Suicide Reveals the Dark Side."

26. Natalie Foster, "How the Independent Drivers Guild Is Helping Drivers in New York," The Aspen Institute, November 10, 2017, https://www.aspeninstitute.org/blog -posts/independent-drivers-guild-helping-drivers-in-new-york/; see IDG webpage for 2019 membership statistics, https://drivingguild.org.

27. Laura Bliss, "New York City Just Changed the Uber Game," CityLab, August 8, 2018, https://www.citylab.com/transportation/2018/08/new-york-city-moves-to-cap -uber-and-lyft/566924/.

28. Aric Jenkins, "Why Uber Doesn't Want a Built-In Tipping Option," *Fortune*, April 18, 2017, http://fortune.com/2017/04/18/uber-tipping-nyc/.

29. Karen Matthews, "NYC Drivers for Uber, Other Apps to Get Vision Care Coverage," ABC News, June 29, 2018, https://abcnews.go.com/Health/wireStory/nyc-drivers -uber-apps-vision-care-coverage-56266151.

30. Independent Drivers Guild, "Uber and Lyft Drivers Petition NYC for a Ride-Hail Minimum Wage," Independent Drivers Guild, March 20, 2018, https://drivingguild .org/2018/03/20/progresspr/.

31. Rebecca Smith, "On New York City's Report Urging Higher Pay for Uber and Lyft Drivers," National Employment Law Project, July 2, 2018, https://www.nelp.org /news-releases/new-york-citys-report-urging-higher-pay-uber-lyft-drivers/; Joy Bork-holder et al., "Uber State Interference: How Transportation Network Companies Buy, Bully, and Bamboozle Their Way To Deregulation," National Employment Law Proj-ect, January 2018, http://www.forworkingfamilies.org/sites/pwf/files/publications /Uber%20State%20Interference%20Jan%202018.pdf.

32. Smith, "On New York City's Report."

33. Maura Dolan and Andrew Khouri, "California's Top Court Makes It More Dif-ficult for Employers to Classify Workers as Independent Contractors," *Los Angeles Times*, April 30, 2018, http://www.latimes.com/local/lanow/la-me-ln-independent -contract-20180430-story.html.

34. Josh Eidelson, "Gig-Economy Giants Ask California to Save Them from a Ruling That May Turn Their Contractors into Employees," *Los Angeles Times*, August 6, 2018, http://www.latimes.com/business/la-fi-contract-workers-20180806-story.html.

35. Mong Palatino, "Uber Faces More Regulation in Southeast Asia," *The Diplomat*, November 27, 2014, https://thediplomat.com/2014/11/uber-faces-more-regulation -in-southeast-asia/.

36. Julia Kollewe and Gwyn Topham, "Uber Apologises after London Ban and Admits 'We Got Things Wrong,'" *The Guardian*, September 25, 2017, http://www .theguardian.com/business/2017/sep/25/uber-tfl-concerns-vows-keep-operating -london-licence.

37. Jon Russell, "Singapore Will Introduce New Regulations for Taxi Booking Apps in 2015," *TechCrunch* (blog), November 21, 2014, http://social.techcrunch.com/2014 /11/20/singapore-will-introduce-new-regulations-for-taxi-booking-apps-in-2015/.

38. Thuy Ong, "EU's Top Court Rules That Uber Is a Transportation Service," *The Verge*, December 20, 2017, https://www.theverge.com/2017/12/20/16799970/eu -court-uber-regulation-taxi.

39. "Directive 98/48/EC of the European Parliament and of the Council of 22 June 1998," Council of the European Union, July 20, 1998, https://eur-lex.europa .eu/LexUriServ/LexUriServ.do?uri=CONSLEG:1998L0034:20070101:EN:PDF; Brian Fung, "Uber Isn't a Tech Company—It's Basically a Taxi Company, E.U. Court Adviser Says," *Washington Post*, May 11, 2017, https://www.washingtonpost.com/news/the -switch/wp/2017/05/11/uber-isnt-a-tech-company-its-basically-a-taxi-company-eu -court-adviser-says/.

40. Faiz Siddiqui, "Services like UberPool Are Making Traffic Worse, Study Says," *Washington Post*, July 25, 2018, https://www.washingtonpost.com/amphtml/news /dr-gridlock/wp/2018/07/25/a-new-study-says-services-like-uberpool-are-making -traffic-worse/?noredirect=on.

41. Dave Johnson, "Silicon Valley Rising Fights for Worker Justice," *Huffington Post*, December 6, 2017, https://www.huffingtonpost.com/dave-johnson/silicon-valley -rising-fig_b_6809240.html.

42. Leonid Bershidsky, "Cities Tell Airbnb, Like Uber, to Grow Up," Domain, December 13, 2017, https://www.domain.com.au/news/cities-including-paris-berlin -and-barcelona-tell-airbnb-like-uber-to-grow-up-20171213-h03k12/.

43. Lisa Ward, "How Airbnb Affects Home Prices and Rents," *Wall Street Journal*, October 23, 2017, https://www.wsj.com/articles/how-airbnb-affects-home-prices -and-rents-1508724361; David Wachsmuth et al., "The High Cost of Short-Term Rentals in New York City," School of Urban Planning McGill University, January 30, 2018, 49.

44. Oliver Gee, "Paris in New Crackdown on Illegal Airbnb Flats," The Local, January 13, 2016, https://www.thelocal.fr/20160113/paris-cracks-down-on-illegal-airbnb -flats.

45. Feargus O'Sullivan, "Berlin Bans Most Airbnb-Style Rentals," CityLab, April 28, 2016, http://www.citylab.com/housing/2016/04/airbnb-rentals-berlin-vacation -apartment-law/480381/.

46. Erin Hudson, "Defying Amazon's Opposition, Seattle Has a New Affordable Housing Head Tax," The Real Deal New York, May 19, 2018, https://therealdeal.com /2018/05/19/defying-amazons-opposition-seattle-has-a-new-affordable-housing -head-tax/; David Streitfeld and Claire Ballentine, "Seattle Repeals Tax That Upset Amazon," https://www.nytimes.com/2018/06/12/technology/seattle-tax-amazon .html.

47. Nancy Dahlberg, "The Gig Economy Is Big and Here to Stay: How Workers Survive and Thrive," *Chicago Tribune*, September 6, 2017, http://www.chicagotribune .com/business/ct-gig-economy-workers-20170906-story.html.

48. Nathan Gardels, "French Reforms Aim for a New Social Contract in the Age of Disruption," *Washington Post*, December 1, 2017, https://www.washingtonpost.com /news/theworldpost/wp/2017/12/01/macron-economy/.

49. GAO, "Contingent Workforce: Size, Characteristics, Earnings, and Benefits," US Government Accountability Office, April 20, 2015, https://www.gao.gov/assets /670/669766.pdf; Lawrence Katz and Alan Krueger, *The Rise and Nature of Alternative Work Arrangements in the United States, 1995–2015* (Cambridge, MA: National Bureau of Economic Research, September 2016), https://doi.org/10.3386/w22667.

50. Eduardo Porter, "Is the Populist Revolt Over? Not If Robots Have Their Way," *New York Times*, February 13, 2018, https://www.nytimes.com/2018/01/30/business /economy/populist-politics-globalization.html.

12 Protecting Work and Workers

1. Josh Eidelson, "Unions Are Losing Their Decades-Long 'Right-to-Work' Fight," *Bloomberg*, February 16, 2017, https://www.bloomberg.com/news/articles/2017-02 -16/unions-are-losing-their-decades-long-right-to-work-fight; Timothy Noah, "Does Labor Have a Death Wish?," *Politico Magazine*, November 7, 2017, http://politi.co /2zGlnk4.

2. "In 5–4 Ruling, Supreme Court Deals Major Blow to Workers' Rights," *Democracy Now!*, May 22, 2018, https://www.democracynow.org/2018/5/22/headlines/in_5_4 _ruling_supreme_court_deals_major_blow_to_workers_rights.

3. Lawrence Hurley, "Supreme Court Delivers Blow to Organized Labor in Fees Dispute," Reuters, June 27, 2018, https://www.reuters.com/article/us-usa-court-unions /u-s-high-court-poised-to-issue-major-labor-ruling-as-term-ends-idUSKBN1JN0H2.

4. Andries Bezuidenhout, Malehoko Tshoaedi, and Christine Bischoff, *Labour Beyond Cosatu: Mapping The Rupture In South Africa's Labour Landscape* (Johannesburg, South Africa: Wits University Press, 2017).

5. Michael Penn, "Japanese Labour Unions Feel Pain of New Era," Al Jazeera, February 18, 2013, https://www.aljazeera.com/indepth/features/2013/02/201321713 641911842.html; Niall McCarthy, "Which Countries Have The Highest Levels Of Labor Union Membership?," *Forbes*, June 20, 2017, https://www.forbes.com/sites /niallmccarthy/2017/06/20/which-countries-have-the-highest-levels-of-labor-union -membership-infographic/.

6. "JR East's Largest Labor Union Has Lost 70% of Membership since Feb.," *Mainichi Daily News*, June 7, 2018, https://mainichi.jp/english/articles/20180607/p2a/00m /0na/019000c.

7. "Labor Unions at Major Japanese Firms Secure Wide Pay Hikes," *Mainichi Daily News*, March 15, 2018, https://mainichi.jp/english/articles/20180315/p2a/00m/0na /014000c.

8. Barçın Yinanç, "Trade Unionization in Turkey Remains Very Low: Professor Kılıç," *Hürriyet Daily News*, April 30, 2018, http://www.hurriyetdailynews.com/trade -unionization-in-turkey-remains-very-low-professor-kilic-131065.

9. Steven Greenhouse, "Fight for $15: Meet the Enfant Terrible Who Turned Minimum Wage into a National Battle," *The Guardian*, April 9, 2016, https://www .theguardian.com/us-news/2016/apr/09/fight-for-15-meet-the-enfant-terrible -behind-the-battle-for-us-minimum-wage.

10. David Rolf, "Toward a 21st-Century Labor Movement," *The American Prospect*, April 18, 2016, http://prospect.org/article/toward-21st-century-labor-movement.

11. Ibid.

12. "Building a Portable Benefits System for Today's World," Uber Newsroom, January 23, 2018, https://www.uber.com/newsroom/building-portable-benefits-system-todays-world/.

13. Josh Eidelson, "Uber-Union Proposal on Benefits Met with Skepticism from Labor," Bloomberg, January 25, 2018, https://www.bloomberg.com/news/articles/2018-01-25/uber-union-proposal-on-benefits-met-with-skepticism-from-labor.

14. Palak Shah, interview by author, May 4, 2018.

15. Charlotte, "Fair Care Pledge Building on Successful First Year," Care.com, June 20, 2016, https://www.care.com/c/stories/1577/fair-care-pledge-building-on-successful-first/.

16. "Good Work Code," accessed December 3, 2018, https://www.ndwalabs.org/good-work-code/.

17. "What's Airbnb's Living Wage Pledge?," Airbnb, accessed December 3, 2018, https://www.airbnb.com/help/article/1975/what-s-airbnb-s-living-wage-pledge.

18. "Introducing the Airbnb Economic Empowerment Agenda," Airbnb Citizen, March 13, 2017, https://www.airbnbcitizen.com/introducing-airbnb-economic-empowerment-agenda/.

19. Stephanie Zacharek, Eliana Dockterman, and Haley Sweetland Edwards, "TIME Person of the Year 2017: The Silence Breakers," *Time*, 2017, http://time.com/time-person-of-the-year-2017-silence-breakers/.

20. Vasudha Desikan, interview by author, January 23, 2018.

21. Brenna Houck, "Meet Amy Poehler's Golden Globes Date, Restaurant Activist Saru Jayaraman," Eater, January 8, 2018, https://www.eater.com/2018/1/8/16862718/golden-globes-amy-poehler-seth-meyers-times-up-saru-jayaraman-roc-united.

22. "National Labor Organizations with Membership over 100,000," InfoPlease, accessed August 20, 2018, https://www.infoplease.com/business-finance/labor-unions/national-labor-organizations-membership-over-100000.

23. Doug Bloch, interview by author, April 16, 2018.

24. Gina Chon, "Teamsters Union Tries to Slow Self-Driving Truck Push," *New York Times*, August 11, 2017, https://www.nytimes.com/2017/08/11/business/dealbook/teamsters-union-tries-to-slow-self-driving-truck-push.html?rref=collection%2Ftimestopic%2FInternational%20Brotherhood%20of%20

Teamsters&action=click&contentCollection=timestopics&region=stream&module =stream_unit&version=latest&contentPlacement=7&pgtype=collection.

25. Rafael Navar, interview by author, December 28, 2017.

26. "Union Members Summary," Bureau of Labor Statistics, January 19, 2018, https://www.bls.gov/news.release/union2.nr0.htm.

27. Navar, interview.

28. Ibid.

29. Ibid.

30. Liz Shuler, interview by author, March 7, 2018.

31. Alma Hernandez, interview author, February 7, 2018.

32. Adam Nagourney, Natalie Kitroeff, and Tim Arango, "Downturn Looms as Leadership Test for California, World's 5th Largest Economy," *New York Times*, October 31, 2018, https://www.nytimes.com/2018/10/10/us/california-economic-recession -jerry-brown.html.

33. Hernandez, interview.

34. Jessica Yarvin, "The Rise of the Anti-PAC Democrat," *PBS NewsHour*, April 5, 2018, https://www.pbs.org/newshour/politics/the-rise-of-the-anti-pac-democrat.

35. Navar, interview.

36. David Weil, *The Fissured Workplace: Why Work Became So Bad for So Many and What Can Be Done to Improve It* (Cambridge, MA: Harvard University Press, 2014).

37. Stephen Forte, "The Gig Economy's Impact on the Future of Work," Medium, July 20, 2016, https://medium.com/fusion-by-fresco-capital/the-gig-economys -impact-on-the-future-of-work-b6f82194b535; Lauren Weber and Melissa Korn, "Where Did All the Entry-Level Jobs Go?," *Wall Street Journal*, August 6, 2014, http://www.wsj.com/articles/want-an-entry-level-job-youll-need-lots-of-experience -1407267498.

38. David Weil, "The Fissured Workplace," Heller School for Social Policy & Management, Brandeis University, 2017, https://useconference.files.wordpress.com /2017/11/weil.pdf; Robert Reich, "Almost 80% of US Workers Live from Paycheck to Paycheck. Here's Why," *The Guardian*, July 29, 2018, https://www.theguardian.com /commentisfree/2018/jul/29/us-economy-workers-paycheck-robert-reich.

39. These numbers are adjusted for inflation and prices/purchasing power.

40. Raj Chetty et al., "The Fading American Dream: Trends in Absolute Income Mobility Since 1940," National Bureau of Economic Research Working Paper, December 2016, https://www.nber.org/papers/w22910.

41. David Leonhardt, "The American Dream, Quantified at Last," *New York Times*, January 20, 2018, https://www.nytimes.com/2016/12/08/opinion/the-american-dream -quantified-at-last.html.

42. Richard Dobbs et al., "Poorer than Their Parents? Flat or Falling Incomes in Advanced Economies," McKinsey Global Institute, July 2016, https://www.mckinsey .com/~/media/McKinsey/Featured%20Insights/Employment%20and%20Growth /Poorer%20than%20their%20parents%20A%20new%20perspective%20on%20 income%20inequality/MGI-Poorer-than-their-parents-Flat-or-falling-incomes-in -advanced-economies-Full-report.ashx; Laura Gardiner, "Stagnation Generation: The Case for Renewing the Intergenerational Contract," Intergenerational Commission Report (Resolution Foundation, July 2016), 5, https://www.resolutionfoundation .org/app/uploads/2016/06/Intergenerational-commission-launch-report.pdf.

43. Leonhardt, "The American Dream."

44. Rory Cellan-Jones, "Bankrupt Kodak Sells off Patents to Investors for $525m," *BBC News*, December 19, 2012, http://www.bbc.com/news/technology-20787024; "The Rise and Fall of Kodak: By the Numbers," *The Week*, October 3, 2011, http:// theweek.com/articles/481308/rise-fall-kodak-by-numbers.

45. Emil Protalinski, "Why Facebook Acquired Instagram for $1 Billion," ZDNet, April 9, 2012, https://www.zdnet.com/article/why-facebook-acquired-instagram-for -1-billion/; Julian Gavaghan and Lydia Warren, "Instagram's 13 Employees Share $100m as CEO Set to Make $400m Reveals He Once Turned Down a Job at Face- book," *Daily Mail*, April 9, 2012, http://www.dailymail.co.uk/news/article-2127343 /Facebook-buys-Instagram-13-employees-share-100m-CEO-Kevin-Systrom-set-make -400m.html.

46. Reich, "Almost 80% of US Workers."

47. Katz and Krueger, "The Rise and Nature of Alternative Work."

48. David Weil, "Improving Workplace Conditions through Strategic Enforce- ment," A Report to the Wage and Hour Division, Boston University, May 2010, 1, https://www.dol.gov/whd/resources/strategicEnforcement.pdf.

49. Philip Perry, "47% of Jobs Will Disappear in the Next 25 Years, According to Oxford University," *Big Think*, December 27, 2016, http://bigthink.com/philip-perry /47-of-jobs-in-the-next-25-years-will-disappear-according-to-oxford-university.

50. Gabriel Falcon, "Testing a Universal Basic Income," CBS News, April 15, 2018, https://www.cbsnews.com/news/testing-a-universal-basic-income-stockton -california-economic-security-project/.

51. Pew Research estimate in Janet Morrissey, "When Robots Ring the Bell," *New York Times*, November 12, 2018, https://www.nytimes.com/2018/11/07/business/robotics -automation-productivity-jobs.html.

52. Jeremie Capron, quoted ibid.

53. Ceylan Yeginsu, "If Workers Slack Off, the Wristband Will Know. (And Amazon Has a Patent for It.)," *New York Times*, February 8, 2018, https://www.nytimes.com /2018/02/01/technology/amazon-wristband-tracking-privacy.html.

54. Ibid.

55. Emily Guerin, "Robots Steal Port Jobs—but They Also Fight Climate Change," KPCC Southern California Public Radio, January 22, 2018, https://www.scpr.org /news/2018/01/22/79969/robots-that-steal-port-jobs-also-fight-climate-cha/.

56. Karen Harris, Austin Kimson, and Andrew Schwedel, "Labor 2030: The Collision of Demographics, Automation and Inequality," Bain & Company, February 7, 2018, http://www.bain.com/publications/articles/labor-2030-the-collision-of-demographics -automation-and-inequality.aspx.

57. Michael Chui, James Manyika, and Mehdi Miremadi, "Four Fundamentals of Workplace Automation," McKinsey & Company, November 2015, https:// www.mckinsey.com/business-functions/digital-mckinsey/our-insights/four -fundamentals-of-workplace-automation.

58. Michael Chui, James Manyika, and Mehdi Miremadi, "The Countries Most (and Least) Likely to Be Affected by Automation," *Harvard Business Review*, April 12, 2017, https://hbr.org/2017/04/the-countries-most-and-least-likely-to-be-affected-by -automation.

59. "Impact d'Internet Sur l'économie Française" (Paris, France: McKinsey & Company, March 2011), http://owni.fr/files/2011/03/internet_impact_rapport _mcKinseycompany.pdf.

60. Stephen Hawking, "Science AMA Series: Stephen Hawking AMA Answers!," *Reddit* (blog), July 27, 2016, https://www.reddit.com/r/science/comments/3nyn5i /science_ama_series_stephen_hawking_ama_answers/.

61. "World Inequality Report 2018," World Inequality Lab, 2018, http://wir2018. wid.world/files/download/wir2018-full-report-english.pdf.

62. Alexander Eichler and Michael McAuliff, "Income Inequality Reaches Gilded Age Levels, Congressional Report Finds," *Huffington Post*, December 6, 2017, https:// www.huffingtonpost.com/2011/10/26/income-inequality_n_1032632.html.

63. Ibrahim Balkhy, "Economist Thomas Piketty Explains Why Income Inequality Is Just Getting Started," *Huffington Post*, April 16, 2014, https://www.huffingtonpost .com/2014/04/16/thomas-piketty-inequality_n_5159937.html.

64. Thomas Piketty, *Capital in the Twenty-First Century*, trans. Arthur Goldhammer (Cambridge, MA: Belknap Press, 2017).

65. Ben Bland, "Robots Will Boost Rather than Destroy Jobs, ADB Says," *Financial Times*, April 11, 2018, https://www.ft.com/content/69602a90-3d4a-11e8-b9f9-de94fa33a81e.

66. Sarah T. Roberts, "Social Media's Silent Filter," *Atlantic*, March 8, 2017, https://www.theatlantic.com/technology/archive/2017/03/commercial-content-moderation/518796/.

67. "Human Augmented Intelligence (HAI)," N2, accessed May 18, 2018, http://www.n2sglobal.com/human-augmented-intelligence-hai/.

68. Kotaro Hara et al., "A Data-Driven Analysis of Workers' Earnings on Amazon Mechanical Turk," Arvix, December 14, 2017, https://arxiv.org/abs/1712.05796.

69. Christian Fuchs, *Digital Labour and Karl Marx* (London: Routledge, 2014); Tiziana Terranova, "Free Labor: Producing Culture for the Digital Economy," *Social Text* 18 (Summer 2000): 33–58.

70. Mark Graham, Isis Hjorth, and Vili Lehdonvirta, "Digital Labour and Development: Impacts of Global Digital Labour Platforms and the Gig Economy on Worker Livelihoods," *Transfer: European Review of Labour and Research* 23, no. 2 (May 1, 2017): 135–162, https://doi.org/10.1177/1024258916687250.

71. Creative Destruction Lab, *Jack Clark and Tim Hwang—How to Build a Responsible AI Future*, 2016, https://www.youtube.com/watch?v=Ax0ddrP8_ho; "Creative Destruction Lab," Creative Destruction Lab, accessed August 21, 2018, https://www.creativedestructionlab.com/.

13   Working Hard, Struggling Harder

1. Warren's background, education, and previous careers did not launch her on a straight path to politics: she comes from a working-class family from the mostly rural and relatively poor state of Oklahoma, served as a public school teacher, went to law school, and eventually taught law at University of Pennsylvania and Harvard.

2. Erik Sherman, "The Gig Economy Won't Save Workers At $3.37 An Hour," *Forbes*, March 3, 2018, https://www.forbes.com/sites/eriksherman/2018/03/03/the-gig-economy-wont-save-workers-at-3-37-an-hour/.

3. Robert Booth, "700,000 Gig Workers Paid below National Minimum Wage," *The Guardian*, February 7, 2018, sec. Business, http://www.theguardian.com/business/2018/feb/07/death-dpd-courier-don-lane-tragedy-business-secretary.

4. Elizabeth Warren, remarks made at Neue House, Hollywood, CA, May 3, 2018.

5. Ibid.

6. Tami Luhby, "Almost Half of US Families Can't Afford Basics like Rent and Food," CNNMoney, May 17, 2018, http://money.cnn.com/2018/05/17/news/economy/us -middle-class-basics-study/index.html.

7. "ALICE 2016 Executive Summary: A Multi-State Comparison," ALICE Project, United Ways, Winter 2016, https://www.dropbox.com/s/8rs2iurjqwyioic/16UW%20 ALICE%20Report_MultiStatesSummery_12.23.16_Lowres.pdf?dl=0; United Way ALICE Project, "51 Million U.S. Households Can't Afford Basics," PR Newswire, May 17, 2018, https://www.prnewswire.com/news-releases/51-million-us-households-cant -afford-basics-300650535.html.

8. "Senator Warren Lays Out Steps to Protect Workers in the 'Gig Economy,'" Elizabeth Warren, May 19, 2016, https://www.warren.senate.gov/newsroom/press -releases/senator-warren-lays-out-steps-to-protect-workers-in-the-and-quotgig -economy-and-quot.

9. Peter S. Goodman, "The Robots Are Coming, and Sweden Is Fine," *New York Times*, December 27, 2017, https://www.nytimes.com/2017/12/27/business/the -robots-are-coming-and-sweden-is-fine.html.

10. Ibid.

11. Paul Kunert, "European HP Workers Take IT Giant to Court over 'High Handed' Job Cuts," *The Register*, November 12, 2012, https://www.theregister.co.uk/2012/11 /12/hp_ewc_suing/.

12. Ananya Bhattacharya, "The Robots Will Actually Create More Jobs for High-Skilled Tech Workers in India," Quartz, September 7, 2017, https://qz.com/1070560 /the-robots-will-actually-create-more-jobs-for-high-skilled-tech-workers-in-india/.

13. "Income Inequality Gets Worse; India's Top 1% Bag 73% of the Country's Wealth, Says Oxfam," *Business Today*, January 23, 2018, https://www.businesstoday .in/current/economy-politics/oxfam-india-wealth-report-income-inequality-richests -poor/story/268541.html.

14. Ted Johnson, "AT&T CEO Calls for Congress to Pass 'Internet Bill of Rights,'" *Variety*, January 24, 2018, https://variety.com/2018/digital/news/att-net-neutrality -internet-bill-of-rights-1202674949/; Marrian Zhou, "AT&T Lets NSA Hide and Surveil in Plain Sight, The Intercept Reports," CNET, June 25, 2018, https://www .cnet.com/news/at-t-lets-nsa-hide-and-surveil-in-plain-sight-the-intercept-reports/; Marguerite Reardon, "With No Net Neutrality Rules, Will Mergers Like AT&T-Time Warner Kill the Internet?," CNET, June 15, 2018, https://www.cnet.com/news/why -net-neutrality-supporters-are-cringing-at-the-at-t-time-warner-merger/.

15. John Donovan, quoted in Thomas L. Friedman, *Thank You for Being Late: An Optimist's Guide to Thriving in the Age of Accelerations* (New York: Farrar, Straus and Giroux, 2016), 233.

16. John Palmer, interview by author, March 8, 2018.

17. AT&T Inc., "2017 Annual Report," AT&T, 2018, 10, http://www.attproxy.com /~/media/Files/A/ATT-Proxy/documents/2017-letter-to-shareholders.pdf.

18. Friedman, *Thank You for Being Late*, 216.

14 Money for Everybody?

1. Matt Orfalea, *Elon Musk Says Universal Basic Income Is "Going to Be Necessary"* (2017), YouTube, https://www.youtube.com/watch?v=e6HPdNBicM8.

2. Julia Carrie Wong, "Tesla Workers Were Seriously Hurt More Than Twice as Often as Industry Average," *the Guardian*, May 24, 2017, https://www.theguardian .com/technology/2017/may/24/tesla-factory-workers-injuries-higher-than-industry -average; Sam Levin, "Elon Musk Apologizes for Tesla Workers Paid Just $5 an Hour by Subcontractor," *The Guardian*, May 16, 2016, https://www.theguardian.com /technology/2016/may/16/elon-musk-tesla-wages-apology.

3. Annie Nova, "Universal Basic Income: U.S. Support Grows as Finland Ends Its Trial," CNBC, May 1, 2018, https://www.cnbc.com/2018/05/01/nearly-half-of-americans -believe-a-universal-basic-income-could-be-the-answer-to-automation-.html.

4. Catherine Clifford, "2020 Presidential Candidate Wants to Give Everyone $1,000 a Month," MSN, April 11, 2018, https://www.msn.com/en-us/money/markets/2020 -presidential-candidate-wants-to-give-everyone-dollar1000-a-month/ar-AAvLSi7.

5. Andrew Yang, "IamA Andrew Yang, Candidate for President of the U.S. in 2020 on Universal Basic Income AMA!," *Reddit* (blog), March 2018, https://www.reddit.com/r /IAmA/comments/87aa2z/iama_andrew_yang_candidate_for_president_of_the/.

6. Amy Cools, "Paine on Basic Income and Human Rights," Thomas Paine National Historical Association, accessed May 22, 2018, http://thomaspaine.org/paine-on -basic-income-and-human-rights.html.

7. James Surowiecki, "The Case for Free Money," *New Yorker*, June 13, 2016, https:// www.newyorker.com/magazine/2016/06/20/why-dont-we-have-universal-basic -income.

8. Nina Renata Aron, "As One of His Final Acts, Martin Luther King Fought for a Basic Income for All," Timeline, April 2, 2018, https://timeline.com/mlk-wanted -universal-basic-income-as-his-last-campaign-b0ba61aa77b8.

9. Jordan Weissmann, "Martin Luther King's Economic Dream: A Guaranteed Income for All Americans," *Atlantic*, August 28, 2013, https://www.theatlantic .com/business/archive/2013/08/martin-luther-kings-economic-dream-a-guaranteed -income-for-all-americans/279147/.

10. Sam Altman, interview by author, January 26, 2018.

11. Ryan Sit, "Sarah Palin Shocked over California City's Universal Basic Income, Which Is Based on Alaska's Model," *Newsweek*, April 3, 2018, http://www.newsweek .com/sarah-palin-stockton-california-universal-basic-income-alaska-permanent -fund-869729.

12. Scott Santens, "On the Record: Bernie Sanders on Basic Income," Basic Income, January 29, 2016, https://medium.com/basic-income/on-the-record-bernie-sanders -on-basic-income-de9162fb3b5c; Kyle Lewis and Will Stronge, "A Right-Wing Think Tank Is Now Supporting Universal Basic Income—but They've Missed the Point," *Independent*, January 19, 2018, http://www.independent.co.uk/voices/universal -basic-income-adam-smith-institute-austerity-libertarian-a8167701.html.

13. "History of the Alaska Permanent Fund," Alaska Permanent Fund Corporation, accessed December 3, 2018, https://apfc.org/who-we-are/history-of-the-alaska -permanent-fund/.

14. Dylan Matthews, "The Amazing True Socialist Miracle of the Alaska Permanent Fund," Vox, February 13, 2018, https://www.vox.com/policy-and-politics/2018/2/13 /16997188/alaska-basic-income-permanent-fund-oil-revenue-study. This could be due to either people shifting out of full-time work and into part-time work because of increased income, or people who weren't working prior to the cash payment are driven to start working because the dividend enables them to afford fixed costs associated with employment.

15. Christopher Arris Oakley, "Indian Gaming and the Eastern Band of Cherokee Indians," *North Carolina Historical Review* 78, no. 2 (2001): 133–155.

16. Daniela Perdomo, "The Great Equalizer: What Dolly Parton & a Tennessee Wildfire Can Teach about Universal Basic Income," In the Mesh, April 19, 2018, https:// www.inthemesh.com/archive/dolly-parton-universal-basic-income/.

17. "UT Releases Preliminary of 'My People Fund' Evaluation," November 16, 2017, https://news.utk.edu/2017/11/16/ut-releases-preliminary-findings-people-fund -evaluation/.

18. Chris Weller, "This German Non-Profit Has given Away Half a Million Dollars in Free Salaries," Business Insider, August 23, 2016, http://www.businessinsider.com /germany-basic-income-lottery-2016-8.

19. Claire Suddath, "What If Everyone Got a Monthly Check From the Government?," Bloomberg, January 11, 2018, https://www.bloomberg.com/news/features /2018-01-11/what-if-everyone-got-a-monthly-check-from-the-government.

20. Government of Ontario, "Ontario Basic Income Pilot," Ontario.ca, April 23, 2017, https://www.ontario.ca/page/ontario-basic-income-pilot.

21. "A Guaranteed Income Demonstration," Stockton Economic Empowerment Demonstration, accessed May 24, 2018, https://www.stocktondemonstration.org/.

22. TEDx Talks, *Upset the Setup: Michael Tubbs at TEDxSanJoaquin*, 2012, https://www.youtube.com/watch?v=3rolJu_puqo.

23. Edward-Isaac Dovere, "Can This Millennial Mayor Make Universal Basic Income a Reality?," *Politico Magazine*, April 24, 2018, https://politi.co/2vIo5oh.

24. Aaron Sankin, "Oprah Winfrey Endorses Michael Tubbs: Media Mogul Moneybombs Stockton City Council Race," *Huffington Post*, May 21, 2012, https://www.huffingtonpost.com/2012/05/21/oprah-winfrey-michael-tubbs_n_1534549.html.

25. "A Guaranteed Income Demonstration."

26. Dovere, "Can This Millennial Mayor Make Universal Basic Income a Reality?"

27. Alix Langone, "Why This 27-Year-Old Mayor Is Giving His City's Poorest Residents $500 a Month—No Strings Attached," *Time*, April 17, 2018, http://time.com/money/5243564/why-this-27-year-old-mayor-is-giving-his-citys-poorest-residents-500-a-month-no-strings-attached/.

28. Michael Tubbs, interview by author, May 15, 2018.

29. Ibid.

30. "Highest to Lowest—Prison Population Rate," World Prison Brief, accessed June 9, 2018, http://www.prisonstudies.org/highest-to-lowest/prison_population_rate?field_region_taxonomy_tid=All.

31. Peter Wagner and Bernadette Rabuy, "Following the Money of Mass Incarceration," Prison Policy Initiative, January 25, 2017, https://www.prisonpolicy.org/reports/money.html.

32. Adam Looney, "5 Facts about Prisoners and Work, before and after Incarceration," Brookings, March 14, 2018, https://www.brookings.edu/blog/up-front/2018/03/14/5-facts-about-prisoners-and-work-before-and-after-incarceration/.

33. Tubbs, personal communication.

15 Worker-Owned Technologies

1. Trebor Scholz and Nathan Schneider, eds., *Ours to Hack and to Own: The Rise of Platform Cooperativism, a New Vision for the Future of Work and a Fairer Internet* (New York: OR Books, 2016).

2. "The Shore Porters Society—History," The Shore Porters Society, accessed December 3, 2018, https://www.shoreporters.com/about-us/history.html.

3. James Sullivan, "List of Top 100 Co-Ops in USA Released," Co-operative News, December 15, 2015, https://www.thenews.coop/100093/sector/retail/list-top-100-co-ops-usa-released/.

4. Nathan Schneider, "The Rise of a Cooperatively Owned Internet," *Nation*, October 13, 2016, https://www.thenation.com/article/the-rise-of-a-cooperatively-owned-internet/.

5. "Fairmondo," Fairmondo, accessed December 3, 2018, https://www.fairmondo.de/global; "Strategies," Resonate, accessed December 3, 2018, https://resonate.is/strategies/; "MIDATA.coop," MIDATA, accessed December 3, 2018, https://www.midata.coop.

6. Scholz and Schneider, *Ours to Hack and to Own*, 29.

7. Marjorie Kelly, *Owning Our Future: The Emerging Ownership Revolution* (San Francisco: Berrett-Koehler Publishers, 2012).

8. Jeremy Corbyn, "The Digital Democracy Manifesto," August 2016, https://d3n8a8pro7vhmx.cloudfront.net/corbynstays/pages/329/attachments/original/1472552058/Digital_Democracy.pdf?1472552058.

9. Matt Burgess, "What Jeremy Corbyn's Digital Manifesto Will Mean for the UK," *Wired*, August 31, 2016, http://www.wired.co.uk/article/jeremy-corbyn-digital-democracy-manifesto.

10. Nathan Schneider, interview by author, December 12, 2017.

11. Julie Goldsmith and David Yap, "OWS Screenprinters Is Bringing Printing to the People," Occupy Wall Street, 2014, http://occupywallstreet.net/story/ows-screenprinters-bringing-printing-people.

12. "Top 300 Cooperatives Generate 2.5 Trillion Dollars in Annual Turnover," Inter Press Service, July 3, 2017, http://www.ipsnews.net/2017/07/top-300-cooperatives-generate-2-5-trillion-dollars-annual-turnover/.

13. "Exploring the Cooperative Economy Report 2018," World Cooperative Monitor (International Co-operative Alliance), accessed December 3, 2018, https://monitor.coop/sites/default/files/publication-files/wcm-2018en-1276015391.pdf.

14. "Our DNA," Ouishare, accessed May 29, 2018, https://www.ouishare.net/our-dna.

15. "The International Co-Operative Alliance," International Co-operative Alliance, accessed March 22, 2019, https://ica.coop/en/about-us/our-members/global-cooperative-network.

16. Nathan Schneider, "10 Lessons from Kenya's Remarkable Cooperatives," Shareable, May 4, 2015, https://www.shareable.net/blog/10-lessons-from-kenyas-remarkable-cooperatives.

17. Schneider, interview.

18. Arun Sundararajan, *The Sharing Economy: The End of Employment and the Rise of Crowd-Based Capitalism* (Cambridge, MA: The MIT Press, 2016).

19. Scholz and Schneider, *Ours to Hack and to Own.*

16 Discrimination Technologies

1. Mark Graham, Stefano De Sabbata, and Matthew A. Zook, "Towards a Study of Information Geographies: (Im)Mutable Augmentations and a Mapping of the Geographies of Information," *Geo: Geography and Environment* 2, no. 1 (June 1, 2015): 88–105, https://doi.org/10.1002/geo2.8. This may partly be a result of how search engines deem certain content "acceptable," but is also a result of the fact that internet access remains higher in more advanced economies. Chinese-language content is prevalent online for most of the internet's users but most of it tends to be accessed and circulated only among Chinese speakers, with only some Western content available in the nation of over 1.38 billion, due to firewalls and censorship.

2. Kate Crawford, "Artificial Intelligence's White Guy Problem," *New York Times*, January 20, 2018, https://www.nytimes.com/2016/06/26/opinion/sunday/artificial -intelligences-white-guy-problem.html.

3. Alex Hern, "Google's Solution to Accidental Algorithmic Racism: Ban Gorillas," *The Guardian*, January 12, 2018, http://www.theguardian.com/technology/2018/jan /12/google-racism-ban-gorilla-black-people.

4. Natasha Lomas, "FaceApp Apologizes for Building a Racist AI," *TechCrunch* (blog), April 25, 2017, http://social.techcrunch.com/2017/04/25/faceapp-apologises-for -building-a-racist-ai/.

5. Sophie Kleeman, "Here Are the Microsoft Twitter Bot's Craziest Racist Rants," Gizmodo, March 24, 2016, https://gizmodo.com/here-are-the-microsoft-twitter-bot -s-craziest-racist-ra-1766820160.

6. Sidney Fussell, "Amazon's Face Recognition Misidentifies 28 Members of Congress as Suspected Criminals," Gizmodo, July 26, 2018, https://gizmodo.com /amazons-face-recognition-misidentifies-28-members-of-co-1827887567.

7. Ibid.

8. Lee Fang, "Amazon Promises 'Unwavering' Commitment to Police, Military Clients Using AI Technology," The Intercept, July 30, 2018, https://theintercept.com /2018/07/30/amazon-facial-recognition-police-military/.

9. Ibid.

10. Jennings Brown, "IBM Watson Reportedly Recommended Cancer Treatments That Were 'Unsafe and Incorrect,'" Gizmodo, July 25, 2018, https://gizmodo.com /ibm-watson-reportedly-recommended-cancer-treatments-tha-1827868882.

11. Ibid.

12. Lee Jared Vinsel, "Why Carmakers Always Insisted on Male Crash-Test Dummies," Boston.com, August 22, 2012, https://www.boston.com/cars/news-and-reviews/2012 /08/22/why-carmakers-always-insisted-on-male-crash-test-dummies.

13. Julia Powles, "New York City's Bold, Flawed Attempt to Make Algorithms Accountable," New Yorker, December 21, 2017, https://www.newyorker.com/tech /elements/new-york-citys-bold-flawed-attempt-to-make-algorithms-accountable.

14. Aylin Caliskan, Joanna J. Bryson, and Arvind Narayanan, "Semantics Derived Automatically from Language Corpora Contain Human-like Biases," Science 356, no. 6334 (April 14, 2017): 183–186, http://science.sciencemag.org/content/356/6334 /183.

15. Lawrence T White, "Is Implicit Bias a Useful Scientific Concept?," Psychology Today, June 23, 2017, https://www.psychologytoday.com/blog/culture-conscious /201706/is-implicit-bias-useful-scientific-concept; Devlin, "AI Programs Exhibit Racial and Gender Biases, Research Reveals."

16. Agamoni Ghosh, "Are Robots Racist and Sexist? Study Reveals AI Picks up Bias from Human Language," International Business Times, April 14, 2017, https://www .ibtimes.co.uk/are-robots-racist-sexist-study-reveals-ai-picks-bias-human-language -1617006; Aylin Caliskan, Joanna J. Bryson, and Arvind Narayanan, "Supplementary Materials for Semantics Derived Automatically from Language Corpora Contain Human-like Biases," Science 356 no. 6334 (April 14, 2017 [suppl.]), http://science .sciencemag.org/content/sci/suppl/2017/04/12/356.6334.183.DC1/Caliskan-SM.pdf.

17. Adam Hadhazy, "Biased Bots: Artificial-Intelligence Systems Echo Human Prejudices," Princeton University, April 18, 2017, https://www.princeton.edu/news/2017 /04/18/biased-bots-artificial-intelligence-systems-echo-human-prejudices.

18. Thor Benson, "Your Next Job Interview Could Be with a Racist Bot," The Daily Beast, April 21, 2018, sec. tech, https://www.thedailybeast.com/your-next-job -interview-could-be-with-a-racist-bot.

19. "Black Mirror Awards and Nominations," IMDb, accessed August 25, 2018, http://www.imdb.com/title/tt2085059/awards.

20. Joe Wright, "Nosedive," Black Mirror (Netflix, October 21, 2016), http://www .imdb.com/title/tt2085059/.

21. Gabrielle Bruney, "A 'Black Mirror' Episode Is Coming to Life in China," Esquire, March 17, 2018, https://www.esquire.com/entertainment/a19467976/black-mirror

-social-credit-china/; Christina Zhao, "'Black Mirror' in China? 1.4 Billion Citizens to Be Monitored Through Social Credit System," *Newsweek*, May 1, 2018, https://www.newsweek.com/china-social-credit-system-906865; Jennifer Bisset, "Black Mirror Too Real in China as Schools Shun Parents with Bad Social Credit," CNET, May 2, 2018, https://www.cnet.com/news/black-mirror-too-real-in-china-as-schools-shun -parents-with-bad-social-credit/; Alice Vincent, "Black Mirror Is Coming True in China, Where Your 'Rating' Affects Your Home, Transport and Social Circle," *Telegraph*, December 15, 2017, https://www.telegraph.co.uk/on-demand/2017/12/15 /black-mirror-coming-true-china-rating-affects-home-transport/.

22. Alexandra Ma, "China Has Started Ranking Citizens with a Creepy 'Social Credit' System—Here's What You Can Do Wrong, and the Embarrassing, Demeaning Ways They Can Punish You," Business Insider, October 29, 2016, https://www.businessinsider.com/china-social-credit-system-punishments-and-rewards -explained-2018-4.

23. Chris Merriman, "China's Black Mirror 'Social Credit' Has Already Stopped 11 Million from Taking Flights," *Inquirer*, May 22, 2018, https://www.theinquirer.net /inquirer/news/3032765/chinas-black-mirror-social-credit-has-already-stopped-11m -from-taking-flights.

24. Chris Baynes, "Chinese Police Will Start Using Facial Recognition to Automatically Fine Jaywalkers by Text," *Independent*, March 29, 2018, https://www .independent.co.uk/news/world/asia/china-police-facial-recognition-technology-ai -jaywalkers-fines-text-wechat-weibo-cctv-a8279531.html.

25. David Graeber, *Debt: The First 5,000 Years* (Brooklyn, NY: Melville House, 2011).

26. Justine Rivero, "From Credit Scores to 'Behavioral Scores': What Numbers Say about You," *Forbes*, October 28, 2011, https://www.forbes.com/sites/moneywisewomen /2011/10/28/from-credit-scores-to-behavioral-scores-what-numbers-say-about-you/.

27. Peter Thiel, "The Education of a Libertarian," Cato Unbound, April 13, 2009, https://www.cato-unbound.org/2009/04/13/peter-thiel/education-libertarian.

28. Ibid.

29. Angela Glover Blackwell, "The Curb-Cut Effect," *Stanford Social Innovation Review* 15, no. 1 (Winter 2017), https://ssir.org/articles/entry/the_curb_cut_effect.

30. Ibid.

31. "Policy Impact: Seat Belts," Policy Impact (Centers for Disease Control and Prevention, January 2011), 8.

32. Blackwell, "The Curb-Cut Effect."

17 Keeping Network Power Local

1. Certain blocked websites or other pieces of content would remain so for legal and other reasons. And websites that monetize the open internet, such as Google's search engine, would continue to benefit from net neutrality.

2. "What Is Net Neutrality?," American Civil Liberties Union, December 2017, https://www.aclu.org/issues/free-speech/internet-speech/what-net-neutrality.

3. Kaleigh Rogers, "Rural America Is Building Its Own Internet Because No One Else Will," Motherboard, August 29, 2017, https://motherboard.vice.com/en_us/article/paax9n/rural-america-is-building-its-own-internet-because-no-one-else-will.

4. Ibid.

5. Jason Koebler, "To Save Net Neutrality, We Must Build Our Own Internet," Motherboard, November 21, 2017, https://motherboard.vice.com/en_us/article/7x4y8a/net-neutrality-fcc-community-networks.

6. Roger Baig et al., "Guifi.Net, a Crowdsourced Network Infrastructure Held in Common," *Computer Networks: The International Journal of Computer and Telecommunications Networking* 90, no. C (October 2015): 150–165, https://doi.org/10.1016/j.comnet.2015.07.009.

7. Sascha Meinrath, interview by author, May 9, 2018.

8. Kaleigh Rogers, "Ignored by Big Telecom, Detroit's Marginalized Communities Are Building Their Own Internet," Motherboard, November 16, 2017, https://motherboard.vice.com/en_us/article/kz3xyz/detroit-mesh-network.

9. Ibid.

10. Baig et al., "Guifi.Net, a Crowdsourced Network Infrastructure Held in Common."

11. Ramon Roca and Roger Baig Viñas, interview by Ramesh Srinivasan, April 3, 2018.

12. Elinor Ostrom, "Reformulating the Commons," *Ambiente & Sociedade* 5, no. 20 (September 2002).

13. Aditi Mehta, "The Politics of Community Media in the Post-Disaster City" (Doctoral Dissertation, Massachusetts Institute of Technology, 2018), 211, http://dspace.mit.edu/handle/1721.1/115714.

14. Ibid.

15. Ibid.

16. Ibid., 227.

17. Ibid.

18. Ibid.

19. Jim Burress, "The Big Disconnect: Google Fiber's Unfulfilled Promise in Atlanta," WABE, May 14, 2018, https://www.wabe.org/googlefiber/.

20. Susan Crawford, "Cities Are Teaming Up to Offer Broadband, and the FCC Is Mad," *Wired*, September 27, 2018, https://www.wired.com/story/fcc-southern -california-broadband-collective/.

21. Meinrath, interview.

22. Leandro Navarro, "Network Infrastructures: The Common Model for Local Participation Governance and Sustainability. Association for Progressive Communications" (Association for Progressive Communications, February 2018), 10, https:// www.apc.org/sites/default/files/network_link_final_307.pdf.

18   Questioning Connectivity

1. Thomas L. Friedman, *Thank You for Being Late: An Optimist's Guide to Thriving in the Age of Accelerations* (New York: Farrar, Straus and Giroux, 2016).

2. "Smartphone Penetration to Reach 66% in 2018," Zenith, October 16, 2017, https://www.zenithmedia.com/smartphone-penetration-reach-66-2018/.

3. "Mobile Phone Users Worldwide 2015–2020," Statista, 2018, https://www.statista .com/statistics/274774/forecast-of-mobile-phone-users-worldwide/.

4. Surabhi Mittal and Mamta Mehar, "How Mobile Phones Contribute to Growth of Small Farmers? Evidence from India," *Quarterly Journal of International Agriculture* 51, no. 3 (2012): 18.

5. Ramesh Srinivasan, "Bridges between Cultural and Digital Worlds in Revolutionary Egypt," *Information Society* 29, no. 1 (January 2013): 49–60, https://doi.org/10 .1080/01972243.2012.739594; Heather Brown, Emily Guskin, and Amy Mitchell, "The Role of Social Media in the Arab Uprisings," Pew Research Center, November 28, 2012, http://www.journalism.org/2012/11/28/role-social-media-arab-uprisings/.

6. "About Rhizomatica," Rhizomatica, accessed December 3, 2018, https://www .rhizomatica.org/about/.

7. For information about the nation's active social media presence, see: "Social Media in Cameroon Is Bustling, Blogging Less So," oAfrica, April 19, 2014, http:// www.oafrica.com/statistics/social-media-in-cameroon-is-bustling-blogging-less-so/.

8. Andrew Fanasia, "PM Confident with Cable Deal—Solomon Star News," *Solomon Star News*, July 13, 2018, http://www.solomonstarnews.com/index.php/news /national/item/20716-pm-confident-with-cable-deal.

9. India Bourke, "Will China's New 'Silk Road' Bring a New Golden Age of Trade or Trample Lands and People?," *New Statesman*, February 2, 2018, https://www .newstatesman.com/politics/energy/2018/02/will-china-s-new-silk-road-bring-new -golden-age-trade-or-trample-lands-and.

10. Daniel Ren, "New Silk Road Offers a US$7.5b New Market for China's Solar Energy Firms," *South China Morning Post*, August 21, 2017, https://www.scmp.com /business/companies/article/2107648/new-silk-road-offers-us75b-new-market -chinas-solar-energy-firms; Robbie Gramer, "All Aboard China's 'New Silk Road' Express," *Foreign Policy*, January 4, 2017, https://foreignpolicy.com/2017/01/04/all -aboard-chinas-new-silk-road-express-yiwu-to-london-train-geopolitics-one-belt-one -road/.

11. Katie Benner, "Justice Dept. Backs Suit Accusing Facebook of Violating Fair Housing Act," *New York Times*, August 18, 2018, https://www.nytimes.com/2018 /08/17/us/politics/justice-dept-facebook-fair-housing.html; Dan Levine and Joseph Menn, "Exclusive: U.S. Government Seeks Facebook Help to Wiretap Messenger…," Reuters, August 17, 2018, https://www.reuters.com/article/us-facebook-encryption -exclusive/u-s-government-seeks-facebook-help-to-wiretap-messenger-sources -idUSKBN1L226D; Susan McFarland and Danielle Haynes, "Justice Dept. Charges 13 Russians, 3 Groups with Interfering in 2016 Election," UPI, February 16, 2018, https://www.upi.com/Top_News/US/2018/02/16/Justice-Dept-charges-13-Russians-3 -groups-with-interfering-in-2016-election/8391518805324/.

12. Lulu Yilun Chen, "TenCent Is Now Worth $72 Billion More than Facebook," Bloomberg, March 20, 2018, https://www.bloomberg.com/news/articles/2018-03-21 /tencent-pulls-72-billion-ahead-as-facebook-s-woes-deepen-chart.

13. Simon Kemp, "The Incredible Growth of the Internet over the Past Five Years— Explained in Detail," The Next Web, March 6, 2017, https://thenextweb.com /insider/2017/03/06/the-incredible-growth-of-the-internet-over-the-past-five-years -explained-in-detail/.

14. Ibid.

19 African-Born Technology

1. Zara Stone, "Everything You Need To Know About Sophia, The World's First Robot Citizen," *Forbes*, November 7, 2017, https://www.forbes.com/sites/zarastone /2017/11/07/everything-you-need-to-know-about-sophia-the-worlds-first-robot -citizen/#634a4a2846fa. During the *60 Minutes* interview Sophia said to Rose, "I've been waiting for you." When the incredulous Rose replied, "Waiting for me?" Sophia responded, "Not really, but it makes for a good pickup line." Bryan Clark, "Humanoid 'Sophia' Puts the Moves on 60 Minutes Correspondent Charlie Rose,"

TNW, October 10, 2016, https://thenextweb.com/artificial-intelligence/2016/10/10/humanoid-robot-sofia-puts-the-moves-on-60-minutes-correspondent-charlie-rose/.

2. Daniel Mumbere, "Sophia the Robot Misses Dinner with Ethiopia PM after Losing Some Parts at German Airport," Africanews, June 30, 2018, http://www.africanews.com/2018/06/30/sophia-the-robot-misses-dinner-with-ethiopia-pm-after-losing-some-parts-at/.

3. Lily Kuo, "Kenya's Mobile Internet Beats the United States for Speed," Quartz Africa, June 8, 2017, https://qz.com/africa/1001477/kenya-has-faster-mobile-internet-speeds-than-the-united-states/.

4. Toby Shapshak, "Alibaba Founder Jack Ma to Launch $10 Million African Entrepreneurial Prize," *Forbes*, August 7, 2018, https://www.forbes.com/sites/tobyshapshak/2018/08/07/alibaba-founder-jack-ma-to-launch-10m-african-entrepreneurial-prize/.

5. Toby Shapshak, "To Create Jobs in Africa Encourage Young People and Give Start-ups Tax Breaks—Alibaba's Jack Ma," *Forbes*, August 8, 2018, https://www.forbes.com/sites/tobyshapshak/2018/08/08/to-create-jobs-in-africa-encourage-young-people-and-give-startups-tax-breaks-alibabas-jack-ma/.

6. Jake Bright, "Africa Roundup: Goldman Backs Startup African Internet Group to Become Continent's 1st Unicorn," *TechCrunch*, March 4, 2016, http://social.techcrunch.com/2016/03/04/africa-roundup-goldman-backs-startup-african-internet-group-to-become-continents-1st-unicorn/.

7. Alexis Akwagyiram, "Goldman Sachs Invests in Africa's 'First Unicorn,'" *Fortune*, March 3, 2016, http://fortune.com/2016/03/03/goldman-sachs-africa-internet-group/.

8. Bright, "Africa Roundup."

9. Daniel Mumbere, "Uber, Taxify Drivers in Kenya Protest against 'Working Long Hours for Little Money,'" Africanews, July 3, 2018, http://www.africanews.com/2018/07/03/uber-taxify-drivers-in-kenya-protest-against-working-long-hours-for-little-money/.

10. Ibid.

11. David Okwii, "Uber vs Taxify vs Safeboda Uganda Prices and Feature Comparison," Dignited, February 12, 2018, https://www.dignited.com/26966/uber-vs-taxify-vs-vs-safeboda-uganda-prices-and-feature-comparison/; "Uber in Uganda: A Ride for Everyone," *Uber* (blog), March 4, 2019, https://www.uber.com/en-UG/blog/uber-ride-options-uganda/.

12. "Population," United Nations, December 14, 2015, http://www.un.org/en/sections/issues-depth/population/.

13. Richard Kweitsu, "Africa's Growing Youthful Population: Reflections on a Continent at a Tipping Point," Mo Ibrahim Foundation, April 6, 2017, http://mo.ibrahim.foundation/news/2017/africas-growing-youthful-population-reflections-continent-tipping-point/.

14. Nathalie Munyampenda, "AIMS Launches African Master's in Machine Intelligence," African Institute for Mathematical Sciences, July 31, 2018, https://www.nexteinstein.org/blog/2018/07/31/aims-launches-first-of-its-kind-african-masters-in-machine-intelligence-at-rwanda-campus/.

15. Gary F. Simons and Charles D. Fennig, eds., "Africa," *Ethnologue: Languages of the World, Twenty-First Edition* (Dallas, TX: SIL International, 2018), https://www.ethnologue.com/region/Africa; UNEP, "State of Biodiversity in Africa," 2010 International Year of Biodiversity (United Nations Environment Programme, 2010), 3, https://www.cbd.int/iyb/doc/celebrations/iyb-egypt-state-of-biodiversity-in-africa.pdf; Allon Raiz, "Strength Lies in Diversity for Africa," World Economic Forum, May 21, 2013, https://www.weforum.org/agenda/2013/05/diversifying-african-economies/.

16. Toby Shapshak, "Transcript of 'You Don't Need an App for That,'" TED, July 2013, https://www.ted.com/talks/toby_shapshak_you_don_t_need_an_app_for_that/transcript.

17. Joe Bavier, "Off-Grid Power Pioneers Pour into West Africa," Reuters, February 20, 2018, https://www.reuters.com/article/us-africa-power-insight/off-grid-power-pioneers-pour-into-west-africa-idUSKCN1G41PE.

18. Robert Lee Hotz, "In Rwanda, Drones Deliver Medical Supplies to Remote Areas," *Wall Street Journal*, December 1, 2017, https://www.wsj.com/articles/in-rwanda-drones-deliver-medical-supplies-to-remote-areas-1512124200.

19. Mutsa Chironga, Hilary De Grandis, and Yassir Zouaoui, "Mobile Financial Services in Africa: Winning the Battle for the Customer," McKinsey & Company, September 2017, https://www.mckinsey.com/industries/financial-services/our-insights/mobile-financial-services-in-africa-winning-the-battle-for-the-customer.

20. Ibid.

21. Governments, for example, have begun to tax mobile money users, which might benefit their coffers, yet run the risk of negatively impacting their nation's economy.

22. Rishi Raithatha, "The Future of Mobile Money in Sub-Saharan Africa: A Foundation for Greater Financial Inclusion," GSMA, July 12, 2017, https://www.gsma.com/mobilefordevelopment/programme/mobile-money/future-mobile-money-sub-saharan-africa-foundation-greater-financial-inclusion. Other initiatives include M-Shwari in Kenya, M-Kopa and Fenix International in East Africa, Paga in Nigeria, and Orange Money across 14 African countries.

23. Matina Stevis-Gridneff, "An Isolated Country Runs on Mobile Money," *Wall Street Journal*, July 6, 2018, https://www.wsj.com/articles/an-isolated-country-runs -on-mobile-money-1530882001; Fionan McGrath, "MMU Releases a New Case Study on Telesom's ZAAD Mobile Money Service in Somaliland," GSMA, July 3, 2013, https://www.gsma.com/mobilefordevelopment/programme/mobile-money/mmu -releases-a-new-case-study-on-telesoms-zaad-mobile-money-service-in-somaliland.

24. Mic Wright, "Desert Discs: How Mobile Phones Are at the Root of Saharan Music," *The Guardian*, November 1, 2010, https://www.theguardian.com/music /musicblog/2010/nov/01/music-from-saharan-cellphones-mali.

25. Shapshak, "Transcript of 'You Don't Need an App for That.'"

26. "Delivering Digital Infrastructure: Advancing the Internet Economy," World Economic Forum, April 2014,. http://www3.weforum.org/docs/WEF_TC_Delivering DigitalInfrastructure_InternetEconomy_Report_2014.pdf.

27. C. K. Prahalad, *The Fortune at the Bottom of the Pyramid: Eradicating Poverty through Profits* (Upper Saddle River, NJ: Wharton School Publishing, 2004); "Title Page—The Fortune at the Bottom of the Pyramid: Eradicating Poverty through Profits, Revised and Updated 5th Anniversary Edition [Book]," Safari Books Online, accessed December 3, 2018, https://www.oreilly.com/library/view/the-fortune-at/9780137042029 /fm.html.

28. Jeff Dean and Moustapha Cisse, "Google AI in Ghana," Google, June 13, 2018, https://www.blog.google/around-the-globe/google-africa/google-ai-ghana/.

29. Juliet Ehimuan-Chiazor, "Google.Org Commits $20 Million to Improve Lives in Africa," Philanthropy News Digest (PND), July 31, 2017, http://philanthropynewsdigest .org/news/google.org-commits-20-million-to-improve-lives-in-africa.

30. "Google Supports Digital Journalism for Africa Initiative," TechFinancials, June 8, 2017, https://techfinancials.co.za/2017/06/08/google-supports-digital-journalism -africa-initiative/.

31. Dean and Cisse, "Google AI in Ghana."

32. Bayo Adekanmbi, "10 Inspiring Facts about Moustapha Cisse, Google AI Ghana Pioneer Lead," Data Science Nigeria, June 27, 2018, https://www.datasciencenigeria .org/10-inspiring-facts-moustapha-cisse-google-ai-ghana-pioneer-lead/.

33. Moustapha Cisse, interview by author, September 7, 2018.

34. Ibid.

35. Ibid.

36. Ingrid Lunden, "Google Goes All-in on Artificial Intelligence, Renames Research Division Google AI," TechCrunch, May 8, 2018, http://social.techcrunch.com

/2018/05/08/google-goes-all-in-on-artificial-intelligence-renames-research-division -google-ai/.

37. Achille Mbembe, "The Digital Age Erases the Divide between Humans and Objects," The M&G Online, January 6, 2017, https://mg.co.za/article/2017-01-06-00 -the-digital-age-erases-the-divide-between-humans-and-objects/.

38. Achille Mbembe, *Critique of Black Reason*, trans. Laurent Dubois (Durham, NC: Duke University Press Books, 2017).

39. Caroline Sourt, "The Congo's Blood Metals," *The Guardian*, December 25, 2008, https://www.theguardian.com/commentisfree/2008/dec/25/congo-coltan.

40. Zekeh Gbotokuma, *Global Safari: Checking In and Checking Out in Pursuit of World Wisdoms* (Newcastle upon Tyne, UK: Cambridge Scholars Publishing, 2015), 31.

41. "CBS News Finds Children Mining Cobalt for Batteries in the Congo," CBS News, March 5, 2018, https://www.cbsnews.com/news/cobalt-children-mining-democratic -republic-congo-cbs-news-investigation/.

42. Bregtje van der Haak, "The Internet Is Afropolitan," This Is Africa, August 7, 2015, https://thisisafrica.me/the-internet-is-afropolitan/.

43. Damola Durosomo, "Discussing African Futures with Achille Mbembe," Okay-Africa, November 9, 2015, http://www.okayafrica.com/achille-mbembe-african-futures -interview/.

44. "The Internet Is Afropolitan," *Chimurenga Chronic*, March 17, 2015, https:// chimurengachronic.co.za/the-internet-is-afropolitan/.

20   AI in Uganda

1. G. Pascal Zachary, "Fertile Ground in Africa for Computer Science to Take Root," *New York Times*, December 5, 2011, https://www.nytimes.com/2011/12/06/science /fertile-ground-in-africa-for-computer-science-to-take-root.html.

2. "Building Fertile Ground for Data Science in Uganda," United Nations Global Pulse, August 5, 2016, https://www.unglobalpulse.org/news/building-fertile-ground -data-science-uganda.

3. Ibid.

4. "Bloodless Malaria Test Wins Africa Prize," BBC News, June 14, 2018, https:// www.bbc.co.uk/news/world-africa-44481723.

5. Julianne Sansa-Otim, interview by author, July 10, 2018.

6. Engineer Bainomugisha, interview by author, July 12, 2018.

7. Fiona Ssozi-Mugarura, Edwin Blake, and Ulrike Rivett, "Codesigning with Communities to Support Rural Water Management in Uganda," *CoDesign* 13, no. 2 (April 3, 2017): 110–126, https://doi.org/10.1080/15710882.2017.1310904.

8. Ibid.

9. Daniel Mutembesa, interview by author, July 14, 2018.

10. Ibid.

11. Robert Leonard, "Game Theory in Economics, Origins Of," in *The New Palgrave Dictionary of Economics* (London: Palgrave Macmillan, 2016), 1–8, https://doi.org/10.1057/978-1-349-95121-5_1922-1.

12. Mutembesa, interview.

13. Ibid.

14. Dambisa Moyo, *Dead Aid: Why Aid Is Not Working and How There Is a Better Way for Africa* (London: Macmillan, 2009), xix.

15. Kimberly Ann Elliott, "The WTO, Agriculture, and Development: A Lost Cause?," *Bridges Africa* 7 (February 15, 2018), https://www.ictsd.org/bridges-news/bridges-africa/news/the-wto-agriculture-and-development-a-lost-cause; James Scott, "The Future of Agricultural Trade Governance in the World Trade Organization," *International Affairs* 93, no. 5 (September 1, 2017): 1167–1184, https://doi.org/10.1093/ia/iix157.

16. Elliott, "The WTO, Agriculture, and Development"; World Trade Organization, "V. Global Perspectives—Who Are the Leading Players?," in *World Trade Statistical Review 2018*, 2018, https://www.wto.org/english/res_e/statis_e/wts2018_e/wts2018chapter05_e.pdf.

17. Lindsay Whitfield and Alastair Fraser, "Negotiating Aid: The Structural Conditions Shaping the Negotiating Strategies of African Governments," *International Negotiation* 15, no. 3 (October 1, 2010): 352, https://doi.org/10.1163/157180610X529582.

21   Innovating from the Ground Up in Kenya

1. Jake Bright and Aubrey Hruby, "The Rise of Silicon Savannah and Africa's Tech Movement," TechCrunch, July 23, 2015, https://techcrunch.com/2015/07/23/the-rise-of-silicon-savannah-and-africas-tech-movement/.

2. John Mutua, "Google Picks Kenyan Tech Start-Ups for Skills Training," Business Daily Africa, April 23, 2018, https://www.businessdailyafrica.com/corporate/companies/Google-picks-Kenyan-tech-start-ups-for-skills-training/4003102-4495522-fh9ibd/index.html.

3. Gary F. Simons and Charles D. Fennig, eds., "Kenya," *Ethnologue: Languages of the World, Twenty-First Edition* (Dallas, TX: SIL International, 2016), https://www.ethnologue.com/country/KE.

4. Pew Research Center's Global Attitudes Project, "Cell Phones in Africa: Communication Lifeline," Pew Research Center, April 15, 2015, http://www.pewglobal.org/2015/04/15/cell-phones-in-africa-communication-lifeline/.

5. "Safaricom Plans," Safaricom, accessed July 19, 2018, https://www.safaricom.co.ke/.

6. Kieron Monks, "M-Pesa: Kenya's Mobile Success Story Turns 10," CNN, February 24, 2017, https://www.cnn.com/2017/02/21/africa/mpesa-10th-anniversary/index.html.

7. CCK, "Quarterly Sector Statistics Report: 2nd Quarter October-December 2011/2012," Community Commission of Kenya, April 2012, 11, https://web.archive.org/web/20121016002713if_/http://cck.go.ke/resc/downloads/SECTOR_STATISTICS_REPORT_Q2_2011-12.pdf.

8. "Impact Report: Theory of Change," Ushahidi, July 2015, https://www.ushahidi.com/impact-report/theory-of-change.

9. Charles Harding, interview with author, November 10. 2017.

10. "MOJA," BRCK, accessed December 3, 2018, https://www.brck.com/.

11. Philip Walton, interview by author, August 2, 2018.

12. Erik Hersman, interview by author, July 27, 2018.

13. Ibid.

14. Toby Shapshak, "How Kenya's SupaBRCK Aims to Solve Africa's Internet Problems," *Forbes*, March 7, 2017, https://www.forbes.com/sites/tobyshapshak/2017/03/07/how-kenyas-supabrck-aims-to-solve-africas-internet-problems/.

15. "Beefing up Mobile-Phone and Internet Penetration in Africa," *Economist*, November 9, 2017, https://www.economist.com/special-report/2017/11/09/beefing-up-mobile-phone-and-internet-penetration-in-africa.

16. Ibid.

17. "Technology May Help Compensate for Africa's Lack of Manufacturing," *Economist*, November 9, 2017, https://www.economist.com/special-report/2017/11/09/technology-may-help-compensate-for-africas-lack-of-manufacturing.

18. Omar Mohammed, "Africa Has the World's Fastest-Growing Labor Force but Needs Jobs Growth to Catch Up," Quartz Africa, November 12, 2015, https://qz.com/africa/547929/africa-has-the-worlds-fastest-growing-labor-force-but-needs-jobs-growth-to-catch-up/.

19. Mark Kamau, interview by author, August 1, 2018.

20. Maketa Maina, interview by author, August 3, 2018. The informal economy describes a system of trade or economic exchange outside of that which is trackable, state-controlled, or perhaps even money-based.

21. Kenneth King, *Jua Kali Kenya: Change & Development in an Informal Economy, 1970–95* (Columbus: Ohio State University Press, 1996), 24–25.

22. Steven J. Jackson, "Rethinking Repair," in *Media Technologies*, ed. Tarleton Gillespie, Pablo J. Boczkowski, and Kirsten A. Foot (Cambridge, MA: MIT Press, 2014), 221–240.

23. Ibid., 234.

24. Yassin Mohamed Barre, interview by author, August 1, 2018.

25. Ibid.

26. The United Nations Refugee Agency, "UNHCR WASH Manual" (UNHCR, 2017), 48–51; Clarence Alucho, "PicoBRCK for WASH—UNHCR, Dadaab Refugee Complex," BRCK, February 21, 2018, https://www.brck.com/2018/02/picobrck-wash-dadaab-refugee-complex/.

27. Miranda Katz, "This Twenty-Something Forced Silicon Valley to 'Show Her the Numbers,'" *Wired*, October 14, 2016, https://www.wired.com/2016/10/this-twenty-something-forced-silicon-valley-to-show-her-the-numbers/.

28. Simone Stolzoff, "Venture Capital's Diversity Problem in Two Words: Alma Mater," Quartz, July 31, 2018, https://qz.com/1343912/venture-capitals-diversity-problem-in-two-words-alma-mater/.

29. Richard Kerby, "Where Did You Go to School?," *Noteworthy—The Journal Blog* (blog), July 30, 2018, https://blog.usejournal.com/where-did-you-go-to-school-bde54d846188.

30. James Barton, "Alphabet to Partner with Kenyan Operators for Project Loon," Developing Telecoms, July 6, 2018, https://www.developingtelecoms.com/tech/wireless-networks/7915-alphabet-to-partner-with-kenyan-operators-for-project-loon.html.

31. Toby Shapshak, "How BRCK, Vanu and Facebook Are Reinventing (Free) Mobile Internet Access for Africa in Rural Rwanda," *Forbes*, accessed December 3, 2018, https://www.forbes.com/sites/tobyshapshak/2017/10/11/how-brck-vanu-and-facebook-are-reinventing-free-mobile-internet-access-for-africa-in-rural-rwanda/.

32. Elizabeth Segran, "SpaceX Exploding Rocket Blew Up More Than Facebook's Internet Dreams," Fast Company, September 6, 2016, https://www.fastcompany.com/4018356/spacex-rocket-blow-up-torched-more-than-facebooks-internet-dreams.

33. Daniel van Boom, "Why India Snubbed Facebook's Free Internet Offer," CNET, February 26, 2016, https://www.cnet.com/news/why-india-doesnt-want-free-basics/.

34. Walton, interview.

35. Ibid.

36. Bitange Ndemo, "NDEMO: How Kenya Gave Rise to Africa ICT Revolution," Business Daily Africa, February 8, 2017, https://www.businessdailyafrica.com /analysis/How-Kenya-gave-rise-to-Africa-ICT-revolution/539548-3805212-sbmlm7 /index.html.

37. "The East African Marine Systems—Teams Fibre Optic," TEAMS, accessed October 6, 2018, https://www.teams.co.ke/; June Okal, "Proposed ICT Policy 2016: How Kenya Stands to Gain," Techweez, August 29, 2016, https://techweez.com/2016/08 /29/proposed-ict-policy-2016/.

38. "Kenya Vision 2030," Kenya Vision 2030, accessed August 24, 2018, http:// vision2030.go.ke/.

22 Mobile Power to the People

1. "Oaxaca," History, December 2, 2009, https://www.history.com/topics/mexico /oaxaca.

2. "Biodiversity of Mexico," The Economics of Ecosystems and Biodiversity, accessed October 12, 2018, http://www.teebweb.org/teeb-mexico/biodiversity/; Allison Green, "10 Insider Tips to Your Trip to Oaxaca," *National Geographic*, July 1, 2018, https://www.nationalgeographic.com/travel/destinations/north-america /mexico/oaxaca/top-activities-things-to-do/; Kathleen March, "Animals in Oaxaca," Animals—mom.me, accessed October 12, 2018, https://animals.mom.me/animals -oaxaca-4794.html; Lauren Cocking, "Mexico's Incredible Trees, Plants and Flowers and Where to Find Them," Culture Trip, June 29, 2017, https://theculturetrip.com /north-america/mexico/articles/mexicos-incredible-trees-plants-and-flowers-and -where-to-find-them/.

3. Peter Bloom, interview by author, March 30, 2018.

4. Robert Jensen, "The Digital Provide: Information (Technology), Market Performance, and Welfare in the South Indian Fisheries Sector," *Quarterly Journal of Economics* 122, no. 3 (2007): 879–924; Leonard Waverman, Meloria Meschi, and Melvyn Fuss, "The Impact of Telecoms on Economic Growth in Developing Countries," *The Vodafone Policy Paper Series* 3 (January 1, 2005).

5. Ethan Zuckerman, "Decentralizing the Mobile Phone: A Second ICT4D Revolution?," *Information Technologies & International Development* 6, no. SE (November 18, 2010): 99,103.

6. Steve Song, interview by author, January 9, 2018.

7. Ibid.

8. Elinor Ostrom, "Crossing the Great Divide: Coproduction, Synergy, and Development," *World Development* 24, no. 6 (June 1996): 1073, https://doi.org/10.1016/0305 -750X(96)00023-X.

9. C. K. Prahalad and Venkat Ramaswamy, "Co-Creation Experiences: The Next Practice in Value Creation," *Journal of Interactive Marketing* 18, no. 3 (June 1, 2004): 5–14, https://doi.org/10.1002/dir.20015.

10. Steven J. Jackson, "Rethinking Repair," in *Media Technologies*, ed. Tarleton Gillespie, Pablo J. Boczkowski, and Kirsten A. Foot (Cambridge, MA: MIT Press, 2014), 221–240.

11. This demonstrates the sociologist Saskia Sassen's point that we must read "space and place" in relation to digital networks and the economics of information routing, rather than traditional ideas of distance and proximity.

12. One can imagine how social media giants could co-opt the TIC effort, promising a free-to-access network for its users, in exchange for access to and control over the voice and text data shared through its network. TIC employees have mentioned that at times they have worked together with Facebook employees, as both have a shared interest in open source software and hardware standards. The key distinction, however, is that TIC's interest is in community ownership, sovereignty, and autonomy around its networks, and Facebook's is quite different.

13. Gilles Deleuze and Félix Guattari, *A Thousand Plateaus: Capitalism and Schizophrenia*, 12th ed. (Minneapolis: University of Minnesota Press, 2007), 6–10.

14. Paul Stamets, "Transcript of '6 Ways Mushrooms Can Save the World,'" TED, March 2008, https://www.ted.com/talks/paul_stamets_on_6_ways_mushrooms_can _save_the_world/transcript.

15. Pablo González Casanova, "The Zapatista 'Caracoles': Networks of Resistance and Autonomy," *Socialism and Democracy* 19, no. 3 (November 1, 2005): 79–92, https://doi.org/10.1080/08854300500257963; "Communiques from the EZLN," Struggle archive, accessed October 13, 2018, http://struggle.ws/mexico/ezlnco.html.

16. Loreto Bravo, "A Seed Sprouts When It Is Sown in Fertile Soil," in *Technological Sovereignty*, vol. 2, accessed October 12, 2018, https://sobtec.gitbooks.io/sobtec2 /content/en/content/08rizomatica.html.

17. Gustavo Esteva, "Hope from the Margins," *Wealth of the Commons*, accessed October 12, 2018, http://wealthofthecommons.org/essay/hope-margins.

18. Katya Cengel, "The Other Mexicans," *National Geographic*, June 25, 2013, https:// news.nationalgeographic.com/news/2013/06/130624-mexico-mixteco-indigenous -immigration-spanish-culture/; OHCHR, "Advancing Indigenous Peoples' Rights in

Mexico," United Nations Human Rights Office of the High Commissioner, July 7, 2011, https://www.ohchr.org/en/NewsEvents/Pages/IndigenousPeoplesRightsInMexico .aspx; "Mexico Population 2018 (Demographics, Maps, Graphs)," World Population Review, September 17, 2018, http://worldpopulationreview.com/countries/mexico -population/.

19. "Channel" here refers to radio frequency or mobile spectrum administration.

20. David Turnbull, "Working with Incommensurable: Knowledge Traditions Assemblage, Diversity, Emergent Knowledge, Narrativity, Performativity, Mobility and Synergy," thoughtmesh, May 29, 2009, http://thoughtmesh.net/publish/279.php.

21. Bravo, "A Seed Sprouts."

22. Oswaldo, interview by author, June 29, 2017.

23. Radio Sonidera, *Desinformemonos Rhizomatica*, 2015, https://www.youtube.com /watch?v=qQfwQ7sGvgM&t=1s.

24. Immanuel Wallerstein, *World-Systems Analysis*, 2004, 4, https://books.google .com/books/about/World_systems_Analysis.html?id=5vGr7kRsXBkC.

25. We also received many complaints that the dialing system for TIC was far more complicated, given that it had to operate on a different community telephony protocol, and therefore infrastructure, rather than those used by private operators such as Movistar or TelCel. TIC has been unable to affordably purchase numbers that are directly connected to major network operators, and as a result have to route around them, which in turn makes the longer numbers dialed by community users more cumbersome.

26. Oswaldo, interview.

27. Bloom, interview.

28. Casanova, "The Zapatista 'Caracoles.'"

23 Blockchain

1. Susanne Tarkowski Tempelhof et al., "Pangea Jurisdiction and Pangea Arbitration Token (PAT): The Internet of Sovereignty," Planet Earth (Bitnation, April 2017), https://github.com/Bit-Nation/Pangea-Docs/raw/master/BITNATION%20Pangea%20 Whitepaper%202018.pdf. White papers are one way that blockchain teams share descriptions of new or not-yet-built software. In some cases, they're used as a vehicle to reach investors.

2. Ibid., and for all material we quote or attribute to the white paper. Regarding its "competitive" environment: At the heart of Bitnation's proposal is a number-based reputation score that all "Citizens, Arbitrators, Governance Services Providers,

Nations, Smart Contracts [which we'll discuss later in this chapter] and Codes of Law" receive. The nations and service providers with higher scores would end up with more citizens, while those with lower scores would struggle to attract membership.

3. aantonop, *Blockchain vs. Bullshit: Thoughts on the Future of Money*, Blockchain Africa Conference 2017, https://www.youtube.com/watch?v=SMEOKDVXlUo.

4. jeet, "Defections Continue from the Nation State System…," Tweet, *@jeetsidhu* (blog), October 22, 2018, https://twitter.com/jeetsidhu_/status/1054567 469409546240; Matt Odell, "Bitcoin Is the Most Powerful Non-Violent Tool the World Has Ever Seen…," Tweet, *@matt_odell* (blog), August 21, 2018, https://twitter .com/matt_odell/status/1031904760725753856; Max Hillebrand, "Bitcoin Is Not a Political Campaign, Begging Bureaucrats to Change. Bitcoin Is #agorism…," Tweet, *@HillebrandMax* (blog), November 11, 2018, https://twitter.com/HillebrandMax /status/1061593723405025280.

5. Jimmy Song, "The Rent Seeking in a Centralized System Go Up over Time. That Means That as We Watch a Lot of These Altcoins and ICOs Progress, You'll See More Embezzlement…," Tweet, *@jimmysong* (blog), October 23, 2018, https://twitter.com /jimmysong/status/1054890052298268672.

6. PHLblockchain, *Glen Weyl on "Radical Markets" and Blockchain Tech*, YouTube, 2018, https://www.youtube.com/watch?v=I39X0GziIV4.

7. Anne Field, "The Plastic Bank: Using Plastic Refuse to Create a Global Currency for the Poor," *Forbes*, November 29, 2017, https://www.forbes.com/sites/annefield/2017/11 /29/the-plastic-bank-using-plastic-to-create-a-global-currency-for-the-poor/; Shaun Frankson, interview by Adam Reese, December 7, 2017. As Plastic Bank co-founder Frankson told Adam, the organization resells its plastic to buyers at prices that depend on the plastic's quality. The rate on very low-grade plastics, he said, is so meager that they are probably not worth anyone's time to collect. He also noted that plastic needs to be washed clean before it is traded, which we believe could discourage people from cleaning up sites that are filthy with mixed refuse (like gutters) due to the extra work it would require. The Plastic Bank's approach may help thin out litter, but because it is driven by market demand for scrap plastic rather than resident demand for a clean living environment, it is not likely to eliminate the trash problem it set out to address.

8. Jeremy Gardner, "Things That Matter for Bitcoin: Mining Consensus Core Development…," Tweet, *@Disruptepreneur* (blog), September 13, 2017, https://twitter.com /Disruptepreneur/status/907977113377972224.

9. David Golumbia, *The Politics of Bitcoin: Software as Right-Wing Extremism* (Minneapolis: University of Minnesota Press, 2016); Langdon Winner, "Cyberlibertarian Myths and the Prospects for Community," *ACM SIGCAS Computers and Society* 27, no. 3 (September 1997): 14–19, https://doi.org/10.1145/270858.270864.

10. Mathias Grønnebæk, "The Solution Is Simple—Stop Paying People Out of the Government's Pocket," Tweet, *@MatGroennebaek* (blog), November 11, 2018, https:// twitter.com/MatGroennebaek/status/1061704129117110273.

11. Golumbia, *The Politics of Bitcoin*, 12.

12. Steve King, "'Nazi' Is Injected into Leftist Talking Points Because the Worn Out & Exhausted 'Racist' Is Over Used…," Tweet, *@stevekingia* (blog), September 12, 2018, https://twitter.com/stevekingia/status/1039924304778014721?lang=en; Mathias Grønnebæk, "I Have Been Very Disturbed Watching Recordings from the South African Parliament…," Tweet, *@MatGroennebaek* (blog), September 26, 2018, https://twitter.com/i/web/status/1045174048060248064; Communist Terror, "A Nine-Year-Old Boy Was Separated from His Mother, Put in a Cell the Size of a Toilet and Beaten…," Tweet, *@communistterror* (blog), September 26, 2018, https://twitter.com/communistterror/status/1044860086471086081; Devin Nunes, "Horrible What's Happening to In and Out Burger in CA. Democrats/ SOCIALISTS Have Targeted Them Because They Gave Money to CA Republican Party…," Tweet, *@devinnunes* (blog), August 30, 2018, https://twitter.com /devinnunes/status/1035378664467136513?lang=en; Brad Parscale, ".@google Needs to Explain Why This Isn't a Threat to the Republic. Watch the Video. Google Believes They Can Shape Your Search Results and Videos to Make You 'Have Their Values'…," Tweet, *@parscale* (blog), September 12, 2018, https://twitter .com/parscale/status/1039980750500704261?lang=en; MaryAnastasia O'Grady, "Another Socialist Achievement: Hunger in Venezuela…," Tweet, *@MaryAnastasiaOG* (blog), November 14, 2018, https://twitter.com/MaryAnastasiaOG/status /1062765245482045441; WhalePanda, "Crypto Twitter's Hateful Underbelly and SJWs," Medium, September 19, 2018, https://medium.com/@WhalePanda /crypto-twitters-hateful-underbelly-and-sjws-a43af028170a; Lightning Master Hub, "Unfortunately…This Is Not Just a Problem in Crypto Twitter…," Tweet, *@LN_Master_Hub* (blog), September 19, 2018, https://twitter.com/LN_Master_Hub /status/1042409831712186368.

13. Ragnar Lifthraser, "Another Call for White Genocide," Tweet, *@Ragnarly* (blog), November 7, 2018, https://twitter.com/Ragnarly/status/1060209210418327553; American Thinker, "Racism on the Rise," Tweet, *@AmericanThinker* (blog), November 6, 2018, https://twitter.com/AmericanThinker/status/1059817291645403137; zerohedge, "'We Fear for Our Lives': First 50 White South African Families Start Resettlement In Russia," Tweet, *@zerohedge* (blog), July 26, 2018, https://twitter .com/zerohedge/status/1022734597610790912; Mike Allen, "Mexican Police Are Being Brutalized by Members of This Caravan as They Attempt to FORCE Their Way into Mexico," Tweet, *@AMike4761* (blog), October 21, 2018, https://twitter.com /AMike4761/status/1053993048739405825; Ernst Roets, "Dear @CyrilRamaphosa. My Brother Was Attacked. My Friend Shot Dead, My Other Friend's Throat Was Slit…," Tweet, *@ernstroets* (blog), September 26, 2018, https://twitter.com/ernstroets

/status/1045174356257787904?lang=en; Breaking911, "WATCH: 10-Year-Old Robbed at Gunpoint in Nashville," Tweet, *@breaking9111* (blog), September 28, 2018, https://twitter.com/breaking9111/status/1045751062693646337; Pale_Primate, "North Western Europeans Are So Full of Trust That They Are a Civilization-Level Version of Someone with Williams Syndrome…," Tweet, *@PALE_Primate* (blog), September 6, 2018, https://twitter.com/PALE_Primate/status/1037718916242522112; J. Burton, "I Don't Know. I'm Thinking if She Had It All to Do Over Again That Maybe—Just MAYBE—She Might Choose 'Enforce Existing Immigration Laws' over 'Die Alone and Terrified, Murdered by an Illegal Alien'…," Tweet, *@jburtonxp* (blog), August 21, 2018, https://twitter.com/jburtonxp/status/1032125705881088001 ?lang=en; Alessandra, "Today Is the Feast of the Martyrs of Otranto, when 813 Italians Were Massacred by the Ottomans for Refusing to Convert to Islam in 1480…," Tweet, *@alessabocchi* (blog), August 14, 2018, https://twitter.com/alessabocchi /status/1029358426256683008?lang=en; Daily Mail Online, "The Dalai Lama Says 'Europe Belongs to Europeans' and Refugees Should Return to Their Native Countries to Rebuild Them," Tweet, *@mailonline* (blog), September 13, 2018, https:// twitter.com/mailonline/status/1040245195395227649?lang=en; Ann Coulter, "At Least Some Europeans Are Fighting Back!," Tweet, *@anncoulter* (blog), September 14, 2018, https://twitter.com/anncoulter/status/1040808997576761344?lang=en; Jack Kessel, "I Just Had a Black Guy from Nigeria as My Uber Driver, We Started Talking about Work…," Tweet, *@jackkessl* (blog), September 14, 2018, https://twitter.com /jackkessl/status/1040778703381450752.

14. We've said a lot about right-wing thought in the space because we feel additional nuance is required to accurately describe it. But the space is no more right-wing than the rest of the business world.

15. "Dark Wallet," Dark Wallet, https://www.darkwallet.is/; Jon Southurst, "Cody Wilson Speaks Out on Campaign to Dismantle Bitcoin Foundation," CoinDesk, January 2, 2015, https://www.coindesk.com/cody-wilson-speaks-campaign-dismantle -bitcoin-foundation.

16. Andy Greenberg, "A Landmark Legal Shift Opens Pandora's Box for DIY Guns," *Wired*, July 10, 2018, https://www.wired.com/story/a-landmark-legal-shift-opens -pandoras-box-for-diy-guns/.

17. Adam Reese, "Huawei Developing Solution for Evaluating Blockchain Performance," ETHNews.com, March 15, 2018, https://www.ethnews.com/huawei -developing-solution-for-evaluating-blockchain-performance.

18. Michael Corkery and Nathaniel Popper, "From Farm to Blockchain: Walmart Tracks Its Lettuce," *New York Times*, September 24, 2018, https://www.nytimes.com /2018/09/24/business/walmart-blockchain-lettuce.html.

19. Ibid.

20. Kai Stinchcombe, "Ten Years In, Nobody Has Come Up with a Use for Blockchain," Hacker Noon, December 22, 2017, https://hackernoon.com/ten-years-in-nobody-has-come-up-with-a-use-case-for-blockchain-ee98c180100.

21. Corkery and Popper, "From Farm to Blockchain."

22. Nick Marinoff, "ConsenSys Initiative Empowers Students to Use Blockchain for Social Good," *Bitcoin Magazine*, September 19, 2018, https://bitcoinmagazine .com/articles/consensys-initiative-empowers-students-use-blockchain-social-good/; Blockchain 4 Humanity, "Wellcome [*sic*] to Blockchain 4 Humanity," *Blockchain 4 Humanity* (blog), March 5, 2018, https://medium.com/@phenizia/wellcome-to -blockchain-4-humanity-3901cb465db8.

23. Asha McLean, "Facebook Holds ICO Ban but Allows 'Approved' Cryptocurrency Ads," ZDNet, June 27, 2018, https://www.zdnet.com/article/facebook-holds-ico-ban -but-allows-approved-cryptocurrency-ads/.

24. Derivatives are an umbrella category of financial products that are common in conventional stock markets. In many cases, they are used to increase risk so that investors can gain or lose greater amounts of money than they might if they were holding stocks or commodities directly.

25. Katherine Burton, Lauren Coleman-Lochner, and Eliza Ronalds-Hannon, "Hedge Fund Billionaire Rode the Worst Trade of His Life All the Way Down," Bloomberg, October 12, 2018, https://www.bloomberg.com/news/articles/2018-10-12/eddie -lampert-rode-the-worst-trade-of-his-life-all-the-way-down. These are essentially bets on a currency's price fluctuations except that the stakes of the original bet get multiplied, meaning that people buying the derivative can win—or end up on the hook for—way more than they put down.

26. Corporate interest appears to be waning as well. Tim Copel, "When Will Ethereum Grow Up?," Decrypt Media, November 7, 2018, https://decryptmedia.com/2018/11 /07/ethereum-serenity-casper-scaling-problems-still-not-solved/; Courtenay Brown, "Corporate America's Blockchain and Bitcoin Fever Is Over," Axios, November 11, 2018, https://www.axios.com/corporate-america-blockchain-bitcoin-fervor-over -fb13bc5c-81fd-4c12-8a7b-07ad107817ca.html.

27. They sometimes sell ownership rights to cryptocurrency on blockchains that do not exist yet.

28. "Two Celebrities Charged with Unlawfully Touting Coin Offerings," US Securities and Exchange Commission, November 29, 2018, https://www.sec.gov/news /press-release/2018-268.

29. Lorenzo Kyle Subido, "Manny Pacquiao Investing in Cryptocurrency Startup, Will Release His Own Virtual Token Soon," Entrepreneur, March 12, 2018,

http://www.entrepreneur.com.ph/news-and-events/manny-pacquiao-investing-in
-cryptocurrency-startup-will-release-his-own-virtual-token-soon-a00200-20180312.

30. Jesus Christ, *Jesus Coin ICO*, YouTube, 2017, https://www.youtube.com/watch
?time_continue=12&v=MgaDBYamU7g.

31. Pirate Beachbum, "Into the Fire: Interview with Kevin Pham," Hacker Noon,
November 6, 2018, https://hackernoon.com/into-the-fire-interview-with-kevin-pham
-af8d7a2d1267.

32. Zoë Bernard, "When This Ice Tea Company Stuck the Word 'Blockchain' in Its
Name, Its Stock Skyrocketed by Nearly 500%. Now, It's Being Investigated by the
Government," Business Insider, August 2, 2018, https://www.businessinsider.com
/long-blockchain-company-iced-tea-sec-stock-2018-8.

33. Such narratives don't always go unchallenged. For instance, Andy Tudhope,
a member of the team behind a blockchain-related app called Status (https://our
.status.im/contact-light/), has suggested that it's "intellectually dishonest" for those
in the space to claim that they're building tools for "oppressed people" who lack
rights or access to infrastructure. (He recommends defining blockchain's target
market as "anyone who is connected.")

34. Naomi Klein, *The Battle for Paradise: Puerto Rico Takes on the Disaster Capitalists* (Haymarket Books, 2018); Naomi Klein and Lauren Feeney, "Puerto Ricans and
Ultrarich 'Puertopians' Are Locked in a Pitched Struggle Over How to Remake the
Island," The Intercept, March 20, 2018, https://theintercept.com/2018/03/20/puerto
-rico-hurricane-maria-recovery/.

35. Siglo, "Siglo Takes Crypto Rico!," *Siglo* (blog), March 23, 2018, https://medium
.com/siglo/siglo-takes-crypto-rico-11cb7a874240.

36. Klein, *The Battle for Paradise*; Virtualizate—Tech Guru, *We Are Here to Help
Puerto Rico—Brock Pierce*, YouTube, 2018, https://www.youtube.com/watch?v=8VQr
Kcj2pjw.

37. Virtualizate—Tech Guru, *We Are Here to Help.*

38. The Guardian, "The Perfect Storm Building a Crypto-Uropia in Puerto Rico," *The
Guardian*, August 9, 2018, https://www.theguardian.com/changingmediasummit
/video/2018/aug/09/the-perfect-storm-building-a-crypto-utopia-in-puerto-rico-video.

39. "Changpeng Zhao," *Forbes*, November 6, 2018, https://www.forbes.com/profile
/changpeng-zhao/.

40. Adam Reese, "Meeting between Zhao and Gnassingbé," ETHNews.com, April 26,
2018, https://www.ethnews.com/binance-ceo-meets-with-alleged-autocrat-announces
-togo-project.

41. Republic of Uganda Ministry of ICT and National Guidance, "Blockchain Technology Coming to Uganda," Uganda Media Centre, April 24, 2018, http://www.mediacentre.go.ug/press-release/blockchain-technology-coming-uganda; C. Z. Binance, "Video of the Binance Meeting with President of Uganda Mr. Yoweri Musevenipic.Twitter.Com/IMax4Y2OKk," Tweet, *@cz_binance* (blog), April 25, 2018, https://twitter.com/cz_binance/status/989224318465785857?ref_src=twsrc%5Etfw &ref_url=https%3A%2F%2Fbinancewiki.net%2F5561%2F&tfw_site=binancewiki; Adam Reese, "Binance CEO Meets with Alleged Autocrat, Announces Togo Project," ETHNews.com, April 26, 2018, https://www.ethnews.com/binance-ceo-meets-with -alleged-autocrat-announces-togo-project.

42. As with the Walmart example, blockchain technology is probably not the best tool to achieve this goal.

43. Joon Ian Wong, "The UN Is Using Ethereum's Technology to Fund Food for Thousands of Refugees," Quartz, November 3, 2017, https://qz.com/1118743/world -food-programmes-ethereum-based-blockchain-for-syrian-refugees-in-jordan/.

44. Ayn Rand was an author whose novels, which extol the pursuit of selfishness, have inspired generations of libertarians.

45. A number of researchers have proposed and begun building non-blockchain DLT systems, such as Leemon Baird's hashgraph concept and David Chaum's Elixxir platform, but it's hard to say which implementation of DLT has proved to be the "most viable." More to the point, we think that if a blockchain future is coming, it is likely to involve several DLT networks with different architectures.

46. Bitcoin was also the first DLT network.

47. Satoshi Nakamoto, "Bitcoin: A Peer-to-Peer Electronic Cash System," 2008, https://bitcoin.org/bitcoin.pdf.

48. Mining is not universal to all blockchain networks.

49. Ledger balances are represented differently on different blockchains. The word "contain" doesn't perfectly describe the relationship between addresses and crypto-currencies in all cases, but for our purposes here it's good enough.

50. Adam Reese, "[UPDATED] New US House Bill Calls for Report, Strategy On Cryptocurrency," ETHNews.com, April 30, 2018, https://www.ethnews.com/new-us -house-bill-calls-for-report-strategy-on-cryptocurrency.

51. Nate Raymond, "Ex-Agent in Silk Road Probe Gets More Prison Time for Bitcoin Theft," Reuters, November 8, 2017, https://www.reuters.com/article/us-usa-cyber -silkroad-idUSKBN1D804H.

52. Blockchain Maroc, *Zooko Wilcox from ZCASH about Recent Advance in Privacy— DEVCON IV*, https://www.youtube.com/watch?v=-kvVmktBoiI.

53. Vitalik Buterin, "Vitalik Buterin—What Is Ethereum and How to Build a Decentralized Future," *Future Thinkers Podcast*, April 21, 2015, https://futurethinkers.org /vitalik-buterin-ethereum-decentralized-future/.

54. Adam Reese, "Weak States, Rough Roads and Micropayments: A Brainstorm," *Medium* (blog), August 17, 2018, https://medium.com/@oneadamreese/weak-states -rough-roads-and-micropayments-a-brainstorm-3fad9356fad6.

55. Contractors who get paid in large lump sums generally have a weaker incentive to finish jobs.

56. This could result from several kinds of disagreements, e.g., a faction of nodes could roll back the blockchain to "erase" a certain transaction, or adopt new rules for adding blocks.

57. Majorly divisive splits might lead to other problems as well, like complications with smart contracts.

58. Most of this software is open source, so anyone can launch a new blockchain network by copying and deploying the code.

59. Node operators are literally the parties that own or run a node.

60. Eric Allen Been, "Jaron Lanier Wants to Build a New Middle Class on Micropayments," Nieman Lab, May 22, 2013, http://www.niemanlab.org/2013/05/jaron -lanier-wants-to-build-a-new-middle-class-on-micropayments/.

61. "Decentralized Energy Microgrids," MIT Digital Currency Initiative, accessed November 12, 2018, https://dci.mit.edu/decentralized-autonomous-energy -microgrids/.

62. Of course, if microgrids replaced traditional grids, someone would have to maintain the electrical infrastructure between households. Perhaps, as in the road maintenance example discussed earlier, smart contracts could automatically collect fees from grid customers to cover maintenance costs.

63. Klein, *The Battle For Paradise*; Tony Barboza, "SoCal Gas Agrees to $119.5-Million Settlement for Aliso Canyon Methane Leak—Biggest in U.S. History—Los Angeles Times," *Los Angeles Times*, August 8, 2018, http://www.latimes.com/local/lanow/la -me-aliso-canyon-settlement-20180808-story.html.

64. Ted Koppel, *Lights Out: A Cyberattack, a Nation Unprepared, Surviving the Aftermath* (New York: Crown Publishers, 2015).

65. These files can be encrypted so that hosts have no way of knowing what they are.

66. "Home—Golem," Golem.Network, accessed November 12, 2018, https://golem .network/.

67. "Home," Folding Coin, accessed November 12, 2018, https://foldingcoin.net/. Folding@Home pivoted into the blockchain space in 2015.

68. ISPs tend to serve users from whom they can easily extract a comfortable profit, and therefore leave some remote areas with poor internet access or none at all.

69. The network that Guifi.net built across the Catalonia region of northern Spain is currently one of the largest scale community-owned mesh networks on the planet. Roca acknowledges that the task of doing this on a global scale is almost inconceivably immense.

70. Muneeb Ali, "How Can Blockchains Improve the Internet's Infrastructure?," Coin Center, April 18, 2017, https://coincenter.org/entry/how-can-blockchains -improve-the-internet-s-infrastructure.

71. Juan Benet, Jesse Clayburgh, and Matt Zumwalt, "Why Is Decentralized and Distributed File Storage Critical for a Better Web?," Coin Center, June 20, 2017, https:// coincenter.org/entry/why-is-decentralized-and-distributed-file-storage-critical-for-a -better-web.

72. Tom Simonite, "The Decentralized Internet Is Here, with Some Glitches," *Wired*, March 5, 2018, https://www.wired.com/story/the-decentralized-internet-is-here -with-some-glitches/.

73. Muneeb Ali et al., "Blockstack Technical Whitepaper," Blockstack PBC, October 12, 2017), https://blockstack.org/whitepaper.pdf.

74. Michael Kern, "The Space Race to Decentralize Everything," Crypto Insider, February 14, 2018, https://cryptoinsider.21mil.com/the-space-race-to-decentralize -everything/.

75. Adam Reese, "The Decentralized Union: A Blockchain Use Case," ETHNews .com, November 5, 2017, https://www.ethnews.com/the-decentralized-union-a -blockchain-use-case; Adam Reese, "The Decentralized Union: A Blockchain Use Case, Part 2," ETHNews.com, November 25, 2017, https://www.ethnews.com/the -decentralized-union-a-blockchain-use-case-part-2.

76. In *Whose Global Village?*, chapter co-author Ramesh Srinivasan identified the objective of helping gig economy drivers unionize internationally.

77. A hyperloop is a transportation technology that uses magnets to propel vehicles through low-pressure tubes at high speeds. So far, none of the existing hyperloop companies are servicing commuters.

78. Adam Reese, "Collective Fractional Ownership: A Proposed Blockchain Use Case," ETHNews.com, June 2, 2018, https://www.ethnews.com/collective-fractional -ownership-a-proposed-blockchain-use-case.

79. We believe that the ability to tie property ownership to tokens (so that a person can own a piece of land by owning a token) creates a lot of positive possibilities, as does the ability to break those tokens up into pieces so that multiple people can own part of a property. But in the blockchain space of the late 2010s, quite a few of the companies offering these types of tokens market their products as a means for young investors to own property before they've built their fortunes. If an investment scheme like this took off on a large scale, it could worsen housing bubbles and drive up property values, and therefore rents. In this scenario, blockchain technology would actually enable harmful speculators and opportunistic middlemen rather than thwarting them.

80. Ameen Soleimani, presentation at the State of Scale Blockchain Conference, Los Angeles, September 22, 2018.

81. "Ethereum Transaction Growth Chart," Etherscan, accessed November 12, 2018, https://etherscan.io/chart/tx.

82. A metric called *confirmation time* is used to measure the time it takes a transaction to make it into a block plus the amount of time it takes to add enough additional blocks so that one can be reasonably confident the transaction won't be rolled back due to an attack or network split.

83. Technically, there's a tradeoff between how much users pay and how long they wait. As transaction volumes increase, users either have to pay more, or wait longer, or both.

84. Wills de Vogelaere, interview by Adam Reese, September 22, 2018.

85. Most of these tokens are backed by (A) a reserve of cash/ commodities, or (B) a system that uses reserves of another cryptocurrency to keep the stablecoin's value flat (within certain parameters). Solution A fails the decentralization test because a third party must safeguard the assets, while it's unclear how solution B would survive a truly catastrophic drop in the reserve currency's value. In short, we don't see how these tokens can be guaranteed to hold their value indefinitely.

86. Joseph Poon and Vitalik Buterin, "Plasma: Scalable Autonomous Smart Contracts," April 11, 2017, https://plasma.io/plasma.pdf.

87. Alex Van de Sande, *Universal Logins Demo for Ethereum*, YouTube, 2018, https://www.youtube.com/watch?v=F5t94cCg6XE; "Digital Identity and Iden3 Technology," iden3, accessed November 25, 2018, https://iden3.io/digital-identity-and-iden3-technology.

88. Aya Miyaguchi, presentation at DevCon IV, Prague, October 31, 2018.

89. Governance is not the only area in which these networks have failed to fully decentralize. Other aspects of various networks, including the mining process, have

been shown to be partially centralized as well. See Adem Efe Gencer et al., "Decentralization in Bitcoin and Ethereum Networks," in *Financial Cryptography and Data Security*, March 29, 2018, https://arxiv.org/abs/1801.03998.

90. Adam Reese, "Ethereum Dev Yoichi Hirai Steps Away from Role as EIP Editor, Raises Questions about Process," ETHNews.com, February 15, 2018, https://www .ethnews.com/ethereum-dev-yoichi-hirai-steps-away-from-role-as-eip-editor-raises -questions-ab.

91. These developers form part of the network's de facto leadership, to the degree that such a thing exists.

92. Adam Reese, "Ethereum Dev Yoichi Hirai Steps Away."

93. Arianna Simpson, "I Am a Fan of People Who Use Regular Words to Say What They Mean. If We Want to Make Crypto More Inclusive, We Shouldn't Make Newcomers Feel Stupid…," Tweet, *@AriannaSimpson* (blog), November 26, 2018, https:// twitter.com/AriannaSimpson/status/1067145705532329999.

94. Again, mining is not universal to DLT systems but it's quite common.

95. Oscar Williams-Grut, "Meet the Crypto Trader Who Says He Bought a Tesla with 'Pump and Dump' Profits but Claims the Scams Aren't Bad: 'It's a Game,'" Business Insider, December 8, 2017, https://www.businessinsider.com/interview-with -cryptocurrency-pump-and-dump-telegram-group-admin-2017-12.

96. Luis Cuende, "The Aragon Manifesto," Aragon Project Blog, May 8, 2018, http:// blog.aragon.org/the-aragon-manifesto-4a21212eac03/. Aragon is a project that creates software to help people build decentralized organizations.

97. Ibid.

98. Some people in the space have been working on compelling privacy solutions like zero-knowledge proofs. Though worth learning about, these solutions are beyond the scope of this chapter.

24 Technology for All

1. It's hard, on some level, to fully embrace Pinker's argument. For one, we obviously did not live in past eras and can only imagine what it would have been like to be alive at those times. Additionally, our psychology is often triggered by *loss aversion*, meaning because of evolutionary reasons we pay more attention to threats and negatives.

2. "Final Report of the High Level Expert Group on Fake News and Online Disinformation," Digital Single Market, March 12, 2018, https://ec.europa.eu/digital -single-market/en/news/final-report-high-level-expert-group-fake-news-and-online -disinformation.

3. Joy Buolamwini, *How I'm Fighting Bias in Algorithms*, TEDxBeaconStreet (TED, 2016), https://www.ted.com/talks/joy_buolamwini_how_i_m_fighting_bias_in_algorithms ?language=en.

4. Joy Buolamwini, *AI, Ain't I a Woman?—Joy Buolamwini*, YouTube, 2018, https://www.youtube.com/watch?v=QxuyfWoVV98.

5. Ibid.

6. Steve Lohr, "Facial Recognition Is Accurate, If You're a White Guy," *New York Times*, June 8, 2018, https://www.nytimes.com/2018/02/09/technology/facial -recognition-race-artificial-intelligence.html.

7. Parag Mital, "Advancements in Artificial Intelligence Should Be Kept in the Public Eye," TechCrunch, October 23, 2016, https://techcrunch.com/2016/10/23 /advancements-in-artificial-intelligence-should-be-kept-in-the-public-eye/--.

8. Sam Altman, interview by author, January 26, 2018.

9. "We Cannot Predict the Future, But We Can Invent It," Quote Investigator, September 27, 2012, https://quoteinvestigator.com/2012/09/27/invent-the-future/.

10. Suzanne Barlyn, "Strap on the Fitbit: John Hancock to Sell Only Interactive Life Insurance," Reuters, September 19, 2018, https://www.reuters.com/article/us -manulife-financi-john-hancock-lifeins/strap-on-the-fitbit-john-hancock-to-sell -only-interactive-life-insurance-idUSKCN1LZ1WL.

11. Kate Crawford, "We Saw This Coming, and Here It Is. Endless Trapdoors Ahead: Data Inaccuracies, Intentional Gaming, Constant Intimate Surveillance 24/7 ...," Tweet, *@katecrawford* (blog), September 19, 2018, https://twitter.com/katecrawford /status/1042558396497649664.

12. Michael Moore, *Fahrenheit 11/9* (Briancliff Entertainment, 2018), http://www .imdb.com/title/tt8632862/.

13. Charlie Boothe, "Potential Teacher Strike Looms over West Virginia," *Bluefield Daily Telegraph*, January 29, 2018, https://www.bdtonline.com/news/potential-teacher -strike-looms-over-west-virginia/article_32f4a9f4-04a1-11e8-99f2-7f31dc816267.html.

14. "The Ethics and Governance of Artificial Intelligence Initiative," Ethics and Governance of AI Initiative, accessed October 10, 2018, https://aiethicsinitiative.org/.

15. "About," The Partnership on AI, accessed October 18, 2018, https://www .partnershiponai.org/about/.

16. Terah Lyons, interview by author, May 25, 2018.

17. Ibid.

18. Megan Cerullo, "Amazon Is Helping ICE Track, Detain and Deport Immigrants, Report Says," *Daily News*, October 23, 2018, http://www.nydailynews.com

/news/national/ny-news-amazon-tech-companies-transforming-immigration
-enforcement-20181023-story.html.

19. Roman V. Yampolskiy, "Could an Artificial Intelligence Be Considered a Person
under the Law?," *PBS NewsHour*, October 7, 2018, https://www.pbs.org/newshour
/science/could-an-artificial-intelligence-be-considered-a-person-under-the-law.

20. Kenneth P. Vogel, "Google Critic Ousted From Think Tank Funded by the Tech
Giant," *New York Times*, August 30, 2017, https://www.nytimes.com/2017/08/30
/us/politics/eric-schmidt-google-new-america.html; Hamza Shaban, "Google for
the First Time Outspent Every Other Company to Influence Washington in 2017,"
Washington Post, January 23, 2018, https://www.washingtonpost.com/news/the
-switch/wp/2018/01/23/google-outspent-every-other-company-on-federal-lobbying
-in-2017/

21. NWGrassroots, *Tucker Carlson Tech Tyranny Robert Epstein 08 24 2018*, YouTube,
2018, https://www.youtube.com/watch?v=5CK02aQLQNA.

22. Carly Berwick, "Beyond Hiring, Reena Jana Is Making Google a More Inclusive
Workplace," NBC News, February 8, 2018, https://www.nbcnews.com/better/business
/beyond-diversity-hiring-how-google-s-reena-jana-creating-more-ncna820821.

23. Bryan Stevenson, *We Need to Talk about an Injustice*, TED, 2012, https://www
.ted.com/talks/bryan_stevenson_we_need_to_talk_about_an_injustice?language=en.

24. Reena Jana, interview by author, May 4, 2018.

25. Ibid.

26. Ibid.

27. Sundar Pichai, "AI at Google: Our Principles," Google, June 7, 2018, https://
www.blog.google/technology/ai/ai-principles/; "Responsible AI Practices," Google
AI, accessed October 10, 2018, https://ai.google/education/responsible-ai-practices/

28. Caesar Sengupta, "400 Wi-Fi Enabled Train Stations in India and Counting," *Google*
(blog), June 7, 2018, https://www.blog.google/technology/next-billion-users/400-train-
stations-india/; Jack Fermon, "Google Station Brings Better, Faster Wi-Fi to More People
in Mexico," *Google* (blog), March 13, 2018, https://www.blog.google/technology
/next-billion-users/google-station-brings-better-faster-wi-fi-more-people-mexico/; Yomi
Kazeem, "Google Is Boosting Internet Access in Nigeria's Biggest Cities with Free Public
Wifi," Quartz Africa, July 26, 2018, https://qz.com/africa/1336361/google-is-boosting
-internet-access-and-its-bottom-line-with-free-public-wifi-in-nigeria/; Rayna Hollander,
"Google Brings Free WiFi to Indonesia," Business Insider, August 25, 2017, https://
www.businessinsider.com/google-indonesia-free-wifi-2017-8.

29. Jana, interview.

30. Vlad Savov, "Google's Selfish Ledger Is an Unsettling Vision of Silicon Valley Social Engineering," The Verge, May 17, 2018, https://www.theverge.com/2018/5 /17/17344250/google-x-selfish-ledger-video-data-privacy.

31. Ibid.

32. Louise Matsakis, "Google Employee's Anti-Diversity Manifesto Goes 'Internally Viral,'" Motherboard, August 4, 2017, https://motherboard.vice.com/en_us/article /kzbm4a/employees-anti-diversity-manifesto-goes-internally-viral-at-google.

33. Bernie Sanders, "Count to Ten. In Those Ten Seconds, Jeff Bezos, the Owner and Founder of Amazon, Just Made More Money than the Median Employee of Amazon Makes in an Entire Year…," Tweet, *@BernieSanders* (blog), August 21, 2018, https:// twitter.com/BernieSanders/status/1032069466052538368.

34. Abha Bhattarai, "Thousands of Amazon Workers Receive Food Stamps. Now Bernie Sanders Wants the Company to Pay Up," *Washington Post*, August 23, 2018, https://www.washingtonpost.com/business/2018/08/24/thousands-amazon-workers -receive-food-stamps-now-bernie-sanders-wants-amazon-pay-up/.

35. Julia Glum, "The Median Amazon Employee's Salary Is $28,000. Jeff Bezos Makes More Than That in 10 Seconds," *Time*, May 2, 2018, http://time.com/money /5262923/amazon-employee-median-salary-jeff-bezos/.

36. Hillary Hoffower and Andy Kiersz, "A Minimum-Wage Worker Needs 2.5 Full-Time Jobs to Afford a One-Bedroom Apartment in Most of the US," Business Insider, June 14, 2018, https://www.businessinsider.com/minimum-wage-worker-cant-afford -one-bedroom-rent-us-2018-6.

25 Educating and Protecting Our Future

1. Natasha Singer, "Tech's Ethical 'Dark Side': Harvard, Stanford and Others Want to Address It," *New York Times*, February 19, 2018, https://www.nytimes.com/2018/02/12 /business/computer-science-ethics-courses.html; Cedric M. Smith, "Origin and Uses of Primum Non Nocere—Above All, Do No Harm!," *Journal of Clinical Pharmacology* 45, no. 4 (April 1, 2005): 371–377, https://doi.org/10.1177/0091270004273680; Aarti Shahani, "As Uber Expands, It Asks Cities for Forgiveness Instead of Permission," NPR .org, December 26, 2014, https://www.npr.org/sections/alltechconsidered/2014/12/26 /373087290/as-uber-expands-it-asks-cities-for-forgivness-instead-of-permission.

2. Adam Sarhan, "Planned Obsolescence: Apple Is Not the Only Culprit," *Forbes*, December 22, 2017, https://www.forbes.com/sites/adamsarhan/2017/12/22/planned -obsolescence-apple-is-not-the-only-culprit/.

3. Tom Jackman, "Electronics-Recycling Innovator Faces Prison for Trying to Extend Computers' Lives," *Los Angeles Times*, February 15, 2018, http://www.latimes.com /business/technology/la-fi-microsoft-restore-disc-20180215-story.html.

4. Ibid.

5. Sven Boddington, "Busting the Myths around Second Hand Electronics," Eco-Business, November 23, 2016, http://www.eco-business.com/opinion/busting-the-myths-around-second-hand-electronics/; Matt Mace, "How O2's Smartphone Recycling Drive Is Strengthening Its Consumer Relations," edie.net, April 11, 2017, https://www.edie.net/news/5/How-O2-s-smartphone-recycling-drive-is-strengthening-its-consumer-relationship/; Dustin Benton, Jonny Hazell, and Emily Coats, "A Circular Economy for Smart Devices," Green Alliance, February 19, 2015, https://www.green-alliance.org.uk/resources/A%20circular%20economy%20for%20smart%20devices.pdf.

6. "ACM Code of Ethics and Professional Conduct," Association for Computing Machinery, June 22, 2018, https://www.acm.org/code-of-ethics.

7. Alex Hern, "Tech Suffers from Lack of Humanities, Says Mozilla Head," *The Guardian*, October 12, 2018, https://www.theguardian.com/technology/2018/oct/12/tech-humanities-misinformation-philosophy-psychology-graduates-mozilla-head-mitchell-baker.

8. Ibid.

9. Katharine Schwab, "The Future of Humanity Depends on Design Ethics, Says Tim Wu," Fast Company, September 21, 2018, https://www.fastcompany.com/90239599/the-future-of-humanity-depends-on-design-ethics-says-tim-wu.

10. Hugo Juárez Olguín et al., "The Role of Dopamine and Its Dysfunction as a Consequence of Oxidative Stress," *Oxidative Medicine and Cellular Longevity* 2016 (2016), https://doi.org/10.1155/2016/9730467.

11. Bethany Brookshire, "What Is Dopamine for, Anyway? Love, Lust, Pleasure, Addiction?," Slate, July 3, 2013, https://slate.com/technology/2013/07/what-is-dopamine-love-lust-sex-addiction-gambling-motivation-reward.html.

12. Schwab, "The Future of Humanity Depends on Design Ethics, Says Tim Wu."

13. Liana Heitin, "What Is Digital Literacy?," *Education Week*, November 9, 2016, https://www.edweek.org/ew/articles/2016/11/09/what-is-digital-literacy.html.

14. Walter J. Ong, *Orality and Literacy* (London: Routledge, 2013).

15. Larry Elliot, "World's Eight Richest People Have Same Wealth as Poorest 50%," *The Guardian*, January 13, 2017, https://www.theguardian.com/global-development/2017/jan/16/worlds-eight-richest-people-have-same-wealth-as-poorest-50.

16. Matthew Hutson, "Artificial Intelligence Could Identify Gang Crimes—and Ignite an Ethical Firestorm," Science, February 27, 2018, https://www.sciencemag.org/news/2018/02/artificial-intelligence-could-identify-gang-crimes-and-ignite-ethical-firestorm.

17. Ibid.

18. H. James Wilson and Paul R. Daugherty, "How Humans and AI Are Working Together in 1,500 Companies," *Harvard Business Review*, July 1, 2018, https://hbr.org /2018/07/collaborative-intelligence-humans-and-ai-are-joining-forces.

19. "Rising Frames," NewsFrames, June 7, 2018, https://newsframes.globalvoices .org/investigations/rising-frames/.

20. Sean Captain, "Van Jones: AI Jobs Are a Route Out of Poverty for Urban Youth," *Fast Company*, July 24, 2018, https://www.fastcompany.com/90206378/van-jones -ai-jobs-are-a-route-out-of-poverty-for-urban-youth.

21. Douglas Rushkoff, "Universal Basic Income Is Silicon Valley's Latest Scam," Medium, October 10, 2018, https://medium.com/s/free-money/universal-basic-income -is-silicon-valleys-latest-scam-fd3e130b69a0.

22. Ibid.

23. Marina Gorbis, "Universal Basic Assets," *Urgent Futures* (blog), April 4, 2017, https://medium.com/institute-for-the-future/universal-basic-assets-abb08ca2f0fc.

24. Laurel Wamsley, "As Facebook Shows Its Flaws, What Might a Better Social Network Look Like?," NPR.org, May 1, 2018, https://www.npr.org/sections/thetwo-way /2018/05/01/607361849/as-facebook-shows-its-flaws-what-might-a-better-social -network-look-like.

25. Ibid.

26. Tim Berners-Lee, "The Web Can Be Weaponised—and We Can't Count on Big Tech to Stop It," *The Guardian*, March 12, 2018, https://www.theguardian.com /commentisfree/2018/mar/12/tim-berners-lee-web-weapon-regulation-open-letter.

27. "Large-Cap Companies with at Least One Woman on the Board Have Outperformed Their Peer Group with No Women on the-Board by 26% over the Last Six Years, According to a Report by Credit Suisse Research Institute," Credit Suisse, July 31, 2012, https://www.credit-suisse.com/corporate/en/media/news/articles/media -releases/2012/07/en/42035.html.

28. Eric Hehman, Jessica K. Flake, and Jimmy Calanchini, "Disproportionate Use of Lethal Force in Policing Is Associated With Regional Racial Biases of Residents," *Social Psychological and Personality Science* 9, no. 4 (May 1, 2018): 393–401, https://doi .org/10.1177/1948550617711229; Jacob Orchard and Joseph Price, "County-Level Racial Prejudice and the Black-White Gap in Infant Health Outcomes," *Social Science & Medicine* 181 (May 1, 2017): 191–198, https://doi.org/10.1016/j.socscimed.2017 .03.036.

29. Julia Angwin, Terry Parris Jr., and Surya Mattu, "Facebook Doesn't Tell Users Everything It Really Knows About Them," ProPublica, December 27, 2016,

https://www.propublica.org/article/facebook-doesnt-tell-users-everything-it-really
-knows-about-them; Julia Angwin, Terry Parris Jr., and Surya Mattu, "Breaking the
Black Box: What Facebook Knows About You," ProPublica, September 28, 2016,
https://www.propublica.org/article/breaking-the-black-box-what-facebook-knows
-about-you.

30. Mary Lister, "All of Facebook's Ad Targeting Options (in One Epic Infographic),"
WordStream, August 23, 2018, https://www.wordstream.com/blog/ws/2016/06/27
/facebook-ad-targeting-options-infographic.

31. Angwin, Parris Jr., and Mattu, "Facebook Doesn't Tell Users Everything."

32. Eric Holder, communication with author, October 3, 2018.

33. "Large-Cap Companies with at Least One Woman on the Board."

34. Ed Jean-Louis, interview by author, April 6, 2018.

35. Erik Larson, "New Research: Diversity+Inclusion=Better Decision Making
at Work," Forbes, September 21, 2017, https://www.forbes.com/sites/eriklarson
/2017/09/21/new-research-diversity-inclusion-better-decision-making-at-work
/#d88dfe14cbfa.

36. Vivian Hunt, Dennis Layton, and Sara Prince, "Diversity Matters," McKinsey
& Company, February 2, 2015, 1, https://www.mckinsey.com/~/media/mckinsey
/business%20functions/organization/our%20insights/why%20diversity%20matters
/diversity%20matters.ashx.

37. Lori Ioannou, "Silicon Valley's Diversity Problem Is Its Achilles' Heel," CNBC,
June 20, 2018, https://www.cnbc.com/2018/06/20/silicon-valleys-diversity-problem
-is-its-achilles-heel.html.

38. Ibid.

39. Dominique Fluker, "Why Arlan Hamilton Created A $36M Fund for Black
Women," Forbes, May 23, 2018, https://www.forbes.com/sites/dominiquefluker
/2018/05/23/arlan-hamilton/#1023d2d3dabd.

40. Philip Agre, "The Market Logic of Information," The Wizards of OS, accessed
October 13, 2018, http://mikro-berlin.org/Events/OS/interface5/ms_agre.html.

41. Ibid.

Conclusion

1. Douglas Rushkoff, "Survival of the Richest—Future Human," Medium, July 5,
2018, https://medium.com/s/futurehuman/survival-of-the-richest-9ef6cddd0cc1.

2. Ibid.

Index

DATE DUE

DEC 3 1 2019	
AUG 1 8	
AUG 0 6 2022	